VENEREAL DISEASE, HOSPITALS AND THE URBAN POOR

Rochester Studies in Medical History

Senior Editor: Theodore M. Brown
Professor of History and Preventive Medicine
University of Rochester

ISSN 1526–2715

Mechanization of the Heart:
Harvey and Descartes
Thomas Fuchs
Translated from the German by Majorie Grene

The Workers' Health Fund in Eretz Israel
Kupat Holim, 1911–1937
Shifra Shvarts

Public Health and the Risk Factor:
A History of an Uneven Medical Revolution
William G. Rothstein

Venereal Disease, Hospitals and the Urban Poor
Kevin P. Siena

VENEREAL DISEASE, HOSPITALS AND THE URBAN POOR

London's "Foul Wards," 1600–1800

Kevin P. Siena

UNIVERSITY OF ROCHESTER PRESS

362.196
S57v

Copyright © 2004 Kevin P. Siena

All Rights Reserved. Except as permitted under current legislation, no part of this work may be photocopied, stored in a retrieval system, published, performed in public, adapted, broadcast, transmitted, recorded or reproduced in any form or by any means, without the prior permission of the copyright owner.

First published 2004
by the University of Rochester Press

The University of Rochester Press
668 Mount Hope Avenue, Rochester, NY 14620, USA
and at Boydell & Brewer, Ltd.
P.O. Box 9, Woodbridge, Suffolk IP12 3DF, UK
www.urpress.com

ISBN 1-58046-148-4

Library of Congress Cataloging-in-Publication Data
Siena, Kevin Patrick.
   Venereal disease, hospitals, and the urban poor ; London's foul wards, 1600-1800 / by Kevin P. Siena.
      p. ; cm. — (Rochester studies in medical history, ISSN 1526-2715 ; 4)
   Includes bibliographical references and index.
   ISBN 1-58046-148-4 (hardcover : alk. paper)
   1. Sexually transmitted diseases—England—London—History—17th century.
   2. Sexually transmitted diseases—England—London—History—18th century.
   3. Urban poor—Health and hygiene—England—London—History—17th century.
   4. Urban poor—Health and hygiene—England—London—History—18th century.
      [DNLM: 1. Sexually Transmitted Diseases—history—London.   2. History of Medicine, 17th Cent.—London.   3. History of Medicine, 18th Cent.—London.
   4. Urban Health Services—history—London. WC 11 FE5
   S572v 2004] I. Title. II. Series.
RA644.V4S495 2004
362.196'951'00942109032–dc22
                                                                                        2003025102

British Library Cataloguing-in-Publication Data
A catalogue record for this book is available from the British Library

Designed and typeset by Mizpah Publishing Services Pvt. Ltd.
Printed in the United States of America
This publication is printed on acid-free paper

CONTENTS

List of Figures	vi
Acknowledgments	vii
Introduction: Institutions and Experiences	1
Chapter 1: The Foul Disease, Privacy, and the Medical Marketplace	30
Chapter 2: The Foul Disease in the Royal Hospitals: The Seventeenth Century	62
Chapter 3: The Foul Disease in the Royal Hospitals: The Eighteenth Century	96
Chapter 4: The Foul Disease and the Poor Law: Workhouse Medicine in the Eighteenth Century	135
Chapter 5: The Foul Disease and Moral Reform? The Lock Hospital	181
Chapter 6: Rethinking the Lock Hospital	219
Conclusion: Poverty and the Pox in Early Modern London	251
Notes	267
Bibliography	329
Index	361

LIST OF FIGURES

1:	Watercolor Drawing of a Lock Hospital Patient, 1849	20
2:	Watercolor Drawing of a Lock Hospital Patient, 1849	21
3:	Watercolor Drawing of a Lock Hospital Patient, 1850	22
4:	Outhouse Patients as a Percentage of All Patients in St. Bartholomew's Hospital, 1622–1696	71
5:	Annual Average Capacity of the Lock and Kingsland Outhouses (combined), 1622–1696	72
6:	Annual Average Number of Venereal Patients Supported by St. Bartholomew's Hospital, 1622–1760	98
7:	Annual Average Number of "Clean" Patients Supported by St. Bartholomew's Hospital, 1622–1760	99
8:	Annual Average Number of "Clean" Patients and Venereal Patients Supported by St. Bartholomew's Hospital, 1622–1760	100
9:	Admissions Fees at St. Thomas's Hospital, 1715–1800	104
10:	Gender Breakdown of Venereal Patients in Four London Parishes	162
11:	Ages of Venereal Patients in the Workhouse of St. Luke's Chelsea, 1742–1769 and 1782–1799	163
12:	Ages of Female Venereal Patients in the Shoe Lane Workhouse, St. Andrew's Holborn, 1776–1781	165

ACKNOWLEDGMENTS

It is strange that a book on such a grim topic could be such a pleasure to write. This was certainly the case thanks to the help and encouragement of friends and colleagues, only a few of whom can be thanked here. Pauline Mazumdar and Barbara Todd assisted me enormously while researching and writing the dissertation out of which this book grew. Donna Andrew shared her great knowledge of eighteenth-century charity with me, read several versions of the manuscript in its entirety, and proposed key revisions. I am just as indebted to Tim Hitchcock. He also read the manuscript, suggested revisions, and shared with me many nuggets he'd unearthed in the archives related to the pox. My friend Ted Brown has had a unique perspective on this project, watching it develop from a seedling of an essay more than a decade ago, and now helping to shepherd it through the process of becoming a book. His careful reading at various stages improved the quality of the manuscript significantly. I am also thankful for the suggestions made by Vivian Nutton, Lesley Hall, the external manuscript reviewers, and the readers who adjudicated the dissertation for the 2001 Hannah-Millennium Dissertation award.

That award was the generosity of Associated Medical Services, Inc., a foundation to which I owe a tremendous debt. AMS supported the project at both the doctoral and postdoctoral levels. This book would not have been written without this support. AMS also enabled me to return to London to conduct further archival research as the 2002 Hannah-Wellcome Traveling Fellow. I wish to thank the Wellcome Trust at UCL for being such gracious hosts. I also received funding from the University of Toronto and its Department of History. I benefited greatly from the support and stimulating seminars of the Lupina Foundation's

Comparative Program on Health and Society. Archival and Library staffs at the Royal College of Surgeons, Royal College of Physicians, London Metropolitan Archives, Guildhall Library, British Library, Corporation of London Records Office, Wellcome Library, Westminster City Archive Centre, Westminster Abbey Library and Muniments Room, Public Records Office, Royal London Hospital, and St. Bartholomew's Hospital all deserve my sincere thanks and admiration.

A few people warrant special mention: Marion Rea, Sally Gilbert, Claire Jackson, Richard Mortimer, Jonathan Evans, Geoff Hudson, Kim Kippen, Nicole Schulman, Keith Walden, Peter Warrian, Margaret Hovanec, Tim Madigan, and John Blanpied. I want also to thank sincerely my colleagues in the History Department at Trent University who supported a teaching release to allow me to finish writing. Finally, I would be remiss if I did not thank Adrianna Bakos who eleven years ago first suggested that I take a look at early modern VD. Sage advice. Despite my good fortune to have received the help of so many, all responsibility for errors or oversights rests with me.

My wife Kate endured the trials of completing this book with immeasurable grace. Her smile makes it all worthwhile. I pledge the same support to her work.

I owe the greatest debt to my parents, Louis and Patricia Siena, whose support means more to me than they will ever know. I can only guess how many times they have had to report to perplexed family and friends that "Kevin is still studying syphilis." They gave me the drive to aim high and the humor to keep things in perspective. I dedicate this book to them, with love.

INTRODUCTION

INSTITUTIONS AND EXPERIENCES

> The Admission of Venereal Patients . . . [is] a Subversion of the Charity, or a Misapplication of the Money given in trust for the Poor . . . the Society [has] constantly rejected Venereal Patients for the very reason of Being Venereal.[1]

So wrote one of the governors of the Westminster Infirmary in 1738. It is clearly a strong invective against allowing so-called "foul" patients into hospitals. Many have presumed that this policy was pervasive in early modern London. It was not.

Considerably more scholarship has explored venereal disease[2] in the modern period. However, there is a growing body of literature on the early modern period that has explored medical treatises, graphic art, and literature, analyzing the various meanings that early modern doctors, artists, and playwrights attached to sexual infection.[3] Yet early modern institutional care has received rather less attention. Robert Jütte has identified the area as a notable gap in the literature and called for further research. This study hopes to add to Jütte's work on Germany and that of Jon Arrizabalga, John Henderson, and Roger French on Italy.[4]

Discussions of institutional care for venereal disease in early modern England have tended to assume that the attitude expressed above by the governors of the Westminster Infirmary was standard throughout the period. Moreover, English scholarship has focused the lion's share of attention on one particular hospital, the London Lock Hospital. The Lock, a voluntary hospital devoted exclusively to venereal disease, was established in 1747. Its appearance in the mid-eighteenth century led many to assume that impoverished venereal patients seeking treatment earlier had nowhere to turn. Historians presumed that the Lock Hospital must have filled some void, that prior hospitals must have excluded venereal patients on moral grounds. With the Enlightenment, in this view, came new tolerance and a new hospital as its manifestation. A kind of whiggishness has colored many discussions of the Lock, portraying it as a progressive step in the march of modernity. To make this case scholars have asserted that early hospital provision for venereal patients was scant or nonexistent.[5] Arguing from a slightly different vantage point, some recent historians of sexuality have advanced a similar picture.[6]

Early London hospital provision for venereal patients was much more impressive than most histories have suggested. As early as 1918 Norman Moore noted early modern venereal care in his massive history of St. Bartholomew's Hospital, as did F. G. Parsons in his 1932 three-volume history of St. Thomas's Hospital.[7] Although neither historian made the pox a prime subject of inquiry, their studies pointed out that both hospitals treated venereal disease as early as the sixteenth century. Despite these early references, treatment of the pox at these two institutions remained largely neglected. Those who have noted this early hospital VD care have tended to mention it in passing on their way to discussing the Lock Hospital[8] or else have presented these findings as an anomaly.[9] Yet both St. Bartholomew's Hospital and St. Thomas's Hospital made treating the disease standard practice from at least the mid-sixteenth century. Moreover, venereal patients loom quite large in both hospitals. Figures from the seventeenth century at St. Bartholomew's and from the eighteenth century at St. Thomas's show that venereal provision at the royal hospitals cannot be described as scant. In many years throughout the two centuries, well more than 20 percent of all patients entering these hospitals went into the so-called foul wards.

Just as contemporary ethics did not keep private doctors from treating the disease, neither did the prevailing morality prevent hospitals from accepting venereal patients. Early London hospitals treated venereal patients as a rule, not an exception.

This study explores treatment of the pox in a range of London hospitals over two hundred years. In many ways it seeks to answer the basic question, how did English society respond to the problem of sexual infection among the urban poor who could not afford to pay for treatment. Because the pox was such a unique condition, it provides a useful litmus test for evaluating London's social welfare system. Some institutions, like the Westminster Infirmary, clearly did exclude venereal patients on moral grounds. Certainly the argument for exclusion was always ready at hand. Moral stain colored the disease throughout the period. Anyone wanting to seize the ethical high ground could easily claim that those infected had brought the disease upon themselves. The attitude that the "foul"—already a delineation dripping with cultural assumptions—deserved what they got and were now feeling the sting of their own immorality was not uncommon. But that attitude notwithstanding, most of London's major hospitals treated the pox.

The hospitals that excluded venereal patients tended to be private charities. However, other institutions were infused with a different sense of civic obligation. The hospitals that treated the foul were not necessarily more tolerant, more enlightened, or more progressive. Rather one gets the sense that they did not have quite the freedom that entirely private charities had to pick and choose who they would treat and who they would not. Exploring how and why certain institutions cared for the venereal poor—certainly the most odious of patients in the period—allows us to think about the sense of civic responsibility that the community invested in particular institutions.

But while this study focuses largely on institutions, it hopes to do more than just provide an institutional history. It aspires to be a patients' history of this disease, one that places the experiences of specifically poor patients at its core. Medical historians have demonstrated the great dividends paid by focusing on patients, in addition to doctors.[10] However, many of the patient-centered histories that we have for the period have tended to document elite illness-experiences, a natural outgrowth of surviving source material. This book hopes to augment that literature through an examination of specifically poor venereal patients. Yet, it must be said that this remains a history based largely on institutional records. Thus the experiences of the poor in these hospitals have to be accessed largely through the lenses of documents created by others, in this case hospital officials, doctors, and churchwardens.

That caveat has become standard operating procedure for those seeking to recover the lives of the pre-modern poor. However, by approaching these institutions from the outside in, by balancing a history of the *provision* of medical welfare with a history of its *use*, and by augmenting institutional records with other sources that shed light on plebeian experiences of illness, I hope to begin to explore what it was like to be both impoverished and infected with a sexual disease in early modern London. Either of those afflictions made life extremely difficult, then just as now. Combined, they represented a dreadful pair, very often deadly in the age before penicillin and the welfare state. Certainly they challenged victims to mount significant campaigns of resistance, marshalling whatever resources they had, and employing every ounce of creativity and resolve they could muster. I hope to contribute to medical and social history of the period by charting how common people reacted to the unique obstacles posed by venereal infection, through exploration of the range of options that they faced and the range of strategies they employed.[11] Ultimately, I hope to show that it is fruitful to study the relationship between poverty and the pox, between sex and social welfare.

Attention to the poor with this disease is also overdue. Some have suggested that the pox was largely an aristocratic disease throughout much of the early modern period. The "modern pattern of infection," which sees wide infection among the general population, allegedly only emerged at the dawn of the eighteenth century. Scholars have argued that before then the pox was rare among common folk. Only after 1690, we are told, do we begin to see the disease spreading widely among plebeian people.[12] Others have argued that the spread of the disease among common people began even later.[13] Seventeenth-century hospital records demand a revision of that picture. The pox was absolutely rife among the London poor long before 1690. There was certainly no dearth of nobles who contracted the disease, and scholars are right to show that one early modern construction of the disease was as a symbol of aristocratic vice.[14] But at the ground the pox raged among the urban poor.

The desire to foreground the patient-experience while still telling a broader story of the social response to the disease led me to organize the material according to institutional type. Following a chapter on private care available on the medical marketplace, two chapters explore the royal hospitals, one chapter looks at parish infirmaries, and two final chapters explore the Lock Hospital. I believe that the specifics of different institutional structures directly affected the scope and format of the care available. Put plainly, different sorts of people relied on different sorts of hospitals in early modern London. Moreover, patients' experiences varied from one hospital to the next. The admissions process provides a good example. What one patient had to do to enter St. Bartholomew's differed in important ways from the process by which someone else secured a bed in a workhouse infirmary or someone else got into the Lock.

While this study presents these different types of hospitals separately, it ultimately hopes to show how they combined to form a system of institutional charitable medicine. Studies of early modern institutional medicine often focus on a single hospital.[15] I believe that studying a single disease in a variety of institutions allows for different perspective. It allows for a cross-section of institutional medicine in the city. Not only do comparisons between institutions become possible, but so also does one learn how these institutions related to one another, how they fit together. These hospitals need to be understood as part of a network for their histories to be fully intelligible. It certainly appeared this way to patients, who frequently moved back and forth between different institutions. Since these hospitals did not exist in isolation, they will not be studied in isolation.

However, it must be said at the outset, that the decisions to explore a broad range of different hospitals and to concentrate attention on patients precluded the attempt to provide a full prosoprographical study detailing the background of hospital administrators. There can be no doubt that this approach to the study of early modern charity is extremely useful[16] and that much more of this type of work needs to be done for English hospitals, as Adrian Wilson has suggested.[17] Thankfully, some of this work has been done for the hospitals studied here, and I have drawn on secondary work wherever possible. However, the desire to focus on the *users* of these institutions as much as possible oriented my research in a different direction. Moreover, I hope to show that ways in which people chose to use institutions could have a significant hand in determining how they functioned on the ground.

Studying how London's hospital network developed over two centuries also reveals a great deal. First, there appears more gradual change over time than is sometimes assumed. The story is not one of absolute prohibition of venereal patients in the seventeenth century followed by enlightened tolerance in 1747. Instead, this is a story of expansion and development. Hospital VD care slowly evolved from its sixteenth-century roots. Over time it took various forms in the large royal hospitals, in transformed leper houses, in workhouses, and in voluntary charities. Indeed, the range of institutions dealing with the pox was much broader in the eighteenth century as the workhouse movement and voluntary hospital movement greatly enhanced the sheer number of medical institutions in the capital. For that reason, the weight of analysis here falls much heavier on the eighteenth century than on the seventeenth century, for there were simply more institutions treating the disease.

In some ways the multi-level healthcare network that came to administer charitable medicine by the eighteenth century reflects the city's own development. The system grew as London grew; the system grew *because* London grew. Over the seventeenth and eighteenth centuries the demand for beds for poor foul patients skyrocketed because London bulged with migrants from rural England. The various responses to provide those beds are revealing. The royal hospitals expanded their services; however, they did so by putting much of the costs of care onto patients themselves, making hospitalization too expensive for many people by the early eighteenth century. From that point forward parish authorities scrambled to come to grips with the problem of the pox among parishioners who could no longer afford hospital care. Their solution was to centralize care inside workhouses. In turn, we then see a group of benefactors establish the voluntary Lock Hospital to provide beds for those migrants with the pox who did not qualify for parochial assistance and thus still fell outside the system.

It is hoped that this study of the multi-faceted response to the problem of venereal disease among the poor will further our understanding of how early modern urban authorities tried to answer some of the pressing social problems caused by rapid urbanization in the late seventeenth and eighteenth centuries.[18] And we will further benefit from viewing the changing system from below. The medical vistas of the poor were colored by a broad range of institutional options as the eighteenth century progressed and an increasing number of hospitals offered treatment. For the prospective patient, each had its own advantages and drawbacks. Each had its own pitfalls. And each had its own admissions process requiring a different strategy on the part of poor applicants.

Those patients' strategies form a central theme that runs through the following chapters. Recent work on the survival strategies of the English poor has advanced our understanding of how early modern institutions functioned.[19] It has been extremely helpful to think about how, why, and under what circumstances applicants for charity and poor relief decided to negotiate welfare institutions. While we have long understood that a power relationship existed between those who administered charity and those who received it, that relationship was never completely one-sided. Recipients were in a weak position within that exchange, to be sure. But they did have some room for maneuver, and they did exhibit agency. The actions of the patients studied here shed light on the institutions themselves. Lists of hospital rules can tell us about how governors hoped their hospitals might function. Attempts by patients to break those rules, to avoid hospital scrutiny, to enter forbidden wards, to pry resources out of local churchwardens, and to force their way into—and out of—hospitals, demonstrate how these institutions actually functioned.

The stigma that attended this disease, and the hope that one might avoid it, gave the patients studied here additional reasons to be careful, even cagey, about where and how they sought public assistance. While it is no surprise that elite people with this disease sought actively to protect their reputations, so, too, did the poor. Plebeian Londoners give every indication that, like their social betters, they were extremely concerned with the ramifications that might follow public exposure of their infection. So frequently they were deeply reluctant to undergo the surveillance and inspection that hospital admission required of them. They only did so when they absolutely had to, and when they did, they were creative in how they applied for admission.

At every step poxed patients leave evidence that they tried to access medical resources in their own ways and on their own terms, guarding their dignity whenever possible. Historians of early modern England have recently begun to apply the insights of James Scott, whose work has proved so fruitful for a wide range of scholars exploring power and resistance. For Scott, the weak generate what he calls a "hidden transcript," a secret discourse in which they express opposition to hegemonic powers that they cannot express publicly.[20] This notion of hidden transcripts has allowed for new insights on relations between weak and strong. In the realm of political history, where Scott's ideas have proven most useful, this has meant master and slave, landlord and serf, colonizer and colonized. Perhaps medical historians can benefit from Scott's insights as well. Both the medical relationship of doctor-patient and the charitable relationship of benefactor-supplicant were power relationships. Perhaps plebeian foul patients had their own hidden transcripts. Being labeled "foul" brought with it a range of ramifications that went well beyond dispensing mercury. At stake were important issues of privacy and reputation, which, if damaged, could bring serious economic results. Explaining his approach, Scott stated "it is apparent that I privilege the issues of dignity and autonomy."[21] So will I. The exchanges between plebeian foul patients and their socially superior doctors, churchwardens, and hospital governors concerned much more than simple material resources.

The cultural constructions of the disease and its links to sexuality have meant that the issue of the moral reformation of patients has always played an important role in histories of the pox in this period. Indeed, the cause of moral reformation is another theme that runs throughout this book. The traditional story of the Lock Hospital suggests that it was founded with a clear reforming mission.[22] It did develop such a mission. However, it only did so at the close of the eighteenth century, and its change in policy marks a larger shift in the history of institutional venereal care. Much of the work on venereal disease and public policy in the modern period has demonstrated that public policy was often aimed at social regulation. Feminist scholars especially have demonstrated how public health measures to control sexually transmitted diseases (STDs) often aimed to regulate working class women's sexuality. Two decades of research illustrate beyond doubt that attempts to control venereal disease in the nineteenth century focused increasingly on attempts to institutionalize poorer women and police their sexuality. Responding to perceptions of new problems in growing urban centers, nineteenth-century medical and state authorities concentrated their efforts on policing working women's sexuality in the belief that plebeian female promiscuity presented a tremendous threat to the social order.[23] And VD hospitals frequently represented the front line of this campaign.

That approach to public health was born in the period studied here. A recent collection of articles on the pox in the eighteenth century has argued that the response to the disease changed in the closing decades of the century, with an increased surveillance of prostitutes marking a signature element of this period.[24] A focused attempt to utilize a VD hospital to incarcerate and reform poor sexually active women emerged only in 1787 when the governors of the Lock Hospital built a sister institution, the Lock Asylum for Penitent Women. Here women recently treated for the pox in the Lock Hospital were confined to undergo a rigorous campaign to save their souls. This proved to be the first in a long series of projects that sought over the next century and a half to merge the hospital with the penitentiary, incarcerating working class women in the name of public health.

However, it is important to point out that this approach only emerged at the very end of the eighteenth century. It is largely absent—surprisingly absent—from the institutions that had been treating the disease during more than two centuries up to the 1780s. The evidence from seventeenth and early eighteenth-century hospitals shows that this gendered penitentiary approach was more properly a modern phenomenon than an early modern one. This does not mean that early modern discourse on the pox was free from attempts to police sexuality. To the contrary, early modern medical writing on venereal disease often aimed at regulating behavior, especially female promiscuity. The clearest evidence of this was an influential gendered theory about the disease's production, which located the generation of the pox in the wombs of promiscuous women and remained popular throughout the late seventeenth and eighteenth centuries.[25] In many ways that theory demonstrates how the underlying anxieties toward unfettered female sexuality that informed later public health campaigns like the Lock Asylum or the later Contagious Diseases Acts can be detected from an early date. Therefore, it is somewhat surprising that such ideas remained confined to medical theory, and that we must wait so long for their clear institutional expression.

However, if attempts at moral reformation were not the primary impetus behind early hospital provision for the pox, we must ask what was. When one confronts the evidence it seems that those directing charitable resources towards treating venereal disease were waging a different battle. Hospitals were not fighting widespread moral turpitude, but rather one of its products, widespread venereal infection. Were our hospitals today so dominated by a single disease—if a single diagnosis accounted for one fifth to one quarter of all hospital patients—the headlines would read daily of one of the worst public health crises in recent memory. I suggest that the pox represented one of the single most pressing public health problems in early modern London.[26] Certainly outbreaks of plague remained, which could strike suddenly and kill many. But the evidence presented below will demonstrate that, unlike plague, which struck occasionally and by surprise, and indeed which faded after the middle of the seventeenth century, the pox formed an ever-present public health problem that only got worse as time went on and the city expanded.

The records show that throngs of poor Londoners applied week after week for a bed in the foul wards. In the eighteenth century new institutions like Guy's Hospital emerged and built venereal wards. Moreover, parish infirmaries cropped up all over London after 1720. Yet despite the increase in hospital beds for venereal patients, there still remained heavy demand for the Lock Hospital in 1747. Even then, that hospital's records leave evidence of lengthy waiting periods still. Notwithstanding the impressive hospital operations presented below, demand constantly outstripped supply when it came to hospital beds for the poxed in seventeenth- and eighteenth-century London. London's hospitals were simply too busy trying to handle the medical problem posed by the pox to devote any serious resources to tackling the complex social problems raised by the disease. At certain points one finds early hospitals trying to take up the larger cause of reforming patients. But those efforts were sporadic during the period before roughly 1780. For example, there is evidence of a campaign to inflict corporeal punishment on venereal patients at St. Thomas's Hospital in the early seventeenth century. And occasionally, governors distributed religious reading material in the foul wards. But these efforts are rather hard to find, and, importantly, they were gender neutral, unlike the later Lock Asylum.

Overall, one is struck instead by the mundane normality of early modern foul wards. Patients went in, patients went out, week after week, year after year, for more than two centuries. More often than not the disciplinary regimen they faced differed little at all from that found in the general wards. I am not saying that early modern foul wards should not be viewed as disciplinary. But what is striking is that the disciplinary aspects of the foul wards tended not to be unique from general hospital wards. Chapter 5 shows just how difficult it was to get a moral reformation program up and running in a hospital. In the end it took quite an effort, and indeed, the governors at the Lock Hospital had to build an entirely separate institution to take on the job. This campaign only got under way after significant effort from a group of Evangelicals that took over that hospitals' administration.

It is furthermore telling that once the Lock Asylum was finally established it faced a steep uphill battle to save the souls of poor women. Evidence from its first years of operation shows that it failed miserably. Women resisted attempts to reform them, demonstrating that the nineteenth-century reformers who would take up the cause and base their efforts on the Lock's model were going to have their hands full. So despite clear evidence that early modern medical authorities did link venereal disease to gendered social and sexual issues, most medical institutions were not able to turn those ideas into clearly formulated moral reformation campaigns. They had enough problems simply administering mercury to the scores of paupers who turned up at their gates week after week.

It is also important that throughout two centuries the administrators at the institutions studied here virtually never turned their backs on these applicants. More than anything else Christian notions of charity probably drove them to continue channeling limited resources to the fight against this endemic and entrenched disease. Chapter 5 shows that the Lock Hospital based its fundraising campaign largely on the Christian appeal that we should love our neighbors, despite their faults. However, the epidemic also threatened to weaken the nation as a whole, and we should consider Donna Andrew's suggestion that national fears of a small and unhealthy population may well have influenced those who administered eighteenth-century charity.[27] Were they saving poxed paupers, or saving England from them? Whatever the reason, the evidence of such a broad response to such widespread infection among the urban poor suggests that the problem was just too big to ignore. The pox was omnipresent in early modern London, and something had to be done about it. Moreover, it is telling to learn which institutions responded to the call. Parish workhouses' responsibilities under the old poor law and the royal hospitals' links to the municipal government determined that they, and not London's essentially private voluntary hospitals, were at the forefront of that response.

Although they provided care, and their efforts to reform foul patients were not as intense as some first assumed, hospitals did treat those labeled foul in unique and revealing ways. For example, foul patients were always segregated. It is certainly telling that the first English venereal wards were transformed leper hospitals. This segregation would continue long after doctors ceased to believe that venereal patients represented a contagious threat to other patients. While the policy of segregation represented a manifestation of the dominating stigma, it simultaneously reinforced that stigma by assuring the population that foul patients *needed* to be segregated.[28] An air of culpability always attended venereal care as well. Hospital administrators show the tendency to hold foul patients responsible for their own infection. For example, they limited foul patients to just one chance; if patients returned to the hospital with the pox a second time they were routinely refused. Increasingly, hospitals also demanded that venereal patients pay admission fees from which other patients remained exempt. These fees increased as time went on, pricing care out of many people's reach. This point alone challenges the picture of an increasingly tolerant hospital system in the enlightenment.

If anything, it became harder and more expensive for poor venereal patients to access hospital resources in the eighteenth century than it had been previously. Governors also clearly privileged "clean" patients over foul ones, and when resources became scarce they steered money away from the foul wards and concentrated their efforts on treating other patients. So while contemporary ethics (medical or otherwise) never prevented hospitals from treating the pox, cultural attitudes colored the form of hospital care at every step. Ultimately, institutional policy reveals enormous tension. One can almost feel the conflict apparent as governors on the one hand felt committed to providing care for the sick and fighting the spread of disease, but on the other they questioned the extent to which foul patients deserved their charity. That uneasy conflict lurks beneath the surface of hospital venereology for more than two centuries.

The stories of these hospitals and their patients demonstrate beyond doubt that class played a fundamental role in the experience of early modern venereal disease. A comparison of hospital patients' experiences with the experiences of those who could pay for private care is implied throughout this book. To enable such a comparison we will first explore the options available to those able to pay for private care. For example, evidence demonstrates that discretion and confidential treatment was a primary concern for medical customers suffering with a sexual infection. In fact, chapter 1 suggests that the presence of the pox may have helped to promote medical privacy as a more pronounced element of the doctor-patient relationship. However, it also stresses that this emerging ethic was a commodity to be bought and sold. Those without means to obtain private care had to rely on institutional care. Ensuing chapters demonstrate how very public that care was. Hospital patients had to disclose details of their condition to a variety of strangers including churchwardens, vestrymen, overseers of the poor, hospital governors, and members of admissions committees. Moreover, by the eighteenth century they found themselves on display before classes of medical students. Therefore there existed a public-private gap between the experiences of the poxed wealthy and the poxed poor. The actions of venereal patients of all classes clearly show that they wished to keep their infection and their treatment a private matter. However, only those of means could do so. It is because of this desire for privacy that we see poorer hospital patients act so shrewdly when they encountered hospital admissions committees.

In their attempts to preserve their privacy, venereal patients initially benefited from the early modern medical exchange, a process of negotiation between doctor and patient in which the patient played an active role in constructing their own diagnosis. In a word, patients were given an important voice within the medical exchange in the period up to the mid-eighteenth century. Venereal patients took advantage of that opportunity and actively contested their diagnosis. Some resisted being labeled "foul." Others used the chance to obscure or rewrite their sexual history. Some lied about the nature of their condition in order to gain access to beds in the "clean" wards, while others maneuvered to arrange beds for themselves as discreetly as possible. Here may be some of medicine's hidden transcripts. However, in the second half of the eighteenth century opportunities for such agency began to evaporate, as there emerged a new form of medical encounter, which afforded patients very little voice at all. The oft-discussed "birth of the clinic" rendered hospital patients voiceless, and this had pivotal implications in the foul wards. In the late eighteenth century, London's two-tiered medical system saw private patients continue to negotiate their diagnoses, while their poorer brethren in hospital foul wards lost that option.

The dynamic of class converged with the force of gender to inform the medical exchange. Chapter 1 also demonstrates that women of means often chose to hire female practitioners rather than expose themselves before a male doctor. The same dynamic and market forces that brought about promises of greater discretion by doctors treating the pox also sustained a sizable number of female practitioners who specialized in treating women with the disease. Again, this service was a luxury. Poor women in hospital foul wards were not afforded the comfort of a female doctor. Despite the important work done by hospital nurses, the majority of people who administered and treated patients in each of the hospitals studied here were male. To obtain the care that they needed, poor women had to expose themselves before a male medical gaze, like it or not.

The concomitant forces of gender and class also contributed to a gender imbalance in hospital foul wards. Gendered realities of early modern poverty guaranteed that the poorest segment of the population was disproportionately female. Thus the workhouse infirmary, the institution catering to London's poorest, was a largely female institution, setting it off in crucial ways from the other hospitals studied here. There emerged in the eighteenth century a gendered landscape of London's hospitals. The royal hospitals, which charged fees, and the Lock Hospital, which required an elite governors' nomination, tended to treat more men than women. By contrast, workhouse infirmaries—London's other hospitals—catered largely to women, who were usually young, always poor, and who could not access care elsewhere. This is one reason why it is crucial to compare the broad range of institutions treating this disease. Different sorts of people relied on different sorts of hospitals. And the resulting illness-experiences could vary tremendously.

Experiencing the Foul Disease

To this point I have spoken of the "foul disease." But what exactly was it? A recent addition to the literature on venereal disease has made an invaluable contribution by reminding us to avoid applying modern diagnostic terms for pre-modern medical conditions.[29] This is particularly important in the case of venereal disease. One single diagnostic term—"the venereal disease"—served to describe a host of conditions that we now separate. There was one recognized venereal disease, the pox, which had a plethora of "symptoms," which included conditions like gonorrhea. It is impossible to say what individual patients "actually" suffered from. Many probably suffered from syphilis, but it would be anachronistic to assume in every case that the early modern pox was the modern condition identified as syphilis. In fact, that term was hardly ever used in the period. Far more common were the terms "the venereal disease," "*lues venerea*," "the pox," "the French Disease" or "the foul disease," all of which stood for the same single disease-concept.

To understand how our predecessors thought about venereal disease, historians must consider it in their terms. David Harley has recently criticized retroactive diagnosis as "deeply misleading" because it requires translating between two different linguistic and conceptual worlds, separated by centuries. Moreover, the process also "privileges supposedly stable modern categories."[30] The assumption that "syphilis" *was* "the foul disease" is a prime example of what Harley criticizes. Therefore, throughout this study I will rely on the terms current in the seventeenth and eighteenth century. Patients will be described as "poxed" or "foul," not as syphilitic. I have considered "venereal patients" all those diagnosed as "poxed," "foul," "venereal," or "clapped" as well as all those who entered identifiable foul wards or the Lock Hospital. This identification is never meant to imply that one suffered from any particular condition. In the pages that follow, reference to individual "venereal patients" merely means that they considered themselves infected with the pox or that their contemporaries so diagnosed them.

So in some ways the pox was imaginary; it was an invented disease-concept. But in other ways it was all too real. Those diagnosed "foul" were often extremely sick people, clearly suffering, and sometimes battling for their very lives. To those fighting illness there was nothing imaginary about the symptoms that ravaged their bodies. Acknowledging that the foul disease is an historically relative disease-concept does not preclude us from discussing the prevailing symptomology or indeed patients' experiences of those symptoms. In the chapters that follow much attention will be paid to the social and cultural meaning invested in venereal disease, which frequently touched on issues of shame, privacy, and stigma. However, focusing on the cultural constructions of the pox is not meant to detract from the very real corporeal struggle that patients often faced. Patients diagnosed foul tended to suffer under particular, often gruesome, symptoms, and any discussion of patients' experiences of the foul disease needs to begin by acknowledging how patients experienced the physical symptoms that signaled they were "poxed."

For starters, how did one come to know, or more precisely, believe, that they had the pox in this period? James Boswell's diary is one of the more infamous descriptions of the self-discovery of venereal infection in the period. For Boswell it was the burning sensation when urinating and the urethral discharge that first signaled his infection. After an evening out on January 19, 1763 Boswell wrote, "When I got home, though, then came sorrow. Too, too, plain was Signor Gonorrhoea. Yet I could scarce believe it. . . ." When he awoke the next morning Boswell complained of the "damned twinges, that scalding heat, and that deep-tinged loathsome matter [which] are the strongest proofs of an infection."[31] Such symptoms seem to have been common among patients' early complaints. Burning when urinating or the presence of a discharge were among the most frequently cited initial symptoms. Court testimony of rape victims who accused their assailants of infecting them often related experiences of noticing a discharge on their linens as the first sign that something might be wrong.[32] One must keep in mind that doctors in this period termed almost any genital discharge "a gonorrhea," frequently using the term in a generic sense to describe what they considered the first stage of the disease. (Although they would at times specify what they considered "non-venereal" discharges, like the "flour albus" or "the whites" in women.[33])

Casebooks support that patients reported some kind of genital symptom far more frequently than other symptoms in the earliest stages. John Pearson, surgeon to the Lock Hospital, kept a casebook of 86 female patients; he recorded the initial symptoms that they reported first noticing, as well as their current symptoms when they entered the hospital.[34] Of these 86 patients, 73 listed their initial symptoms. All but three cited "gonorrhea." That this complaint is cited so frequently by women is especially significant because some scholars have suggested that women with gonorrhea may have remained less aware of their infection,[35] and thus would have been less likely than men to cite gonorrhea as the first signal of their infection.

Some patients listed more than one initial symptom; the remaining initial symptoms listed in Pearson's casebook give a rough early symptomology as recognized by the patients themselves. A variety of dermatological manifestations were also listed, confirming the useful description of symptoms offered by Phillip Wilson.[36] Fifteen women complained of "buboes," ten of "chancres," eight of "ulcers," three of "eruptions," three of "blotches," one of "a swelling," and one of "spots." Some of these may have been genital ulcers, but Pearson too rarely gives the location of these sores. On only five instances did he specify inflammation of the labia. While most sufferers first took notice of genital symptoms, some patients also complained of such sores on other parts of their body at an early point.[37]

Contemporaries recognized a progression of the disease and a changing symptomology that usually followed. In the most basic terms, doctors believed that the disease proceeded in two stages: first a clap, and then a pox. A clap represented the first stage, the genital symptoms: localized sores, pain when urinating, and usually a discharge. Doctors commonly proposed that they could treat the disease more easily at this early stage, and warned that if patients did not act quickly the disease would progress to the next stage, the pox. Essentially, doctors conveyed a picture of the disease's "poison" entering the body in the genitals (or in the mouths or nipples in cases of infant-nurse transmission) and creeping slowly inward. During the initial stage of infection the disease stayed in the genital region, but if it were not eradicated while localized it would move inward, infest the internal organs and eventually the entire body. This progression from clap to pox became a central theme in medical attacks on quackery, as doctors lamented that poorly treated claps evolved into much more serious pox cases. In 1693 Charles Peter warned his patients this way: "for as the Whore gives the Clap, here the Quack gives them the Pox."[38]

Indeed, after a few weeks patients did complain of a wider range of non-genital symptoms. The patients in Pearson's casebook entered the Lock Hospital on average about twenty-six weeks after they reported first noticing symptoms. They still complained of gonorrhea and other genital symptoms. However, they now also complained of other symptoms, including some of the classic symptoms of secondary syphilis. For example, many suffered from pains in their long bones, especially their legs.[39] Women like Ann Cook complained of a "swelled tibia." Patients seem to have found these pains especially bad at night, something often mentioned in contemporary treatises.[40] For example, a patient named Phoebe Ballard complained of "nocturnal pain in the thighs," while Elizabeth Knight suffered from "nocturnal pain in the chin bone." Such symptoms were not just painful, but could be debilitating and severely restrict mobility. When an eighteen year-old apprentice named Valentine testified in a larceny trial in 1766, he told of his limited mobility, describing himself as "lame." When the court asked "what is the nature of your lameness?" Valentine replied that he "had got the venereal disease" and that he usually only "hobbled along."[41]

Often patients who had carried the disease to this point also now had the symptoms for which the disease had been named, the pox.[42] Pearson noted that several patients had skin eruptions all over their bodies. He described Jane Vero as having "Scabby Blotch covering the back & upon thighs, [and] sides" in addition to a swelled right foot and another unlocated "ulcer." Poor Ann Miller believed she had carried the disease for about twenty-four weeks by the time she entered the hospital with "eruptions around her face arms & leg" to go with her persistent "gonorrhea." If patients had listed ulcers or sores among their initial symptoms, these had now often worsened over the course of the few months before they entered the hospital. Pearson frequently described initial sores as deteriorating; buboes have now "supported" or broken open, and formerly "inflamed" labia he now described as "tumified."

Graphic evidence survives to help show what the disease could look like in its advanced stages. Sixteenth-century woodcut depictions of pox sufferers often highlighted sufferer's "pox," their dermatological symptoms. Historians of syphilis have frequently relied on such images.[43] Later in the eighteenth century, Hogarth is one of the best-known artists who continued to depict the pox as a dermatological disease, the pox-mark serving in his moral tales to mark characters who slip from the path of righteousness, growing in direct proportion to the character's depravity.[44] But Hogarth's images, and sixteenth-century woodcuts, are akin to a kind of cartoon. Perhaps more useful for our purposes is a set of sketches commissioned by the surgeon at the Lock Hospital in the mid-nineteenth century.[45] Judging from the images it seems that the house surgeon ordered a series of sketches of patients to document symptomlogy and progression—either of the disease or of a cure. This seems the case because portraits of the same patients appear at earlier and later stages, their symptoms having worsened or receded, depending on which was sketched first. Still it is clear that the surgeon and artist posed the patients so as to highlight particular symptoms. Some are of genitals; some are taken from behind. Some have patients posing with their mouths open to expose their throats. Many of these images are full-body portraits depicting open sores covering large portions of the patient's body.

Figure 1 is such a sketch. The disrobed male patient is sketched from behind to illustrate the wide spread of skin eruptions all over his trunk. Figure 2 (which depicts the same man) and figure 3 illustrate how the pox could be disfiguring as well, scarring the face and scalp with large open sores. And as is well known, extreme cases witnessed the collapse of the nose.[46] Early diarists like Samuel Pepys noted such facial disfigurement as a particularly frightening aspect of the pox's wrath. Pepys called it a "miracle" to hear that an acquaintance had been cured "after it had come to an ulcer all over his face." On another occasion he reported hearing how Prince Ruppert suffered, "the horrible degree of the disease upon him, with its breaking out on his head."[47] "Foul" is a loaded term, at work on many levels in the seventeenth and eighteenth centuries. But as a mere description of the symptoms it works rather well. Ravaged, the flesh corrupted. The skin began to rot, and the body oozed noxious matter. Regardless of whether the patients depicted in these sketches had syphilis or some other condition that contemporaries called the pox, the kinds of sores borne by those diagnosed foul were deeply painful and frequently disfiguring. When considering the experiences of those diagnosed with this disease—before analyzing issues such as shame, the construction of what the pox signified, or how people strategized to cope with it, all of which we will explore in later chapters—it is worth remembering

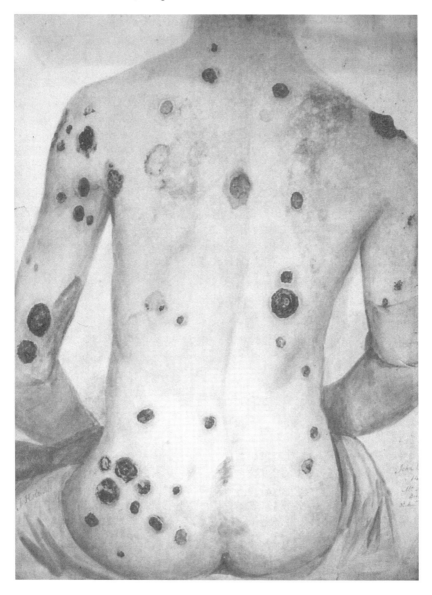

Fig. I: Watercolor Drawing of a London Lock Hospital Patient, J. Holt, 1849. Reproduced by Kind Permission of the President and Council of the Royal College of Surgeons of England.

Introduction: Institutions and Experiences 21

Fig. 2: Watercolor Drawing of a London Lock Hospital Patient, J. Holt, 1849. Reproduced by Kind Permission of the President and Council of the Royal College of Surgeons of England.

22 Venereal Disease, Hospitals and the Urban Poor

Fig. 3: Watercolor Drawing of a London Lock Hospital Patient, J. Holt, 1850. Reproduced by Kind Permission of the President and Council of the Royal College of Surgeons of England.

that many of the patients for whom the diagnosis foul meant more than just a short-lived urinary tract infection experienced a great deal of pain throughout their bodies.

Doctors utilized several types of mercury treatment to combat the disease. The most common, and odious, was mercurial salivation. Promoting salivation was merely one of many avenues by which early modern doctors tried to rid the body of infectious matter.[48] Humors or "poisons" could be evacuated via any form of excretion: sweat, urine, vomit, stool, menses, saliva, or age-old bleeding. Doctors at one time or another attempted to purge the pox by all of these routes. For whatever reason, salivation became the evacuation of choice with sweating and purging (i.e., through stool) also common. These treatments involved the application of mercury, usually through mercurial ointments rubbed into the skin, mercurial fumigation, or internal consumption of mercurial pills. Patients were then confined and kept warm, while physicians monitored the amount of saliva or sweat that they produced.[49] Given in significant doses, mercury causes patients to salivate profusely; patients often spit several pints per day. Thomas Sydenham, for example, insisted on "about four pints in the twenty-four hours."[50] This treatment tended to last at least four to six weeks.[51] The toxicity of such prolonged regimens of a heavy metal usually produced dreadful side effects. Patients frequently suffered internal pain, intense nausea, and permanent damage to their mouths including loss of teeth, gum damage, and the complete loss of the uvula (which may account for the Lock Hospital sketches highlighting patients mouths and throats).[52] Allegedly, patients undergoing mercurial salivation also suffered embarrassingly wretched breath.

Mercury treatment always accompanied other forms of preparatory and post-salivation care. These usually aimed either to assist the mercury in its task or to ease patients into and out of the excruciating process and mitigate some of its side effects. However, as Herman Boerhaave's influential treatise demonstrates, this could often make matters worse. He held that the venereal poison situated itself in human fat. Therefore, a successful cure demanded the elimination of patients' body fat by extreme dieting, bordering on starvation. He stated: "[W]e are to judge that the cure is not rightly performed, unless the person looks pale as death, and is so emaciated like a skeleton; and . . . thro' the whole course of his cure he has been careful to avoid fat food." He warned doctors that "unless these admonitions be regarded . . . you'll repent you have tortur'd him to so little purpose."[53]

Though this rough treatment earned trust, it simultaneously instilled fear. This is hardly surprising when one considers the difficult conditions that doctors readily admitted. Consider Nicholas Robinson's description of the procedure.

> Under a salivation [the patient] is ever dejected, low spirited and his stomach quite off him. . . . In a salivation he can scarce ever close his eyes but the noisom slaver glides down into his stomach, makes him sick, apt to reach [wretch]; and subject to puke; so that often when he awakes, he finds his stomach disturbed; and cannot be easy till he has disburthened his stomach of that load of saliva, that was swallowed during his sleep. . . . Under a salivation palsies, numbness, deafness, contractions of the jaws, apoplexies, are often the unhappy consequences. . . .[54]

At points court records capture the details of those forced to testify while undergoing the procedure. Thomas Cinnamon dragged himself into court in 1719, "appearing . . . in a very weak and helpless Condition as if not able to walk, nor help himself, and with a Rug hanging over his Shoulders, under pretence of being in a Salivation, and his Tongue so swelled he could not speak; and being arraigned, [he] supported himself upon, the bar."[55] The court did not believe Cinnamon, assuming that these were antics aimed at excusing him from testifying. Even so, that salivation could provide one with such an excuse, feigned or real, suggests that its debilitating effects were widely recognized. Robinson admitted: "I must acknowledge, I am at a Loss to discover the true reason why the patient is so readily brought to submit to a salivation."[56]

In fact, many patients tried to avoid salivation. Some simply refused the procedure altogether. Such was the case reported in 1710 of a woman whose symptoms returned following her first salivation. When her surgeon advised a second course of mercury the suggestion "so affrighted her, the last she underwent being fresh in her Memory, that she resolved the contrary taking her leave of him, said she would rather chuse to dye."[57] Generally, contemporaries believed that mercury could eventually work. But any less gruesome alternative that promised a cure held great attraction throughout this period. One man put it in plain terms in 1721 when he claimed "he would sooner lose his Nose than take Physick."[58]

So the pox and its treatment could involve a grave and painful corporeal struggle. But just how bad was the experience of the disease and its "cure"? Some rich material contained in coroners' inquests offers particularly useful insight on some of the more dire experiences of those who suffered under the foul disease. It demonstrates just how serious, indeed how dismal, this diagnosis could be in this period. Despite doctors' continuing confidence that mercury could cure the disease, contemporaries recognized that the pox could kill, as it did a soldier named William Urin, who succumb to it on December 26, 1771. The coroner's jury reported that William had "for a Long Time laboured under a Grievous Disease of the Body, to Wit, the Venerial [sic] Disease," and ruled the pox as the cause of death. Describing Urin's last days, George Sparrow, a domestic servant and one of Urin's neighbours, testified that he found Urin in a poor state of health on December 9th, claiming that William was bedridden and had trouble communicating. He "could not speak plain," said Sparrow, who reported that William was "very bad [having] been afflicted with the foul disease for some time." Urin reportedly told Sparrow that he had seen a doctor some time ago, and he feared that he must undergo salivation. Sparrow's employers, the Becketts, returned to check up on Urin two weeks later. By then they had to have a parish beadle break down the door because William failed to answer their calls. Inside they found him dead in his bed.

Sparrow described that William's illness ate him away to almost nothing, reporting that William appeared "very much wasted, having very little flesh upon his bones." William's poverty did not help matters. Like many of the patients that we will see in coming chapters, his poor finances complicated his poor health. On November 3, William had written to his regiment explaining that his illness prevented him from working. He complained that he had not been paid since October 14, and requested to be discharged "for he is not fit to go to the Regiment." He wrote to an acquaintance, Lewis Watters, asking for money but to no avail. In his deposition, neighbor John Beckett concurred that William's poverty contributed to his deteriorating health, reporting that Urin was starving. Beckett offered to arrange to send him to the parish, but he again cited Urin's inability to communicate in his weakened state, claiming that William "made him no answer." Beckett gave his opinion to the jury that "the Disease the Deced. had and the want of necessaries were the occasion of his Death."[59]

In William's case it seems he was unable to conquer the disease, spending his last days in bed wasting away. So the pox could kill. But would anyone really rather "chuse to dye" than face the disease and its treatment, as the woman cited a moment ago claimed? Was it merely hyperbole when George Robertson, testifying at his trial for theft and described as "very ill with the foul disease," claimed "I wished I had been dead" because he had neither friends nor money to help him in his illness?[60] Perhaps not. For there is strong evidence that scholars must take this reaction to the pox and salivation very seriously indeed. Some patients did not wait to waste away like William Urin. Instead, they chose to take their own lives before the disease did it for them. When considering how patients reacted to the pox it is revealing to explore this the most tragic of responses.

The Westminster Coroner's Inquests provide rather dramatic evidence that in a surprising number of instances patients considered their suffering so intense, their treatment so painful, or their prospects for renewed health so dim, that they took matters into their own hands and ended their suffering through suicide. One such case was that of Richard Miller, who, like William Urin, appears to have faced a particularly difficult illness-experience due to his poverty. Deponent Mary Roberts had known him for three years. She described Richard as "greatly distressed" claiming that he "had not money to pay for a Lodging." Nor, it seems, could he pay for treatment. A domestic servant named John Smith testified that he had seen Richard the night before his death in an alehouse. They spoke, but Richard was too sick to drink anything. Smith reported that Richard was scheduled to enter the Lock Hospital to begin salivation the next day. Perhaps he feared the notorious operation that Nicholas Robinson had described as so gruesome, or perhaps he despaired that the treatment would not work. He was far from optimistic, to say the least. Smith described Richard as "melancholy and uneasy on account of his Distemper" on his last night alive. The following day, rather than report to the Lock Hospital, Miller hanged himself in a barn.[61]

If Miller indeed feared salivation, it was with some good reason. Soldier John Jackson had twice endured the procedure in a military hospital. Ann Turfrey, the hospital matron, testified that he had left the hospital about a fortnight earlier and seemed severely affected by the mercury. Robert Wilson, a sergeant from whom Jackson rented a room, concurred that Jackson emerged from the hospital in bad shape, both physically and emotionally. He reported that upon leaving the hospital Jackson physically "looked worse than he used to do." Moreover, he testified that Jackson now seemed dejected; he was "low spirited and declined all conversation." Jackson shot himself at home on the morning of March 31, 1771, roughly two weeks after leaving the hospital.[62]

Mary Elson appears to have endured her treatment no better. In fact, she was right in the midst of salivation in the workhouse of the parish of St. Clement's Dane when she leaped to her death from a third story window. Both house surgeon John Andree, and nurse Mary Jones depicted Elson's mood as frequently gloomy; both described her as "very sullen and melancholy" throughout her mercury treatment.[63] A man named David Hughes similarly jumped to his death, on May 5, 1767. Hughes' co-lodger James Wear testified that David had been in particularly bad health from the pox. He had been under an apothecary's treatment for some time, with little success. His strength and mobility had diminished severely. Near the end Hughes was unable even to climb stairs. Two men had to carry him up three flights to his bedroom on his last night alive. Witnesses on the street below reported that Hughes went in and out of his third story window several times before he sat down on the ledge for a few minutes and finally pushed himself off, tumbling down to the pavement. James Sharp's story is just as sad. He had suffered from the foul disease for some time. At the end of his life a woman named Elizabeth Turner nursed him in his home. She described his physical condition as "very bad." She also described him as emotionally dejected. On the morning of May 27, 1789, Sharp ended his struggle with the disease by drawing a razor across his own throat.[64]

These are just some of the suicides linked to the pox that emerge from the Westminster Coroner's Inquests. We will explore others in later chapters. At the outset, they surely raise many more questions than they can ever answer. There can be no doubt that each case represents complex event that no historian can hope to decipher completely more than two centuries later, especially on such fragmentary evidence. Still, it is striking that so many venereal patients reached the same desperate conclusion when they faced later stages of the disease. Among the factors that have been identified as contributing to early modern suicide, two bear directly on the patients of this study—poverty and shame.[65] I hope that the twin effects of those factors on venereal patients' experiences will become clearer in coming chapters.

However, at the outset we should add to those factors the overall impact of physical illness itself, because further review of coroner's inquests reveals no dearth of instances in which other sick people similarly chose to end their suffering by taking their own lives.[66] For example, Mary Simpson hanged herself after she was turned out of St. George's Hospital and declared incurable. Deponents reported that a woman named Mary Harris had been ill and had complained of constant pains in her head over five long years before she jumped to her death from a window. Charles the Duke of Bolton, who had fought long bouts with illness for several years, suffered from particularly bad internal pains and a fever during the final month of his life before he shot himself. Just a review of a single year, 1765, uncovers eight instances in which the testimony in cases of suicide record that the deceased complained of physical pain or illness immediately before they killed themselves.[67] Further work on the complex relationship between illness and suicide would make an invaluable contribution. But the evidence that illness generally could mark a contributing factor in early modern suicide suggests that the sociocultural adversity specific to the pox, which we will explore in coming chapters, only contributed to the more general despair that could beset those suffering from serious illness in this period.

That venereal patients were prevalent among the early modern ill who reached such a desperate decision may begin to give some indication of just how serious this diagnosis was in the eighteenth century. While no one can know just why someone like John Jackson or Mary Elson killed themselves, can there be any doubt that their illness had something to do with it? By all accounts they faced a disease that brought painful symptoms and painful treatment. The records certainly suggest that they themselves, and those who knew them, believed that the disease was at the heart of their stories, because they highlighted the pox when trying to explain their deaths. These peoples' actions, while difficult for the historian to decode, betray stark hope of a cure. For those who had already endured mercury only to see their symptoms persist or return, the optimism may have been especially dim.

Of course, these experiences of the pox exist at the far end of a spectrum that also includes Boswell's bouts with the clap, which seem by contrast quite tame. Young Boswell was confined to his room for five weeks. He went through periodic spells of anger, fear, and depression. He swore at his bad fortune—"What a cursed thing this is! What a miserable creature I am!" But he followed his surgeon's directions and emerged from his bedchamber seemingly none the worse for wear.[68] Still, there can be no denying that for others not so fortunate as to harbor a manageable urinary tract infection, the pox could be met with fear, anxiety, and dread. I hope that as we explore the options available to the poxed and how the urban poor experienced this disease, we may learn a bit more about why some of the people diagnosed foul came to believe that their prospects were so very dire. It will be important to remember that those illness-experiences, which will include the strategies for accessing medical care, the intricacies of the doctor/patient relationship, the realities of hospital life and more, were first and foremost experiences of pain. The symptoms that fell under the umbrella diagnosis of the foul disease, and the main therapeutic responses were, by all accounts, deeply painful to endure. The patients we will witness throughout the ensuing pages, therefore, often faced a palpable, ongoing, and grim struggle.

1

THE FOUL DISEASE, PRIVACY, AND THE MEDICAL MARKETPLACE

People of means did not lack for medical options when they contracted the "foul disease" in early modern London. This book is not about people of means. However, an understanding of the experience of poverty and the pox demands a point of comparison. For that reason we should first consider the place of the foul disease within the oft-described medical marketplace of Stuart and Georgian London. If one had money to spend, what were their options? There is no doubt that there were plenty of customers for anyone who could relieve the suffering just described. There was a rich living to be made in the trade in pox-cures, and the powerful economic force of demand indeed spurred vibrant activity in London's medical market. Few corners of the early modern medical market could challenge VD care for the range of therapies or providers. The range of the market response to the pox was driven, not just by the widespread incidence of the disease, but also because of the unique nature of the pox. Patients diagnosed foul tended to have particular demands that others who suffered from more banal ailments did not. Attempts to provide this huge customer base with services for these unique demands resulted in a rich range of medical services. Thus there were a wide array of options for the poxed, but they were frequently not cheap. And there is the rub. For early modern London had a two-tiered medical system, to use modern parlance; private fee-based care of the market existed alongside publicly funded care for the poor. The bulk of this book will explore the forms that such public care took. But to make the most sense of it we must first explore VD care in the private sector, and this means first studying care for middling and well-off Londoners.[1]

In addition to the desire to ease their physical suffering, which all medical customers sought, venereal patients also made unique demands springing from the socio-cultural reactions specific to the pox, namely its stigma. There has been some disagreement regarding the relative stigma attached to the disease over the long period. Some scholars have asserted that early modern sensibilities regarded venereal infection with cavalier light-heartedness, particularly during the Restoration.[2] The evidence from medical texts does not support this view. Shame, not frivolity, dominated early modern medical discussions of the circumstances surrounding venereal disease and its cure.

It is worth noting the prevalence of the term "shame" in medical discussions throughout the period. In 1660, for example, John Wynell commented on people's reluctance to seek treatment, claiming "I was led to believe, that either ignorance or shmefac'dnesse to discover it, made [patients] carry it about them too long."[3] Restoration practitioners highlighted patients' "shame" and "tenderness of reputation," Charles Peter lamenting in 1673 that "People . . . either scorn to own that they have it, or those who being ashamed to confess their condition, rather venture to undergo it, than wisely to seek for help."[4] The dermatological manifestations of the pox illustrated earlier were well known and recognized. This made it difficult for foul patients to hide their condition from public view, try as they might. In 1670 John Archer noted the effect of such visible symptoms, calling the pox "hell on earth for man's punishment, bringing at once with a sting of discontent, a cursed pain with loathsome and shameful symptomes."[5] Restoration medical writers stressed the problems caused by the social stigmas attached to the disease, especially people's reluctance to seek treatment. Playwrights may have employed the pox in satires, but clearly doctors, especially in England,[6] did not consider this disease in comic terms.

Neither did laymen, as evidenced by Samuel Pepys. Pepys came to suspect that his brother, Thomas, might be infected in March 1664. As Thomas lay delirious and near death Pepys found himself consumed with anxieties about his brother's health (as well as his debts, which he might have to assume). But he also feared public embarrassment for his family. Commenting on those who discussed his brother's condition too freely in mixed company, Pepys proclaimed that "the shame of this very thing, I confess, troubles me as much as anything."[7] When Thomas's doctor declared him free of the disease, Pepys expressed great relief that Thomas, and by extension his family, would be free from "that reproach which was spread against him." Despite his seeming relief, Pepys still insisted on inspecting his delirious and dying brother's genitals to verify for himself that "he was as clear as ever he was born."[8] Analyzing Pepys's reaction, Raymond Anselment noted how Pepys was struck deeply by the personal shame he might face for any association with the disease.[9]

That reading is undoubtedly right. For Pepys recorded that he even threatened his brother's doctor if he dared breathe a word about Thomas's supposed pox. "I threatened him that I would have satisfaction if I heard any more such discourse."[10] Pepys's reaction may have been typical. John Graunt related that families often bribed the officers who recorded deaths for the Bills of Mortality to register a different diagnosis when a loved one died of the pox, a reaction that Claude Quétel notes in France as well.[11] Pepys's account demonstrates how shame and satire might indeed be opposite sides of the same coin. For Pepys himself often enjoyed poking fun at *others* who contracted the pox, as he did once in a conversation with one Mrs. Turner, criticizing "the baseness of Sir W. Penn and the sluttishness of his family—and how the world doth suspect that his son Lowther, who is sick of a sore mouth, hath got the pox."[12] However, Pepys discovered that when it struck closer to home everything changed. Anselment makes a convincing case that the seventeenth-century satire is best understood not as evidence of flip or light-hearted attitudes towards the disease, but as a device for confronting and relieving the quite profound feelings of fear, guilt, and shame that accompanied it.[13]

A look at a piece of English satire allows us to see just how biting some representations of pox sufferers might be. For example, the ordeal of mercurial treatment made potent material for satirists.[14] One suggestive example touching on the side effects of salivation comes from the anonymously penned eighteenth-century poem "The Phlegm Pot," which highlights the torments of patients suffering through mercury.

> A world of joys, I'll now complete, a winters frost and summers heat,
> With Kibey heels or sweaty feet, but whoring brings the pox;
> Each brick each blood that shines so fine, each showy fop or grave divine
> May brighter than the day light shine, yet have the same mischance.
>
> There's many a fine and flashy beau that's taking pills for what you know.
> They are all beshit from top to toe, tho' they strut St James's Mall;
> There's many a lady at Vauxhall to go to stool must have a call,
> Thus up in a corner she lets it fall whence it stinks as strong as hell.
>
> And when this way you have done your best, and your rotten cups wont let you rest
> At St Thomas's hospital you are carest, where you spit the phlegm pot ore';
> And as the phlegm begins to rise, and the fever works out at your eyes
> And the death stool from your arse gut flies, oh, then you feel no more.[15]

Here the poet moralizes at the expense of those enduring mercury therapy. He gives the impression in the first stanza that indeed polite Londoners might try to conceal their treatment of the disease, attending social events publicly, while purging themselves with mercurials privately. The poet offers his grim prediction for the success of these attempts at self-remedy, offering a marked departure from the frequently optimistic, self-congratulatory boasts that color many doctors' treatises. In a three-stanza fall from grace such dandies are driven into poverty and despair in one of the lowly hospitals that we will explore in the coming chapters. Notably, the author shifts to the second person in the final stanza, addressing his moral warning directly at the reader. We can assume that those who came across this kind of rough song had a good laugh and speculated that such folk as those depicted here got what they deserved. Yet simultaneously here was a warning of what one might have in store if they came to need this treatment. They would suffer. And to make matters worse they, too, might find themselves the target of this kind of ridicule.

The song also begins to show that the attitudes that Anselment demonstrated for the seventeenth century remained prominent into the eighteenth century. Like his predecessors half a century earlier surgeon Joseph Cam complained that patients were often "so hinder'd by a clownish Shame" that they avoided pursuing medical treatment.[16] Physician Solomon Sawrey expressed compassion for his patients who tried to hide their symptoms: "fearing that the knowledge of them might injure their future prospects, they mostly conceal them. If they find their complaints durable, they are often more anxious not to be disgraced by them."[17] John Burrows showed that Pepys's anxieties remained strong a century later, describing the horrors of the symptoms "not only on account of the[ir] danger, but the shame with which they are attended."[18] Finally, we also find discussions of shame within case studies. In these the conditions of patients come through and we begin to see the effects that the pox could have on common peoples' lives. Consider the twenty-eight year old woman Joseph Plenck described in 1768. She was a poor woman who sold poultry whom he treated out of charity. She was covered in pustules, and likely resembled the patients we encountered in the Lock Hospital sketches. Plenck indicated how her appearance affected the way she presented herself—"being very much ashamed, she covered her face over with a cloth." "Nobody would buy her poultry from her," Plenck continued, "on account of her being of a nasty appearance."[19] Sawrey claimed that "humanity shudders to view" such advanced cases of the disease.[20] The visible symptoms like the ulcers, the collapsed nose, or the baldness associated with the disease were well known, associated in the popular mind with such endeavors as prostitution, and connected all those who bore them to the dreaded French Disease.[21] The extreme disfiguring capabilities of the pox contributed greatly to the personal effects of the disease, which went well beyond the physical and scarred on a more intimate level. This was especially the case considering how the ideal of the polite body came increasingly to be symbolized in this period by clean expanses of white skin, which the propertied classes strove to obtain through ever more application of powder and cosmetics. With its pustules, ulcers and scabs, the pox stoods as the antithesis to this unblemished ideal.[22] Even the contemporary names for the disease—the "foul disease" or the "secret disease"—indicate its stigma as well as patients' desire to conceal their infection. This factor helped to make the foul disease unique among early modern maladies. This is not to say that no other early modern medical condition might be bring some measure of embarrassment. But finding a contemporary condition worthy of comparison, in which shame is noted so regularly as a common element of the illness-experience, proves difficult.[23] The stigma attached to the disease meant that privacy would be a central concern of venereal patients.

However, we must bear in mind that concepts such as shame or privacy are not fixed, but culturally relative and shifting. Indeed, a range of important studies have placed the issue of the changing relationship between the public and the private at the very heart of our understanding of the seventeenth and eighteenth century.[24] A variety of explorations of the issue have suggested that the period witnessed a turning point in the history of the relationship between public and private, even if these authorities have not entirely agreed on how exactly these terms are best employed. To a great degree the debates focus on Habermas's view of the Enlightenment with the birth of the public sphere at its center.[25] But other scholars, such as Phillipe Ariès, have tended to conceptualize the public and the private as shifting along a slightly different axis that explores the individual's relation to the community.[26] Here issues like personal intimacy are explored through such features as diary-keeping, personal correspondence, architecture and family life, to argue for a growth in individualism in the eighteenth century, which brought with it heightened concerns about privacy and intimacy at the personal level.

In articulating his notion of increasing sociability, Ariès proclaims a "new modesty" for the eighteenth century. His focus on modesty, intimacy, and manners is not unlike many elements of Nobert Elias's depiction of the civilizing process, which placed modesty and personal manners at the heart of his major theoretical framework, upon which Ariès draws.[27] Elias describes an advancing "embarrassment threshold" or "shame frontier" in the early modern period, as self-restraint gradually became a central feature of polite culture. In a word, shame may have been on the rise as our period progressed. Elias explores commentary on manners, bodily functions, and sexuality to argue for the growing value put on controlling individual drives as a component of increasingly influential codes of politeness, which demanded that issues like personal sexuality and bodily hygiene be kept out of public view, largely through the mechanism of shame.[28]

Debates about public and private are in full swing and have yet to be settled, but a constant throughout much of this literature is that the late seventeenth and eighteenth century may have witnessed a shift regarding issues of privacy. The implications for intimacy and individual privacy that these lines of scholarship suggest may bear directly on the developments under examination here. For if Ariès and Elias are correct, we may well witness an increasing tension surrounding public revelations of venereal infection. Indeed, there is reason to believe that we may witness an increasing demand for medical privacy in our period, especially as it related to the pox. It is revealing that we find discussions of the doctor-patient relationship and of venereal disease itself woven into both Ariès's and Elias's discussions of early modern discretion. Ariès links his new modesty directly to the doctor-patient exchange, citing increased tension surrounding male surgeons' examinations of female patients as a prime example of what he means,[29] while Elias spotlighted the pejoratives attached to venereal infection as one of his many examples of the advancing shame attached to sexual transgression.[30]

This advancing "shame threshold," or "new modesty," probably only intensified what was already an established element of the experience of the pox from an early date. Certainly, Anselment's, Schleiner's, and Fabricius's studies go a long way towards demonstrating the entrenched nature of the disease-stigma well before the eighteenth century. Moreover the evidence from the considerable work done on sexual slander supports the point. For example, one finds reports of people invoking venereal infection to augment insults from a very early date. Court records show that neighbors often called one another not just a "whore," but a "pocky" whore, employing the kind of pollution taboos that Mary Douglas explored so long ago.[31] The pox was rich material for insults because it linked to the common sexual forms of slander further accusations of impurity. It hit on two levels, accusing simultaneously of sexual deviancy and of filth, making it a rich pejorative for public shaming. Thus one finds such evocative examples in sixteenth-century court records as the woman who proclaimed that her neighbor "berth feyre [fire] in her ars for every man to light his candyll att."[32]

These share similarities with court records from more than a century later which continue to show how the slur of venereal infection could be used as a character assassination to undermine credibility in court. In a 1731 bigamy case, Andrew Robinson testified as a character witness and undercut Mary Moore thus: "I had the Misfortune to be acquainted with this base Woman, Mary Moore (and it has been many a Man's Misfortune, as well as mine) and she has given me the foul Disease twice." While in 1735 Mary Proctor similarly testified against the claimant of a rape, Sarah Evans: "I have heard that she is a wicked vile base lying Girl . . . and that a Barber gave her the Pox."[33]

The foul disease was pejorative from virtually the moment it walked onto the European stage in the 1490s. This is evident in one of the most commonly noted phenomena in the pox's history, the tendency to use it to smear entire nations and other peoples, whether the French, the Spanish, the Jews, or the Amerindians.[34] The coincidence of broad national smearing and local personal smearing indicates how venereal disease functioned to slander on both the macro and micro-levels. It was a widely used and powerful slur from its earliest points, bringing with it the defensive reaction of all who faced it. Realizing that "foul" was a weighty insult heaped on any and all enemies caused people of all stripes to strive to avoid the designation. The advancing embarrassment threshold that Ariès and Elias describe most likely served only to exaggerate that reaction. However, even if the divide between eighteenth-century constructions of the pox and earlier allegedly "light" attitudes may have been drawn slightly too starkly, scholars who characterize eighteenth-century attitudes as particularly intense find support in the general picture of cultural drift that Ariès and Elias suggest.

Regardless, privacy remains a slippery issue for the scholar. It demands, for example, that one consider the kinds of expectations for privacy that members of former cultures might have had, a question about which debate continues.[35] Moreover, one needs to refrain from applying modern notions of privacy retroactively. When thinking about how to approach the difficult task of studying privacy, Linda Pollack has offered a useful way forward: "We should be investigating what people themselves wanted to keep private as opposed to what historians think they should have been keeping private."³⁶ Using such a guideline, the evidence is strong that throughout the seventeenth and eighteenth century people sought to keep certain delicate medical details—especially details related to possible sexual infection—to themselves. Pepys supplies another fruitful example that demonstrates the utility of Pollack's approach (and which incidentally bears clear likeness to reactions recorded a century later, suggesting again notable similarities between seventeenth- and eighteenth-century responses³⁷). In November 1663 Pepys's wife, Elizabeth, suffered from a vaginal infection, for which she consulted a doctor named Hollyard. Even though she had no reason to believe that the infection was a symptom of the pox,³⁸ she worried that others might make the assumption. Elizabeth fretted that gossip might spread that she was foul. She refused to allow a nurse to tend to her, being "loath to give occasion of discourse concerning it." In the end Samuel and Elizabeth convinced the doctor to prescribe a treatment that Elizabeth's maid could apply under the guise that "it be only for the piles." In other words, they conspired, husband and wife, doctor and patient, to obscure the nature of her medical details, even though her condition was "nothing but what is very honest."³⁹ Facing the possible stigma of the pox, undeserved though it may have been, they maneuvered to guard Elizabeth's medical privacy.

While the lion's share of work on privacy in the period has concerned elite folk like Pepys, we should not make the mistake to assume that the well-heeled had a monopoly on reputation.⁴⁰ The pox was a slur in plebeian culture, too.⁴¹ When Susannah Carter accused Elizabeth Lewis of transmitting disease in 1743, Lewis responded "You B---h, I never gave the Carman the Clap"⁴² The accusation started a fight that resulted in Lewis's death and Carter's conviction for murder. It is well known that "pox" itself had long been a common swear word among plebeian folk. "What a-pox do ye do out of your own Room?" demanded Mary Jones's neighbor when she discovered that money Jones was holding for her had been stolen: "Ah, pox o' your Pad-lock! the Money's all gone."⁴³

Much more importantly, quite tangible ramifications could follow disclosure of infection for plebeian Londoners. Just like slanderous accusations about sexual probity, speculation that someone was poxed might well cost them their job,⁴⁴ as it did apprentice William Frazier, who sought a certificate of clean health from an apothecary in order to convince his master that the servant maid accusing him of infection was lying. William poached a few shillings and snuck away in search of an apothecary, only to be caught and end up in the Old Bailey facing a felony theft charge. The trial transcript described his defense thus: "That the Reason of his going away so suddenly was, that the Servant Maid had told his Master that he had the foul Disease, and therefore he would turn him away, unless he brought a Certificate from the Apothecary, that he had no such Distemper."[45] Even the hospitals that agreed to treat the disease did not make the most forgiving employers. Elizabeth Staling, a nurse at St. Bartholomew's, found herself fired when the surgeon discovered that what she claimed was a fever was actually the pox. Not only did she lose her job, but the hospital refused to treat her as a patient.[46] The evidence suggests that plebeian Londoners had good reason to share their elite neighbor's anxiety about public exposure of sexual infection—or, in both wealthy Elizabeth Pepys's and poor William Frazier's cases, the mere *speculation* of infection.

I am not trying to suggest that plebeian sexual culture mirrored patrician sexual culture. Rather I agree with Anna Clark that while plebeians may not have accepted the notions of bourgeois respectability that elite reformers tried to impose on them, they certainly recognized the ramifications of not *appearing* to conform to those ideals.[47] Even if they did not internalize quite the same sense of shame as their wealthy neighbors, they understood quite well the consequences they might face if their infection became public knowledge. It is useful here to consider plebeian women's reaction to rape or out-of-wedlock pregnancy. It was not uncommon for working-class women similarly to conceal the details of these incidents. Randolph Trumbach's exploration of the topic illustrates how the similar damage done to working women's sexual reputation frequently cost them their jobs.[48] This was especially the case in domestic service, where, because they were members of bourgeois households, employers paid great heed to the ramifications of their servants' reputations on their own—hence William Frazier's trouble with his master. Domestic servant Mary McGowen found herself out of work when her pregnancy became known, even though her mistress admitted that she was an exemplary worker.[49] Such pressures could force working women to hush up sexual assaults, to conceal their pregnancies for as long as possible, or in some cases give up their children to the Foundling Hospital in order to maintain appearances of chastity to keep their jobs. As Laura Gowing succinctly put it, "[t]he maintenance of a single woman's reputation had a concrete economic importance: survival in the parish, in service or out of it, depended on keeping a good name as long as possible."[50] We must add venereal infection to pregnancy and sexual assault on the long list of things that plebeian folk must have tried to conceal from their social betters, especially their employers, in the name of economic survival. And we should furthermore not overlook the importance of the marriage market. Even the suspicion of venereal taint could be enough to thwart young people's marriage prospects. So for the rich and for the poor, the pox could seriously endanger a reputation; and whether rich or poor, that damage could bring serious economic results. Despite class differences, Londoners of all ranks often exhibited great sensitivity about medical disclosure of sexual infection.

It was just such sensitivity that set venereal patients apart from other early modern medical consumers, and led to specific demands on their doctors. This affected medical services in a variety of ways. For example, it helped fuel the opposition to salivation. Of course, patients who resisted salivation did so because the procedure was, by all accounts, excruciating. However, other factors contributed. One of the main complaints about salivation was that it was conspicuous.[51] It was not a treatment that could be easily kept secret. Patients feared that by undergoing the procedure they risked exposing the nature of their illness to public scrutiny. This was so because mercury treatment tended to be long; it required confinement for a minimum of four to six weeks. Such a period of absence was not so easily explained to family, neighbors, or business associates.

Doctors recognized the concern. John Profily, for example, claimed that the "shame of a suspicious retirement" kept many patients from submitting to salivation. He reported that patients commonly reacted to the suggestion of a course of mercury thus: "[M]y employment requires my attendance almost every day, and I cannot absent myself five or six weeks on any consideration." He acknowledged that patients feared the physical pain of the treatment. But Profily depicted that some considered salivation's cost for their honor an even worse side effect: "[B]ut what I think is worse is that I should, by going under a course of

Salivation, lose my reputation by a suspicious absence. . . ."[52] Other medical writers stressed that such concern was of particular importance to wealthy patients, claims which overlooked the effect of the disease on plebeian honor. Regardless, Bouez de Sigogne's treatise, translated and published in London in 1724, described salivation as undesirable to well-off patients because

> The time it requires to perform it, is so long, and a patients [sic] withdrawing himself for the space of forty to fifty days during which time he must keep his chamber is so troublesome and often so suspicious, specially for persons whose stations oblige them to have a more than ordinary regard for their reputation.[53]

The market responded to such widely held concerns for discretion, assuring a constant supply of alternative treatments that promised to cure the pox without resorting to lengthy, painful, and potentially embarrassing salivation. A generation of scholarship has explored the medical scene through the prism of the marketplace, demonstrating how the force of patient-demand helped to craft the contours of early modern medicine.[54] That model continues to prove useful, for the particular demands unique to foul patients and the intensity of the competition to provide those services made VD care one of the most vibrant and competitive corners of the London medical scene.

Before charting further how such demand affected medical provision, it is worth considering just how widespread that demand was. The pox was simply far more rife than scholars have assumed. Margaret Pelling was undoubtedly right when she claimed that "venereal disease remains seriously underestimated" as a force in early modern England.[55] Evidence from doctors' publications and promotional literature begin to demonstrate the point. The sheer number of publications gives some indication. Consider that *The Short Title Catalogue* contains more than three hundred titles on venereal care printed in London between 1650 and 1800. Advertisements also abounded. Ads for pox cures littered the pages of periodicals such as the *Gentleman's Magazine* and literally dominated other forms of early modern medical advertising.[56] For example, within three collections of medical advertisements dated ca.1660–ca.1715 held in the British Library, which together total 512 advertisements,

284, or 55 percent, advertised treatment for venereal disease.[57] Such ads covered all aspects of early modern health and medicine, ranging from cure-all elixirs, to bone-setters, general practitioners, specialists, even dentistry and skin-care products. Yet more than half of all the practitioners who advertised in this way treated venereal disease. Such a predominance of printed material begins to suggest very high rates of infection needed to support what seems like something of a venereal industry at work in the city.

Hospital records confirm that suggestion beyond any doubt. Put simply, the foul disease was omnipresent in contemporary hospitals. In most years for which records survive that lone diagnosis accounted for more than 20 percent of all hospital patients, a phenomenon which coming chapters will explore in greater detail. Such figures indicate extremely widespread infection—or at least widespread *diagnosis*—throughout seventeenth- and eighteenth-century London. However we may account for it, it remains true that anyone living in the city would have perceived that the disease was all around them. Doctors certainly did. For the presence of such a large clientele could hardly be ignored. Swift movement to tap into this clientele instilled the market with some of the stiffest competition witnessed in the London medical scene.

Thus the stakes in the debates over salivation were extremely high. Challengers to traditional mercury therapy did not mince words. They denounced the practice as dangerous, ineffective, and unnecessary. And importantly, they appealed directly to patients' concerns for discretion. A host of doctors heralded their new therapies with great fanfare, although many of these "new" cures actually still employed mercury as the main ingredient. Many doctors criticized catharsis via the salivary glands and offered allegedly safer alternatives. Pierre Desault, for example, administered mercury but advocated stool evacuations,[58] while Sigogne advised combinations of evacuations via stool, perspiration, and urine.[59]

Such practices began to catch on in London and posed a significant threat, in part because they came from France.[60] Desualt and Sigogne were, in fact, typical of the eighteenth-century French movement away from salivation. Doctors at Montpellier devised a system of mercurial ointment rubs that did not raise salivation, which they trumpeted as a safer alternative.[61] This new system claimed that salivation acted too fast to be totally effective. According to Montpellier theorists mercury was swept out of the body before it had cleansed the poison "from the finer vessels or inmost Recesses of the Body."[62] Francois Chicoyneau, in a treatise detailing the Montpellier method published in English in 1723 and entitled *The Practice of Salivating Shewn to be of no Use or Efficacy in the Cure of Venereal Disease*, not only sharply criticized the London medical establishment but he invited patients to come to Montpellier to receive their cure. There is evidence that patients made the trip. Nicholas Robinson for example, made specific reference to foul Londoners who made the medical pilgrimage to Montpellier.[63]

Astute businessmen, the anti-salivationists stressed the benefits of alternative mercury regimens to London's mercantile class. They crafted their ad copy to the mercantile concerns about the lengthy period of debilitating confinement that salivation demanded. London merchants could not simply take four to six weeks off without seriously hurting their business as well as their reputation. Of course, as recent work on early modern "credit" has shown, the two were inextricably linked.[64] Sigogne clearly targeted mercantile Londoners when he specified those "whose stations oblige them to have a more than ordinary regard for their reputation."[65] Desault's translator pitched his attack on salivation even more directly to London's merchant class. He promised to "entirely secure the reputation of the patient" and went on to brag:

> you save him the shame of a suspicious absence, since he may do the business in his own house without any body's knowing it, and he may even receive visits without raising suspicion. His tongue and speech are free, he takes nourishment without pain and with pleasure. One may even keep in the dark persons of the profession.[66]

Merchants (and especially patients who could afford a trip to France for treatment) were not clients London doctors would give up without a fight. Surgeons Joseph Cam and Daniel Turner took up the cause of responding to the challenge and defending salivation. Turner was one of the most active defenders of orthodox practice in eighteenth-century London.[67] Turner challenged one Montpellier doctor, calling him "ignorant" and mocking him for offering "to save our People the Charge and Trouble of a Journey to Montpellier."[68] Joseph Cam responded to the initial publication of the Montpellier method with *The Practice of Salivating Vindicated: In answer to Dr. Willoughby's Translation of Mons. Chicoyneau's Pamphlet against mercurial salivations*. His counter-attack responded on many levels. First, Cam made the case for mercurial salivation, claiming that mercury is dangerous and therefore must evacuate the body as quickly as possible. "I say mercury must not lodge within, as soon as it has performed its office, it must be thrown out again."[69] But Cam's treatise also illustrates that the tension between Montpellier frictions and London salivation was both economic and even nationalistic in nature. When he came to the issue of patients traveling to France for treatment he challenged Chicoyneau:

> So that, sir, if our English travellers have no other views in repairing to Montpellier, than to be rubbed with six-penny worth of quicksilver, and take up with confinement; two or three hundred pounds given for these favors is such *bon marchee* as we generally have from the French; and then their practice (as well as their goods) ought to be reckon'd *contraband*.[70]

Despite their insistence on salivation's benefits, doctors like Turner had to admit that their patients resisted the operation for exactly the reasons that their competitors suggested. Turner recorded that patients frequently requested an alternative regimen because their "[a]ffairs . . . [did] not favour . . . such recess from Business."[71] Regardless, London salivationists were entrepreneurs, protecting their business from foreign trade. To them, salivation came increasingly to be seen as a distinctly English, and VD care thus deserves its own small place in the larger history of eighteenth-century Anglo-French competition. The Montpellier doctors continued to make inroads in the London market, converting an increasing number of London practitioners to the benefit of mercurial frictions.[72] With so much at stake, the debate raged on. However, Phillip Wilson's examination of the debate has shown that Turner and Cam were fighting a lost cause. Montpellier style frictions slowly but surely gained the upper hand over traditional salivation, and in the second half of the century, hardened proponents of salivation like Turner would come under attack as dangerous quacks.[73] Salivation would continue to be viewed as a therapy that ultimately worked, but it grew to be exceedingly unpopular, and alternatives prospered.

Nonmercurial cures also challenged orthodox salivation treatment. Some sort of vegetable matter provided the active base ingredient in many of the most common nonmercurial cures, usually pills or drops. Such remedies were popular from the early sixteenth century. Guaiac, a wood from the New World was by far the most popular alternative to mercury, although doctors also prescribed sasparilla, china-root, and a host of other herbs.[74] Marie E. McAllister has given the most complete analysis of this branch of medicine in the eighteenth century.[75] Her work on the "Vegetable Wars" examines the fierce competition between establishment doctors and proponents of vegetable remedies, as well as between various vegetable proponents themselves. McAllister's biographical sketch of one of these men, John Burrows, also further blurs the lines between establishment and fringe practitioners, demonstrating that Burrows came from a traditional medical background before embarking on his career selling "Velno's Vegetable Syrup."

Just as the Montpellier doctors had attacked salivation, so the proponents of vegetable-based alternative strategies also stressed the inconvenience and conspicuous nature of confinement, alongside the pain of salivation's side effects and suggestions that it just did not work. Such was the approach of John Leake, whose "Lisbon Diet Drink" proved to be one of the most widely promoted nonmercurial venereal cure of the eighteenth century.[76] These drugs were not cheap, as Boswell found out, spending half a guinea per bottle on Kennedy's Lisbon Diet Drink, a treatment that required patients to drink two bottles per day.[77] These practitioners relied on a variety of techniques to marshal legitimacy for their forms of alternative medicine. For example, some obtained royal patents to attract customers and create the impression of proven efficacy.[78] Others advertised the results of experiments they claimed to have performed.[79]

But it was their assurances of privacy that made the best marketing tool of all. As surgeon John Marten's partner I. F. Nicholson put it "the Distemper is Ignominious, [that] it requires secrecy in the Cure."[80] The aforementioned collections of medical advertisements in the British Library illustrate nicely how anti-salivationists trumpeted the benefits that their alternative therapies held for patients concerned about their reputations. Consider first the sheer number of pox ads that specifically mentioned medical privacy. Promises of "private" and "secret" treatment abound in their promotional literature; in one of the collections alone over 32 percent of all the advertisements for treatment of the pox specified discreet or confidential treatment, while not one of the remaining ads (which did not mention the pox) made a similar claim.[81]

Such advertisements did not merely mention confidentiality in passing. They often began their handbills with clear promises of privacy. In fact, the promise of discretion seems so important that doctors prominently situated it at the very top of their ads right next to other crucial selling points, namely that their therapies were reliable, affordable, and quick. Quite typical were ads like the following, which began "At the Golden-Ball in Princess-Street, near Stocks market you may have A Certain, Safe, and Private Cure for any Venereal Pox or Clap . . ." or, "At the Second house on the left hand in Peterborough Court in Fleet street You may have a private, cheap, and speedy cure for the Venereal Pox. . . ."[82] John Case was a particularly active marketer of his services; fifteen different handbills by him survive. Many begin "A most infallible, and Sure, Cheap, Secret, Safe, and Speedy Cure for a Clap. . . ."[83] Eight other handbills survive by a different practitioner that similarly begin, "A Perfect, Safe, and Private cure for the Clap, Pain, Heat, Scalding in making water, Running &c. . . ."[84]

Most commonly, anti-salivationists stressed that their treatment was brief and did not require confinement. Thus practitioners like the "chymical physician" named G. Dean plainly advertised a "Speedy and Absolute Cure for the French Pox, without Fluxing or Confinement," again linking the issue of confinement to reputation.[85] Targeting the mercantile class they often promised that patients would experience no "hindrance of business" while undergoing treatment. In fact, no fewer than fifty separate surviving handbills made that promise in those very terms. Quite common were assurances like the following from an anonymous Covent Garden practitioner, who

> has a Method that Cures the French Disease. . . . And for those that have lately been in the Garden of Venus, and have got a Clap or Running of the Reins, I promise to Cure in six or eight days without hindrance of Business.[86]

The handbills virtually never indicated anything substantive about the method of cure, so it is impossible to know which alternative therapy these practitioners advocated. Regardless, they all made the same point patently clear. They would treat the pox quickly and discretely, without resorting to salivation.

Such practitioners did not merely make vague promises of confidentiality; they employed several specific tactics to ensure their patients' discretion. First, doctors specializing in the foul disease frequently arranged the location and layout of their offices to allow patients to visit them without being noticed. As Porter noted, pox doctors often advertised that they resided in out of the way places and employed backdoor entrances for shy patients. Gilbert Anderson was one such practitioner who eased his patients' worries by assuring them "you may enter which way you please."[87] Similarly in the 1680s, an anonymous doctor who treated the clap in Bartholomew Close assured patients that "his House is so private, that no notice can be taken of you coming to him."[88] It also seems that doctors treating the pox kept late hours specifically to allow patients to come under cover of darkness. Such was the case at the home of the practitioner who informed patients that "There is a Lamp lighted at his door in the Evening, by which he may be readily found in the Night time without Enquiry."[89] At least one commentator suggested that some patients concealed their identity with masks when they called on doctors.[90]

Since so many doctors complained that embarrassment kept patients from going to doctors, some practitioners tried to render a face-to-face exchange unnecessary and advertised that patients might have the option to avoid an office visit altogether. To do this some practitioners utilized some of the new methods by which doctors began to market and distribute their medicines.[91] One way to avoid the necessity of coming to the doctor in person was to utilize messengers or the penny-post. In this way patients could describe their circumstances and symptoms through the anonymity of a letter and receive both diagnosis and medicine without ever having to admit their condition in person to anyone. For example, Robert Grubb solicited business through the mail, inviting patients either to "apply in person or by letter (post paid) to . . . his place in the 'Old Bailey', London."[92] Patients could even employ pseudonyms if they wished. The author of the 1728 treatise, *The Practical Scheme of the Secret Disease,* suggested that his patients might do just that.

> So as soon as the *Post-man* brings your letter, you shall presently have what you write for BROUGHT (carriage-free) to you, and with the utmost privacy, if you live anywhere not exceeding 12 miles out of London, by this Servant, who is employ'd on purpose to go with these things. And your charge of *Postage* will be repaid you again. Or what you *Write for*, shall be *Left till called for* (if you like it so better) at any house *Privately* for you under any feigned name, as you shall order in your letter.[93]

This practitioner made it clear as well that he would ship his medicines by all possible modes of transport to any location further outside the city to rural Englishmen who lived in small communities where the risk of exposure was greater than in the populous and anonymous metropolis of London. The rural poxed could thereby "cure themselves of the Venereal Distemper, in the private Family where they live in the Country, without making your case known to others."[94]

In addition to shipping medicines directly, some pox doctors sold medicines and instructions through retail agents. These represented still another avenue for those patients too embarrassed to visit the doctor. The advertisements indicate that booksellers were the most common agents through whom pox doctors sold their products. Some practitioners sold their medicines exclusively through one agent, but often these distribution networks were quite elaborate. For example, Stephan Freeman sold his pox medicine with instructions at ten different London locations. In addition to these sites within the city he also advertised an impressive list of "Printers and Booksellers" in sixteen other British towns, and undisclosed agents in "Jamaica," "Ireland and America."[95] Bookshops and other stores, then, became a sort of anonymous medical space for the London poxed. Sick men and women could enter the shop under the guise of browsing a merchant's innocent wares and, largely out of the public eye, obtain medicine and information on how to administer it.[96] When discussing why he preferred to sell his treatments at a bookseller's rather than at an apothecary's shop, one eighteenth-century practitioner claimed that a medical man like an apothecary might "know the Occasion" for the medicine, "and so be more inquisitive than the Patient be pleased with. . . ."[97]

Handbills also advertised that pox medicines were sold at coffee-houses and ale-houses.[98] Attempts to secure privacy in these spaces bordered on covert. Consider the cloak-and-dagger exchange a doctor named Vanforce arranged. His handbill promised that he had "ordered it to be sold at these places hereunder mentioned, being Ale-houses, where you may go into a Box, call for a pint of Ale, and have it delivered by the Landlady, without the least Notice or Suspicion, sealed up with my Cypher and Directions. Price Two Shillings."[99] Of course, they were hardly the only doctors to sell medicine through the mail. But the stigma attached to the pox made self-medication and mail-order medicine particularly attractive to foul patients. Subsequently, pox doctors became prevalent among those early practitioners to experiment with these emerging forms of promotion and distribution, marketing techniques that were well suited for protecting privacy.

It is telling that these early attempts to promise and provide for patient confidentiality were so prevalent in the treatment of venereal disease. Other branches of early modern medicine do not exhibit similar tendencies to stress secrecy when advertising. For example, none of the practitioners who treated ailments besides VD specified privacy in their handbills. Thus the actions of pox doctors may well have been innovatory. Confidentiality would eventually become codified in western systems of medical ethics. However, it must be borne in mind that in the seventeenth and early eighteenth centuries this was not yet the case. Studies of early modern medical ethics show that doctor-patient confidentiality is a relatively modern concept. Of course privacy is specifically mentioned in the Hippocratic Oath. However, too often have some assumed that this ensured patient confidentiality throughout history.[100] Robert Baker has termed this blanket application of the Hippocratic Oath "the myth of the Hippocratic footnote," and demonstrated that the influence of the oath on early modern medical ethics was minimal.[101]

Studies that explore seventeenth- and eighteenth-century medical debates support Baker and show that Hippocrates was seldom, if ever, cited by early modern doctors debating ethical issues.[102] Rather than assume an inherited tradition of medical morality, historians such as Baker and Laurence B. McCullough have argued that "the ethical concept of the professional-patient relationship had to be invented" in the late eighteenth century.[103] The "inventor" was Scottish physician John Gregory, whose *Lectures on the Duties and Qualifications of a Physician* (1772) applied the enlightened philosophy of Hume to medicine and marked the birth of modern medical ethics. Gregory did specify that doctors should respect the privacy of their patients. And Thomas Percival carried this forward the following century. His *Medical Ethics* (1803) drew on Gregory's ideas to offer the first codified system of modern medical ethics.

Studies of medical ethics in the two centuries prior to these publications illustrate that patient confidentiality was not yet a primary ethical concern of early modern doctors. Commenting on early modern diaries Porter noted the frequency of complaints that physicians commonly "blabbed about their patients."[104] Andrew Wear has shown that early modern debates over medical morality tended to concentrate, not on privacy, but rather on accusations of quackery and malpractice.[105] Similarly, Schleiner has examined the concept of confidence in early modern medicine. He found, however, that debates over confidence did not relate to patients' rights to have their conditions kept secret, but to doctors' responsibility to instill patients with confidence in their skills, and the lengths to which they could go (including lying) to ensure that confidence.[106] Moreover, the culture of medicine itself seems not to have placed primacy on the kind of individual patient-privacy that later medical systems would stress. Pelling has made a number of salient points, demonstrating on the one hand that in their shops barber-surgeons consulted with patients freely in the presence of others, but that even at home one could still expect to consult with a doctor with others present. She considers nineteenth-century assumptions about the importance of private medical exchanges in our period misplaced, suggesting that medical privacy developed significantly only after 1800.[107] Certainly many Stuart and early Georgian doctors practiced discretion out of basic respect and/or good business sense, but a confidentiality pact between doctor and patient such as we know today was not yet universally agreed upon.

In fact, significant tensions surrounded delicate medical exchanges, as illustrated by an incident recorded by Pepys in 1669. Pepys spoke with a surgeon named Pierce, and found him willing to discuss the treatment of one of his female patients, a Mrs. Churchill.

> But here Mr. Pierce, I asking him whither he was going, he told me as a great secret that he was going to his Maister's mistress, Mrs. Churchill

[with] some physic; meaning for the pox I suppose, or else that she is with child; but I suppose the former, by his manner of speaking about it.[108]

The exchange is revealing on several levels. On the one hand, we note the lack of discretion on the part of Pierce, who spoke openly about his patient's care with a third party. On the other hand we sense a heightened sensitivity and a need for modesty arising from the nature of Churchill's condition, which Pierce clearly felt. He considered the information he imparted to Pepys a "great secret." However, his "manner of speaking about it," to say nothing of the fact that he spoke about it at all, still betrayed confidences. The issue of medical privacy was thus imperfectly settled. It was perhaps for this reason that doctors treating the pox had to specify privacy in their promotional literature. It was not a service that was entirely assumed, but one that some doctors began to offer more than a century prior to John Gregory's *Lectures*. By trumpeting private care in their treatises and advertisements in a period when other medical men seem less concerned about privacy, and when confidentiality was not yet the primary concern of medical ethicists, these early modern practitioners may have been quite progressive indeed.

This is not to suggest that pox doctors were single-handedly responsible for giving the world the gift of medical confidentiality. But in their own way they played a role. We would do well to conceptualize the evolution of doctor-patient confidentiality not exclusively according to a top-down model, situated in university curricula and based on the intellectual heirs of Hippocrates. The evolution of medical confidentiality was also driven by forces emanating from below. The doctors who produced the advertisements studied here drew on neither Hippocrates nor John Gregory when maneuvering to ensure patient privacy. They were inspired by their patients' demands. Such a perspective is particularly useful considering the wide array of professional healers who did not learn their trade in a university milieu. In the main, Oxford-trained physicians did not resort to the kind of advertising explored here. Most of the practitioners who produced these handbills did not have a university background and would have been much more likely to feel the influence of immediate economic concerns than ethical arguments made by members of the intellectual elite.

Nor is this to suggest that this literature—both the ads as well as published treatises—worked only one way. The running discourse about embarrassment and privacy that pervades the genre might well have stemmed as much from attempts to *generate* shame as from attempts to respond to it. A view of this literature that privileges the commercial activities surrounding the production of these texts has to consider the mercenary benefits which would have come to doctors who convinced prospective patients that they *should* be ashamed of their infection, and that they *should* seek private care. As Porter put it, many pox doctors were indeed "trading on fear."[109] So it is difficult to come down on one side or another. To say that these practitioners were sensitive to patients' concerns is certainly not to champion them as ethically superior to other contemporary doctors. In fact, to argue that their ethics stemmed to a great extent from attention to their own personal gain goes some way towards casting them in a dubious light. There can be no doubt that some of this discourse was carefully constructed to help foster a culture of shame from which doctors with backdoor entrances might then benefit. Just like the satirists who poked sharply barbed fun at venereal infection, or the moralists who railed against the "wages of sin," medical writers contributed significantly to the prevailing culture of embarrassment that surrounded the disease.

In practice, the actual provision of medical privacy demonstrates the profound influence of gender. Contemporary medical ethics did not compel doctors to inform patients' sexual partners of their possible infection, nor to convince patients to do so themselves. In fact, doctors' assurances of privacy went so far as to promise treatment so discreet that such disclosure would be unnecessary. Thus, just like business associates and neighbors, sexual partners—especially wives—could be kept in the dark about one's infection and treatment. In advertisements these two promises were usually placed side by side. One practitioner in Holborn promised to cure "with Speed and Secrecie, and so much ease, that [patients] may follow their occasions, and not the nearest Relation take notice of the Cure. . . ."[110] Similarly, a separate advert promised patients treatment for the pox "without hindrance of any business, and . . . so privately, that a friend in bed shall not know it."[111]

It is clear that this discretion was extended to married patients. Peter Maris's advertisement makes this point clear when it promised prospective customers treatment "so private that no person; nay the wife or husband shall not [sic] be able to perceive your being in cure."[112] While Maris offers this service to both men and women, in practice doctors were far from egalitarian when guaranteeing patient confidentiality. Male practitioners went to much greater lengths to assure the privacy of their male patients, even when this meant lying to their female patients about their conditions in order to protect the reputation of the men under their care.[113] Several case studies demonstrate the point. One early eighteenth-century couple consulted the London surgeon John Marten, each complaining of pain when urinating. After examining them Marten was sure that the infection was venereal, but he withheld this diagnosis while the wife was present. As Marten described:

> Finding his wife ignorant as to the real cause of their indispositions, and believing it to be his fault, I enquired not so strictly, nor told them what my opinion was as to their distemper when they were with me together, lest I should stir up difference between them; but the next day when he came for the medicines, I plainly told him their cases were venereal. . . .

The husband denied this diagnosis at first, but when pressed by Marten he admitted his extra-marital exploits. Marten described how he promised the man his "secrecie" and went on to brag of how by this act he saved their marriage. The surgeon treated them "with such success that both of them were very well cured . . . and to this day his wife knows no otherwise . . . [and] they live very kind and lovingly together."[114]

The case is telling on many levels. First, it is worth noting that Marten took it upon himself to keep the diagnosis a secret from the wife before the husband even had a chance to ask him to do so. In other words Marten made this important decision *on behalf* of his patients. By making that decision and then calling the husband in alone to discuss the couple's condition, Marten clearly indicates that he by no means offered his services equally to men and women. Second, one must consider the manner in which Marten justifies his actions. There is a sense of ethics in this story, but it does not concern the wife's right to know. Instead Marten seems influenced by some higher obligation to the institution of marriage. To Marten his lie is petty and justified in the name of that crucial early modern institution, the family. Finally,

Marten was also well known for self-promotion.[115] His treatise (like many VD tracts) was part medical theory, part advertisement. Considering this, this case study, which Marten placed very near to the beginning of the book, has to be seen not only as the story of an actual case, but as an advertisement for similar services to other infected married men.

Sadly, Marten was not an anomaly. Henry Wastell similarly described the case of a woman whose husband he had treated for the clap previously. She was seven months pregnant and complained of pain when urinating. "As her husband had been my patient some months before, I concluded it was venereal, without mentioning my suspicions to her." Like Marten, he chose not to tell the woman of her condition, despite the obvious risk to her child.[116] Similarly, the author of *The Tomb of Venus* (1710) bragged of how he upheld the reputation of a wealthy male patient by withholding the diagnosis from his wife. The gentleman "was so well pleas'd that by all means his lady must go into the same course . . . (not knowing for what reason, or for whom) [she] did so for about a fortnight, under the notion of sweating her blood. . . ."[117] William Saunders was no better, hiding the nature of a woman's ailment in one case study, and conspiring with a husband in another: "He was desirous of the cure being kept secret of his wife and I therefore prescribed Mr. Plenck's medicine. . . ."[118]

As mentioned, Schleiner's study of early modern medical ethics indicates that doctors often justified lying to patients. At times, doctors knowingly misdiagnosed patients, administered placebos, or overestimated the outlook for the terminally ill for the sake of the patient or their family. They justified themselves by claims of acting for a higher good often buttressed by references to classical ethical philosophy.[119] The evidence from VD literature indicates that this tradition was alive and well into the eighteenth century and came to conjoin with the emerging medical ethic of privacy. Clearly this came on behalf of male sufferers and at the expense of women.[120]

Contemporaries, it seems, were not oblivious to the important gendered features of the medical exchange. In light of the work on early modern slander and the importance of women's sexual reputation, it is certainly true that women felt the stigma of venereal infection even more acutely than early modern men (although revisionist work on masculinity is showing that men were also anxious to sustain a reputation for sexual probity, a point worth bearing in mind[121]). John Profily, who it has been suggested may have targeted a female readership in his treatise,[122] relates how the sting of the pox's stigma made some women reluctant to submit to the male medical gaze.[123] "[Y]ou know how scrupulous Women are of exposing themselves and that modest women would rather suffer death, than be the subject of . . . censorious acquaintance."[124] Solomon Sawrey concurred, noting how "modest women . . . will either not mention their disease, or not suffer to be examined."[125] Again we see that practitioners maneuvered to provide services that met the unique demands of foul patients, in this case female patients who preferred not to consult with male doctors when infected with a sexual disease.

It is now well documented that many early modern women earned a living by healing professionally (i.e., for money).[126] The evidence found in handbills demonstrates that a significant number of London women treated the pox, specializing in treating modest women. In fact, they may show that treating the pox constituted one of the more common and important services that female healers provided. Women largely excelled in that field because they provided a popular service that responded to the gendered disparity of doctor-patient relations. The advertisements of nine different women who treated the pox survive from the period ca.1660–ca.1715. Their ads are quite similar to men's, following the same format, offering roughly the same range of services, and providing directions to their homes with specified hours. Just like men women bragged about their exotic training and some offered services through the mail. Men and women used similar language when advertising, frequently employing the same common phrases. For example, no less than thirty-four handbills for the pox distributed by men invited male patients who had been infected "in the garden of Venus."[127] At least one female practitioner employed this same imagery, but stood it on its head by making the female Venus the victim of the male god Mars. She invited female patients thus:

> And if Venus should Misfortunately be wounded with a Scorponious Poyson by Tampering with Fiery Mars, to her own Sects [sic] it is she brings Comfort and Relief and by her Antidote expels the Poyson Jove-like, though never so far gone without Fluxing.[128]

The fundamental similarity between men's and women's advertisements suggests that there was not necessarily a battle of the sexes in the medical sphere. In fact, medical men and medical women frequently cooperated and practiced medicine together in seventeenth- and eighteenth-century London. Usually this cooperation took place within the family, between husbands and wives, and/or parents and children.[129]

Women's ads differed from men's in at least one telling way. Women went to much greater lengths to highlight their training. Female healers who advertised based their claims to medical expertise almost exclusively on their familial connections, where, essentially, they described apprenticeships. Barred from university, women instead bragged about their long training under the tutelage of experienced doctors, usually fathers or husbands.[130] Ann Laverenst and Willemina van Soeburgh advertised that their parents were their primary teachers.[131] Elizabeth Maris, Sarah de Heusde, and Joyce Lucas all cited the joint training by their parents and their husbands, alongside whom they practiced for many years.[132] Heusde, for one, proudly proclaimed the medical tradition in her family, lauding herself "Sarah Cornelius de Heusde, Widdow of Dr. Sashout, and Grandmother of the Doctor that had his Stage upon the Great Tower-Hill, and did so many cures before the Fire." Given that apprenticeship continued as an acceptable form of training in so many fields, there is not reason to believe that these claims did not resonate.

In fact, the evidence from men's advertisements indicates that these women brought in clientele that men might not attract on their own, and thus could enhance men's practices by providing the unique service of discreet care for poxed women. Ten male doctors who advertised to treat the pox made specific reference to women, usually their wives, who shared in their medical practice. In all cases they stressed the issues of privacy and female modesty and the advantages that their practices held for women afflicted with the embarrassing foul disease. Consider the following examples. One typical "physician" who practiced in Holborn treated the foul disease "without the least Hazard of Loss of Time, and with that Secrecy, that the most intimate Friend shall not be able to observe the same." These standard promises were also afforded to women because "His Wife likewise for the more modest Accommodation of the Female Sex (being well skill'd in all Distempers incident to Women) gives her Attendance."[133]

Male doctors lamented that infected women were reluctant to discuss their conditions to them. Cornelius Tilbourn added a note to this effect at the close of his advertisement: "Note. If any Gentlewoman hath gotten a Rupture or any foul Venereal Disease; or is subject to obstructions, and is ashamed to discover her Distemper to me; they may speak to my Wife and she shall relieve them very secretly."[134] The empiric who advertised "An Herculeon [sic] Antidiote Against the Pox" assured women discreet treatment in the same fashion, promising that "the Female Sex that have any private Distemper, may be accommodated by my Wife."[135] Gerardts Gonsale, W. Langham, and four other unnamed practitioners all advertised that their wives practiced medicine with them and would treat the pox in any women too ashamed to disclose their conditions to them.[136] Abraham Souberg promised that his sister treated shy women; it is possible that his sister may have been Willemena van Soeburgh, mentioned above.[137] It is unclear who was the female assistant to the "physician" who treated the pox at "In Carthusion-street, in Charter-House-Yard." He advertised, "If any Gentlewoman be unwilling to disclose her Distemper to him, she may Relate her Grievance to one of her own Sex, of great Experience. . . ." This leaves open the possibility that this female practitioner was not related to the him, though from the weight of the evidence this would be an exception to the general rule.[138] Finally, it is worth pointing out that these claims about female modesty do not seem like mere advertising copy. In fact, they parallel statements made by women during rape trials. When Ephemia Wiesenthall was asked to testify about sensitive details of an alleged rape she responded, "I am ashamed. . . . I should like better to speak it to a woman; it is not fit to be mention'd in the hearing of a man."[139]

The case of one of these husband/wife medical teams illuminates both the role of the family in medical training and the unique nature of these partnerships. There are three surviving advertisements printed for Peter and Elizabeth Maris. Peter's name appears on two of them,[140] and Elizabeth put one out herself following Peter's death.[141] They resided at the same address in Holborn "next Door to the Old King Charles's-Head" at the time of the first two handbills. In both advertisements Peter claims to "perfectly cure the Morbus Gallicus" along with the host of other common ailments. However, he devotes the majority of space to his discussion of the pox. Interestingly, he only mentions the availability of consultation with his wife in one of the two handbills. In language typical of other advertisements, Maris claims "If any Woman be unwilling to speak to me, they may have the conveniency of speaking with my Wife who is expert in Womens Distempers."[142] If we recall those doctors who promised treatment so discreet that sexual partners could not detect it, we might remember that Peter Maris stood out for his egalitarian promise of discretion for both married men *and* married women. It is now apparent that this unique promise should be considered in the context of Maris's shared practice with his wife. Medical practices like that of Peter and Elizabeth Maris stand out not only because a man and women practiced medicine together but because they offered unique services as a result of that partnership. Women infected with the foul disease who visited traditional doctors had to submit to the male medical gaze and were not assured of the same discretion as male patients. In some cases doctors even lied to them. On the contrary, at the home of Elizabeth and Peter Maris infected women could consult with an experienced practitioner of their own sex and obtain the privacy they craved regardless of their gender.

Following Peter's death Elizabeth continued the business herself, a common occurrence.[143] She moved to a new address, although she stayed in Holborn. Now advertising under her own name, Elizabeth had to account for her training. Like other women practitioners she cited her family. She indicated that she was knowledgeable in "physick" before she married Peter, mentioning the tutelage of her parents. In her own handbill Elizabeth now elaborated on the various female maladies that her husband's advertisements only mentioned in passing. She did not, however, shy away from mentioning venereal disease, promising to cure the pox "with ease, speed and secrecie." The pattern of medical training through familial connections is further reinforced when Elizabeth mentions her son. Despite the loss of Peter, she did not wish to alienate potential male clientele afflicted with the pox. Just like the female healer who reversed the Venus/Mars imagery, Elizabeth turned a standard male advertising technique on its head.

> If any Gentleman has any Distemper not fit to be discourst of to a Woman, he may speak to my Son, who hath practis'd Physick above

twenty years with good success. He hath great experience in curing any Venerial and all other Distempers . . . [144]

Thus medical training in the family comes full circle.

Taking into account that some doctors produced multiple advertisements (as many as seventeen different handbills can be attributed to a single doctor), the three collections of advertisements studied represent 154 separate practitioners who treated venereal disease in London ca.1660–ca.1715. Nineteen different women can be identified in that group who either advertised themselves or who treated the pox alongside their husbands. Thus women represented a significant portion, 12.3 percent, of those London pox doctors who advertised in this fashion. That figure possibly could have been higher. Many other men advertised that their wives treated female patients, but they did not specify venereal disease among their services.[145] Similarly, several other female practitioners who advertised on their own failed to specify treatment of the pox, although there is evidence that some may well have treated the pox, but chose not to specify it in the text out of modesty. One "gentlewoman," a widow who practiced with her husband for over twenty years in the Strand, was purposefully vague when she promised to treat "all other weaknesses incident to her Sex, not proper to mention here."[146] In the same vein, "The Doctor's Wife in Dean's Court" also practiced modesty when she claimed to treat "many Infirmities in Women not fit to be mentioned here."[147] Whether these claims, or those like them, should be interpreted as veiled advertisements for pox treatment must remain an open question. Regardless, it is clear that the sensitive nature of the doctor/patient relationship changed when confronted with sexual infection, and this change manifested itself in patients' demands for same-sex practitioners, demands which the market scrambled to meet.

Foul patients with money to spend, therefore, faced a wide range of options. These medical consumers could choose from various types of practitioners who offered various types of treatment, under varied circumstances. However, private care, multiple therapeutic options, or a same-sex doctor were luxuries. Fissell and Porter have both noted that patent medicines like the ones trumpeted in handbills and newspapers were usually priced well beyond the reach of working people.[148] Moreover, treatments for the pox have been shown to be among the most expensive medicines of the period.[149] London's plebeian population often found themselves excluded from the benefits that the market offered.

This was especially so in the case of the pox, for we know that the symptoms that attended early modern venereal ailments, especially in cases of actual syphilis, often abated and returned in an ongoing, frustrating cycle, as James Boswell discovered, battling his recurring infection throughout much of his adult life. Considering this, the importance of wealth to the patient experience increases even more. The luxury of sustained private treatment over several years, if not a lifetime, required significant resources. A portion of the venereally infected were *perpetual* medical consumers. This augmented the impoverishing dynamic of the pox, and needs to be borne in mind when considering the care available to those Londoners who could not afford to *continue* paying for private care.

Thus many people could not afford to take advantage of the market. Sometimes they tried, nonetheless. For many people accessing private medicine meant staying out of a hospital, guarding their privacy, keeping their job, or avoiding the horrors of mercury. Some plebeian Londoners left evidence that they did what they could to continue utilizing private medicine for as long as possible. At times these efforts became desperate, as when the poor were reduced to theft in order to pay for medicine. Domestic servant Sarah Bloss confessed to stealing money from her boss, bookbinder Thomas Barlow. When asked what she did with it she explained that "she gave [one guinea] to a doctor that had partly cured her of the foul disease." Soldier Phillip Allen stole money and goods from the house of his landlord, Joseph Terry. When arrested and interrogated, Allen confessed that "he had got the Foul Distemper, and he got that Money on purpose to get him cur'd." Domestic servant Sarah Coates stole thirty shillings, claiming that she needed it for treatment of scrofula, the so-called "King's Evil." But as her former employers testified "it was more her own Evil than the King's" for she used the money to pay for a salivation. Henry Pinnock eeked out a living running errands for a watchmaker named William Addis. In 1758 he was tried for stealing one of the watches. Explaining his actions at trial Pinnock similarly cited his illness as his defense: "I own myself to be in a fault, I had the foul disease four years ago, and I am so bad that I can hardly crawl along." The debilitating physical effects of mercury also had real economic effects and could make a major difference when it came to plebeian attempts to survive. Desperation seems to have led barber John Palmer to steal thirty shillings' worth of hair. When apprehended authorities discovered that he owed his doctor forty shillings. He confessed "that he was just come out of a Salivation, and that mere Necessity drove him to take the Goods."[150]

Those who failed in their scramble to sustain themselves when infected would have had to rely on institutional care. People forced by need to enter a hospital almost always faced physically debilitating salivation. As late as 1784 surgeons like Peter Clare still criticized hospitals for continuing to employ the "old methods[,] when it is known and lamented, that many poor unfortunate people perish through the great abuse of mercury in their hospitals."[151] The divergent illness-experiences of rich and poor that this book hopes to demonstrate was openly admitted by such eighteenth-century commentators as surgeon John Andree, who stated that "the method practiced in Hospitals, of salivating for the cure of most venereal disorders, cannot be adopted in private, as very few patients would submit to so severe a mercurial course."[152] Historian Phillip Wilson concurs, noting that salivation remained standard treatment for hospital patients throughout the eighteenth century, even though "salivations by public (i.e., non-hospital) surgeons under whom patients had more say about their treatment distinctly lost . . . support [after 1740]."[153]

The poor could not choose between various treatment strategies; they did not have a say about the circumstances in which their treatment took place; and hospitals did not provide female practitioners for the benefit of female patients. Thus, just as there were important gendered differences between the experiences of men and women with this disease, so we will see there were crucial class differences between the experiences, options, and strategies available to the poxed wealthy and the poxed poor.

2

THE FOUL DISEASE IN THE ROYAL HOSPITALS: THE SEVENTEENTH CENTURY

Many Londoners could not avail themselves of the services that the market provided. This posed a problem. What can be done, lamented surgeon Charles Peter in 1693, for "those poor unhappy wretches where the Pox and Poverty are complicated"?[1] In a sense, that is the central question of the remainder of this book. Such folk were not entirely without options in the seventeenth century, as scholars have sometimes presumed. One of their main options lay in one of the two royal hospitals that offered venereal care, St. Bartholomew's and St. Thomas's.[2] However, contemporaries did not always see these as a desirable choice. An anonymous commentator criticized hospital foul wards, claiming that "more rude Ignorance, and slighter Management in curing this disease, has not been heard of, than in these places." He went on to claim that "some that have undergone their common unctions there, and come forth uncured, have protested they had rather chuse to dye, than to return thither again."[3] In 1696 a doctor named Wall similarly described the "Despised Hospitals and Lock Nursery" that represented the only resort for paupers who had been taken advantage of by ruthless quacks who took their meager pennies and sold them phony cures.[4]

Wall's depiction of seventeenth-century hospital VD care is notable for its lack of a clear opinion; it is a description that, though brief, captures well the tension and difficulty inherent in trying to sum up early modern hospital provision in simple terms. On the one hand, Wall offered rather little hope to the "captives" who resorted to the foul wards, where they might find themselves "reduc'd . . . to a dribling Condition" by the substandard care and "slack attendance" that characterized hospital therapeutics. Yet despite that grim picture he also acknowledged that the foul wards still "furnish[ed] out more Mercy to the Afflicted" than the dog-eat-dog market from which paupers were excluded. Wall's depiction is generally accurate, exaggeration notwithstanding. There can be no question that hospital patients faced inferior care when compared on many counts to the care available to paying patients. However, before gauging that care we must first acknowledge the absolutely impressive scope of both hospitals' provision for venereal patients, which has been largely unrecognized. The royal hospitals steered quite significant resources towards tackling the complicated problems of poverty and the pox, and saw venereal patients as worthy objects of that significant charity. Ultimately, Wall's ambiguity seems well placed: despite significant "mercy" these hospitals indeed remained "despised."

London hospitals admitted foul patients fully two centuries before the famous Lock Hospital was even a glimmer in its founder's eye. The first instances of poxed patients treated in Bart's and St. Thomas's Hospitals come from the earliest administrative records, which for Bart's survive from 1549 and for St. Thomas's from 1556. For example, the governors of St. Thomas's agreed to treat a poxed woman named Eme Danyall on March 10, 1556.[5] The earliest surviving records of St. Bartholomew's also contain similar entries. Financial records indicate that surgeons received gratuities for treating patients with the pox during the fiscal year 1547/48.[6] Considering that venereal patients emerge from the earliest surviving records, it is likely that both hospitals may have accepted venereal patients even earlier. One of Bart's surgeons, William Clowes, proclaimed as early as 1579 that the hospital was practically overrun by such patients. According to Clowes, over one thousand venereal patients had received cures in St. Bartholomew's between 1574 and 1579. He stated, "among every twentye diseased persons, that are taken in, fiftene of them have the pocks."[7] Clowes may have exaggerated his figures, but hospital records indicate that his claim must still be taken seriously. According to the gratuities granted to surgeons for specified cures in 1547/48 almost one quarter (21 of 87) were for the pox.[8]

While this evidence alone already runs counter to the traditional histories of early English institutional VD provision, research indicates that the English case was hardly unique in Europe. Jon Arrizabalaga, John Henderson, and Roger French have examined how various *Incurabili* hospitals also treated venereal patients throughout Italy during the first half of the sixteenth century, and institutional care has been explored in sixteenth-century Germany as well.[9] The continental context shows that various charitable organizations responded to provide relief during the raging first phase of the pox, which, by all accounts, seems to have been horrible.[10] The new disease spread rapidly and killed many, causing lay Italians to define the disease as "incurable" and develop special hospitals to cater to the poor so afflicted. As mercury and guaiac became increasingly popular over the century this belief that the pox was incurable faded, but the *Incurabili* hospitals continued to provide the important service of housing and treating the infected poor. The Italian hospital records pre-date the London ones, but it is likely that the chronology of the English response parallels the case that these scholars describe in Italy, and we should not be surprised to find English institutional VD care in the first decades of the sixteenth century as well.

The governors at St. Bartholomew's hit upon one solution to the problem presented by the poor foul Londoners who came weekly to the hospital in search of care. They sent venereal patients to one of the various leper houses under the hospital's control, a response similarly recorded in Germany and Norwich.[11] The City of London transferred control of six lazar houses to Bart's in 1549.[12] As indicated, a number of venereal patients were recorded in the hospital just the previous year. There may well be a connection between the significant number of foul patients in the hospital in 1547/48 and the transfer of the six lazar hospitals in 1549. Although no explanation has been given for just why Bart's took control of these six hospitals, it is probable that the prevalence of venereal patients and their demands on Bart's limited ward space may have encouraged hospital governors and/or City officials to hit upon the scheme of transforming leper hospitals into "outhouses" for venereal patients.[13]

It is likely that venereal disease may have provided the original impetus behind the transference of the lazar hospitals, because it is clear that leprosy itself was no longer a significant medical concern by the mid-sixteenth century. By all accounts the disease faded rapidly after 1500,[14] and lazar houses all over England began to shut throughout the first half of the century.[15] While the decision to house venereal patients in lazar houses may seem unique, it hardly required a great stretch of the imagination. Popular and medical beliefs about the pox and leprosy had linked the two diseases in a variety of ways from an early date. First, some medical thinkers argued that they were actually the same disease.[16] Well into the eighteenth century this notion continued to be a cornerstone of the argument in favor of the antiquity of the disease (against the Columbianists, who argued that it was new to Europe). In the late seventeenth century, medical writers like Stephan Blanckaert continued to remark on the striking symptomatic similarities between leprosy and the pox, concluding that "there is all the probability in the world that men are frequently deceived . . . and that the Venereal Distemper has been taken for the Leprosie."[17] This belief remained alive and well into the eighteenth century, when William Beckett produced the most extensive exploration of the subject, publishing three letters on the antiquity of the venereal disease in the Royal Society's *Philosophical Transactions*.[18] Beckett marshaled evidence from various classical sources on the symptoms, transmission, and treatment of leprosy to claim that "leprosy" was merely a different name for the pox. The longstanding belief in the sexual transmission of leprosy also helped connect the two diseases.[19]

Many medical writers disagreed, but some nonetheless forged theoretical links between *lues venerea* and leprosy. The most influential of these was Paracelsus. He proposed a novel explanation of the French Disease's origin, which medical authorities would debate throughout the following two centuries. It is interesting that the Paracelsian theory exhibits a theme common to theories about early venereal disease production: namely, that a disease transmitted by sex must be originally produced by sex. This idea manifested itself in various explanations for venereal disease's origin, including theories about promiscuous women, miscegenation, and even bestiality.[20] In Paracelsus's case, he argued that the pox arose from the following event.

> The French Disease derives its origin from the coition of a leprous Frenchman with an impudent whore, who had venereal bubas, and after that infected everyone that lay with her; and thus from the leprosy and venereal bubas, the French disease arising, infected the whole world with its contagion, in the same manner as from coition of a horse and ass the race of mules is produced.[21]

Metaphorically speaking, leprosy gave birth to the pox. Those who believed that the venereal disease *was* leprosy were a minority. But there were many others who believed that, although distinct, the two were very closely related.[22]

The suspected etiological connection between the foul disease and leprosy helped inform Bart's decision to house venereal patients in former leper hospitals. But practical issues also played a key role. Contemporary medical writers believed strongly that the pox spread by various forms of casual contact. Sex was not the only means of contagion that worried sixteenth-century doctors. Bart's surgeon, William Clowes, illustrates the point. In addition to sex and nursing, the two most commonly discussed means of transmission, Clowes warned readers that they might catch the pox by eating from the same dish as an infected person, sleeping on bed linens after a poxed person, sharing their clothing, or using the same privy.

Clowes believed so strongly in the dangers of casual contagion that he advised people to avoid all contact with foul patients. For example, he warned them to stay clear of places like alehouses, an environment filled with "rogues and vagabonds," where innocent people might easily catch the disease.[23] Such fear of casual contagion made a strong case for segregation. To early doctors like Clowes the poxed posed a significant health risk that demanded quarantine measures. In this way, again, the pox paralleled leprosy. In fact, it may be that the fear of casual transmission represented another aspect of the pox's cultural inheritance from leprosy. Regardless, when the governors of St. Bartholomew's Hospital faced the problem of large numbers of infected Londoners seeking treatment, members of the medical staff like Clowes must have warned them of the dangers that these people posed to other patients who might easily share bedding, dishes, or clothing. Luckily, in 1549 the City of London put at the hospital's disposal a series of institutions specifically designed to segregate such patients. Thus as foul patients assumed the discursive and imaginative space left in the wake of disappearing lepers,[24] so they also took up the physical (segregated) space that they once inhabited.

Throughout the second half of the sixteenth century Bart's sent most of its venereal patients to one of these six lazarettos situated on the outskirts of the city, where they would be treated by a surgeon called a "guide."[25] But they also seem to have continued treating foul patients in wards on the main hospital grounds, a phenomenon first noticed by Margaret Pelling. In the 1570s, well after the City had granted the lazarettos to Bart's, two of its eight wards, "Cloister Dorter" and "Garden Dorter," and possibly a third, the "Sweating Ward," still received venereal patients.[26] But Bart's strategy became much more uniform in the early seventeenth century. In 1622 the governors decided to close down all of the former lazar houses, save the Lock in Southwark (not to be confused with the eighteenth-century voluntary hospital of the same name, which will be explored in coming chapters), and the Kingsland near Hackney. The hospital focused its venereal operations entirely in these two institutions, and stopped accepting foul patients at the main hospital; from this point forward foul men went to the Lock and foul women went to the Kingsland.[27]

Evidence indicates that VD provision became the primary function of both these institutions. To begin with, the records do not indicate that any foul patients were admitted into the main house at any point after 1622. Nor did any of the house surgeons or physicians receive gratuities for treating such patients at any point in the seventeenth century, as they had in the sixteenth century. A 1633 set of rules drawn up by William Harvey, Bart's chief physician, also indicates that Bart's had by now focused its VD operations in these institutions. Harvey stressed that all patients diagnosed "uncurable & scandelous or infeccous shal be putt out of the said howse [Bart's], or to be sent to an out-howse."[28] The scholars who have examined this rule have interpreted "scandelous" patients as venereal cases.[29] They are undoubtedly right. For it is likely that Harvey's comments on "uncurable & scandalous or infeccous" patients actually refer to two categories of patients, not three: "uncurable & scandalous" patients on the one hand and "infeccous" patients on the other. Harvey, as is well known, studied in Padua; he very likely brought back from Italy the Italian term *incurabili*, which Arrizabalga, Henderson, and French have shown was the dominant term for the poxed there.[30] Harvey's clumsy term "uncurable" seems likely an anglicized version of the Italian term, which he then linked to the moralizing term "scandalous."

The records leave little doubt, as the board restated their rule on excluding foul patients throughout the century. They named the disease in question in 1687, when they declared that all patients with the "French Pox" would be sent to the outhouses.[31] Harvey's comments indicate that they also sent "infeccous" patients to the outhouses on occasion, notably when plague struck. However, such instances became increasingly rare as the seventeenth century wore on.[32] The seventeenth-century discussions of the outhouse focus almost exclusively on the pox. That said, it is worth noting that it would not have been entirely rare for patients in the Lock or Kingsland to find themselves lying next to someone with small pox or some other infectious "fever" that the governors or medical staff considered too contagious for the main wards. This undoubtedly gave cause for serious and well-placed anxiety on the part of paupers who had to utilize these institutions.

Whereas Bart's kept its foul patients at a safe distance by utilizing old leper hospitals, St. Thomas's treated foul patients within the main hospital itself. Situated not far from the Lock outhouse, the presence of St. Thomas's Hospital meant that the majority of seventeenth-century hospital beds for foul patients lay in Southwark, a fact that may have contributed to that neighborhood's black reputation as a center of vice. But Southwark's longstanding reputation may have made situating foul wards there more acceptable to early modern Londoners who hoped to confine these dubious institutions to a section of the city already tainted by notions of immorality, vice, crime, and contagion.[33]

Unfortunately, the seventeenth-century records for St. Thomas's do not yield quite as much as those of Bart's concerning venereal disease.[34] It is not clear when St. Thomas's established its foul wards, but it is likely that the 1560 reference to the "Sweate ward" marks the earliest reference.[35] We do know that St. Thomas's, lacking its own lazarettos, occasionally availed itself of Bart's, as it did in February 1556 when it sent two patients to the Lock.[36] Though the records do not specify the diagnoses of such patients, and it seems clear that by 1600 the hospital concentrated its venereal care within its own wards.

As at Bart's, St. Thomas's made it standard practice to segregate venereal patients. The governors mention the segregated foul wards frequently, and it is clear that they limited venereal patients to these specified spaces. For example, in 1633 the board specified that "no patient of the fowle Disease be admitted into this house till there shall be room in the fowle wards."[37] The prospect of sending foul patients to some distant house certainly had its appeal, for during the ensuing debate over how many patients the hospital could accommodate the board considered emulating Bart's strategy in order to free up space for "clean" patients. On March 10 of that year the board considered a proposal "for an outhouse or spittlehouse" to isolate foul cases.[38] However, the proposal never made it beyond this stage, perhaps because the financial straits which prompted the discussion of limiting patients in the first place precluded any grand design to build or buy a separate institution for venereal patients. Thus St. Thomas's would continue to house its venereal patients within the walls of the main hospital for the rest of the century.

It is worth exploring the scope of these two operations. At Bart's the outhouses represented a significant part of the hospital's overall operations. The hospital's annual ledgers provide key data on the number of patients treated at the outhouses that helps to show how venereal care fit into St. Bartholomew's larger scheme. Each year the debit or "payments" section of the annual ledger began by listing the monthly payments for patients' food. The ledger's prescribed format listed the main hospital's "diet" payments separately from those of the outhouses. They combined the figures for the two outhouses, making the exact figures for each house impossible to calculate. Still, since these figures are complete for the entire century to 1696, simple division provides the average number of patients treated in both the hospital and the outhouses throughout each year. The diet figures cannot indicate the exact number of patients treated. The figures represent the number of beds that were kept full throughout the year, not the number of *different* patients who filled those beds. It would have cost the hospital the same amount to keep the outhouses filled with the same forty patients all year long as for a new group of forty patients replacing the old ones each month or each week. The same rate, four pence per patient per day, applied to both the hospital and the outhouses. Dividing the annual total payments by 365 produces the average amount that the governors paid to feed patients each day throughout the year. Dividing that figure by four converts it into the average number of patients that could be found in either the hospital or the outhouses on any given day that year. In short, these figures yield what we might call an average annual capacity for both the main hospital and the outhouses.

The patterns are telling. In the initial year in which they focused all venereal patients in the two outhouses the governors laid out £1105.17.3 to feed patients within the main hospital and £333.18.10 to feed outhouse patients. This translates into an average capacity of 182 hospital patients and 55 outhouse patients, meaning that on an average day in 1622 one could expect to find 182 patients in the main hospital and 55 patients in the two outhouses combined. Put another way, outhouse patients represented almost one quarter, 23.8 percent, of all the patients Bart's received that year.[39] Far from refusing venereal patients, Bart's made treating them one of its primary services.

The graph in figure 4 demonstrates the portion of Bart's total patient population in the outhouses for the period 1622–1696. This initial pattern remains fairly constant for a decade. During the ten-year period 1622–1631 the outhouses had an annual average capacity of just below 57 patients, representing just over one fifth (20.7 percent) of all of St. Bartholomew's patients. Throughout this period the hospital had hoped to maintain a twenty patient limit for each outhouse.[40] However, the figures indicate that this limit was routinely exceeded, each outhouse treating closer to thirty patients at a time throughout the decade.[41] Figure 5 represents essentially the same data, expressed in terms of the average capacity of the two outhouses—the number of patients treated at a time—rather than as a percentage of the total population. After 1631 the number of outhouse patients increased steadily. Within ten years the outhouses combined to treat an average of over one hundred patients at a time. This accompanied a steady rise in the number of patients admitted to the wards of the main hospital and thus represents a general increase in Bart's overall services. However, the growth of the outhouses actually outpaced that of the hospital. The £640 paid to feed outhouse patients in 1643 now represented just under 30 percent of

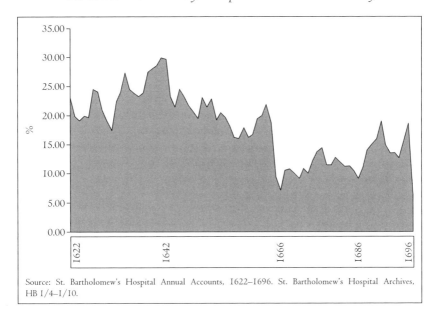

Fig. 4: Outhouse Patients as a Percentage of All Patients in
St. Bartholomew's Hospital, 1622–1696.

Bart's total population. The governors' minutes do not explain this increase. One might attribute the high rates of admission in the early 1640s to casualties of the civil war, but such a simple answer does not explain the steady increase during the previous decade and the similarly high rates in the years prior to the outbreak of hostilities.[42] Nor do the records explain the drop off in the number of outhouse patients in 1644. But the total number of clean patients also drops off simultaneously, and the outhouses continue to represent about one fifth of Bart's total clientele throughout most of the 1650s. While outhouse patients represent a slightly smaller portion of total clientele during the period 1658–1662, they still made up about one fifth of all patients in the mid-1660s.

Because St. Thomas's did not set up independent VD operations that generated their own records, we cannot make such precise calculations of the percentage of foul patients at that hospital. Still, it is possible to make estimates, and these suggest that foul patients represented a similar, only slightly smaller, portion of the St. Thomas's clientele. For example, one

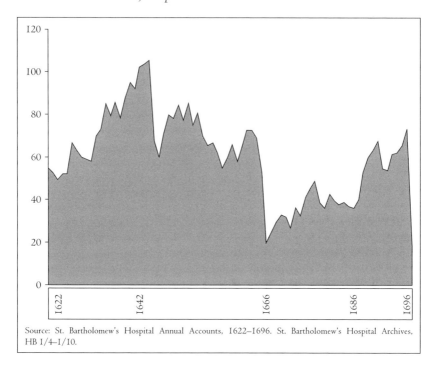

Fig. 5: Annual Average Capacity of the Lock and Kingsland Outhouses (combined), 1622–1696.

can relate the ward space reserved for venereal patients to St. Thomas's total ward space. This method is not without risk since the governors realigned the wards several times over the course of the century. But according to F. G. Parsons, during the early seventeenth century St. Thomas's housed its foul patients in two wards (one for each sex), at a time when the hospital had a total of fourteen wards. Parsons estimates that each ward had twenty beds.[43] By this rough estimate, venereal patients made up almost 15 percent of all patients. Or, more exactly, St. Thomas's reserved nearly 15 percent of its beds for venereal patients during the first half of the century. However, the forty beds in these wards meant that the number of VD patients that St. Thomas's budgeted to treat was in the same general range as Bart's.

Sometime in the middle of the century St. Thomas's expanded its operations and committed an additional two wards to venereal care. By the 1660s, Job and Lazarus wards were devoted to foul men and Judith and Susannah wards housed foul women. An admission register gives the number of beds in each ward,[44] allowing for a more exact account of the proportion of beds allotted to pox patients. The four foul wards now had 52 beds combined. The register specifies the number of beds in all but two of the clean wards, Isaac and Abraham. These combined to total 189 beds. Assuming that the two missing wards had an average number of beds (18.5), St. Thomas's had approximately 278 total beds: 52 for venereal patients and 226 for clean patients. Thus, St. Thomas's devoted nearly 19 percent of its ward space to foul patients by the 1660s, quite in line with Bart's operation at the same time.

As the graphs suggest, the situation changed dramatically after 1666. This was a direct result of the Great Fire. That event had terrible economic repercussions for both hospitals, which venereal patients felt almost immediately. Both hospitals and the two outhouses survived the blaze. However, the royal hospitals relied heavily on rental incomes. Bart's and St. Thomas's had each been endowed with hundreds of properties around London, the rents of which provided the vast majority of their operating budget. The fire destroyed a great many of these buildings, sapping the vital income that they provided, causing an immediate fiscal crisis at both hospitals. St. Thomas's estimated its annual losses at £600; Bart's estimated its at £2000.[45] The administration at Bart's scrambled to react to the crisis and decided quickly to cut costs by closing the outhouses. On October 27 the board ordered the outhouse surgeons to "discharge all theire patients in either house att or before xmas day next." Both houses would have to remain empty until "by some providence the Hosp[ital] revenues shall be enlarged."[46] As the diet figures show, they stuck to their words. The £20.1.4 paid to the outhouses in early October represents approximately 43 patients in the two outhouses when they were ordered closed. Four weeks later this dropped to 37. By mid-December there were 23 patients left, and by the end of the month a mere 11. On December 29 the governors made a final payment of £2.14.0, after which payments stopped, and the outhouses were empty.

St. Thomas's governors also felt the crunch. In the years following the fire they set up the "Committee for Burnt Houses" to deal with the hospital's financial woes. They may have followed suit and immediately cut back on accepting venereal patients, but again the issue is not as clear in the records as it is at Bart's. The records do show that St. Thomas's fiscal problems continued for several years. In June of 1670 the Committee for Burnt Houses was still at work and made a number of cost-cutting recommendations that affected VD provision. Most importantly, they cut operations in half, closing down two of the four foul wards.[47]

Such decisions indicate a great deal about attitudes towards venereal care. Certainly, when resources became tight the hospitals were clearly willing to sacrifice the care of venereal patients in order to keep providing care to others. The data from diet payments at Bart's shows that after a brief dip immediately following the fire, the number of clean patients treated in the main hospital rebounded and remained basically constant. In other words, hospital cuts almost entirely targeted foul patients. Moreover, no evidence indicates that anyone opposed the proposals to close the outhouses and foul wards, or even that any debate attended those decisions. Venereal patients clearly occupied the lowest rung on the hierarchical ladder at both institutions.

However, it is just as notable that the hospitals quickly tried to resume treating foul patients. Tellingly, Bart's did not keep the outhouses closed for very long. When they closed them Bart's authorities indicated that the surgeons could continue to reside on the premises and continue to "make use of the bedding and goods in the several houses" for the "present advantage for patients."[48] So venereal patients could still resort to the outhouses, but the hospital would not pay for their treatment. Presumably patients had to strike private agreements with the surgeons. But even this arrangement did not last long. By March the governors had already acquiesced to admit patients to the outhouses and support them "upon necessity."[49] The outhouses, therefore, remained closed a total of only three months; diet payments resumed in April.

So despite dire financial straits, and the low status of venereal patients, the hospitals still remained committed to providing at least some level of venereal care. Considering the pressing financial needs, Bart's might have sold or leased the two outhouses. There was certainly no shortage of people looking for shelter for their homes or businesses in the wake of the fire. Instead, the board allowed the surgeons to continue treating venereal patients at no profit to the hospital, they covered the overhead, and resumed supporting foul patients as soon as they could. By June, within eight months of the fire, there were close to twenty patients in the two outhouses combined.

Despite attempts to continue providing care, there is no doubt that the overall scope of the operations at the outhouses diminished after the 1666, as figures 4 and 5 indicate. Through the mid-1680s the average capacity of the outhouses now generally remained below forty patients. Notably, during the same period the number of clean patients in the hospital gradually increased from an average of 222.7 in 1665/66 to over 350 in 1684/85, a rise of 57 percent. This meant that for the next three decades Bart's applied a much smaller portion of total hospital resources to the treatment of venereal patients; the outhouses usually received well below 15 percent of the hospital's clientele.

The fire had one other major legacy. After 1667, hospital treatment of the pox was no longer free. For the first time hospital patients had to start paying for a portion of their care. And it is telling that the first patients to do so were those in the foul wards. When it resumed sending patients to the outhouses, Bart's governing board now stipulated that the hospital would provide only food; they would not pay the surgeons a salary. The patients, their friends, or families would have to pay the surgeons an unspecified fee for their cure.[50] In 1668 the governors also allowed the surgeons to accept private patients in the outhouses, with whom they could make their own financial arrangements.[51] Of course, this development means that the actual number of patients in the outhouses was somewhat larger than the diet figures indicate, helping to explain the apogee of the graph in the late 1660s and 1670s, when the figures fail to account for patients who paid their own way. It also opens the question of whether the quality of treatment differed for private patients than for those supported by the hospital. The governors allowed surgeons to continue accepting fees until sometime before 1679. In May of that year the records stipulate that no hospital employee may receive any payment from outhouse patients, reverting to the former policy of treating all patients for free. They reiterated the order in 1682.[52] Within a few years the number of patients sent to the outhouses (and supported by the hospital) finally began to approach the levels seen during the first half of the century. By the late 1680s the outhouses generally treated more than 60 patients at a time, and venereal patients represented more than 15 percent of the total patient population.

But like a bad episode of *deja vu*, another fire decimated hospital properties in 1696, again disrupting the care of foul patients. Just as they had thirty years earlier, hospital governors faced difficult decisions in the face of the financial impact of disaster. The fire could not have come at a worse time, because the hospitals already suffered from separate fiscal difficulties. This crisis resulted from the effects of England's wars on the continent, during which the royal hospitals were traditionally obligated to receive soldiers. This strained hospital resources and often left few available beds for other patients.[53] The years of war that followed the Glorious Revolution put significant strain on both hospitals. Soldiers poured into St. Thomas's in 1690, and by 1691 governors had to shut down all venereal operations, commandeering all the beds in the foul wards for military men.[54] Bart's found that regiments soon stopped paying to support hospitalized soldiers, as was customary, and their accounts fell seriously into arrears.[55]

These were the circumstances when fire struck in the summer of 1696. Financial officers estimated that it destroyed eighty-three of Bart's rental properties, depriving the hospital of about £880 per annum. In August Bart's formed a committee to assess the impact of the fire and submit suggestions for coping with the fiscal crisis.[56] Citing the destruction wrought by fire and the debts owed by various regiments, which they now estimated at above £900, the committee reported that it did not see how the hospital could meet its debts and ordered that expenses be pared significantly.[57] The committee proposed the following:

> Thereupon this committee doth think fitt and are of opinion that there be no more than twenty patients att any one time admitted into each of the outhouses (Viz) The Lock and Kingsland and that the Churchwardens that recommend the said patients or their friend for the time to come should before their admittance into either of the said places become bound to pay 4d p. diem to the surgeon and guide of those houses for each patients dyett during his, her, or their continuance in those houses they being admitted only for cure of their respective infirmitys.[58]

Just as in 1666, venereal patients were the first to feel budgetary cuts. Despite the respite from 1679 to 1696 when the governors resumed paying the full cost for their treatment, venereal patients now had to pay for their care once again, while "clean" patients still received free treatment.

This time it would be permanent. The hospital cemented a two-tiered fee structure that would continue throughout the next century, and which ensured that foul patients would always pay more for hospitalization than any other patients. Again, the hospital remained committed to providing some level of care. It kept the outhouses open and paid the surgeons' salaries. But through the end of the century, the governors refused to pay for any foul patients. After 1696/97 the annual ledgers do not include diet payments to either of the two outhouses. From this point onwards, foul patients had to pay for their hospitalization. And importantly, the hospital reiterated the strict limit on the number of patients they would accept, making it much more competitive for patients to get a bed.

The decisions governing the VD operations at such points of fiscal crisis indicate a great deal about contemporary attitudes towards foul patients. There is no denying that both hospitals devoted significant resources towards treating poor foul patients in the seventeenth century. The pox was not so maligned that it "barred even charity itself," as Peter Lewis Allen has suggested.[59] The sheer numbers tell the story: such patients represented roughly one-fifth of all patients at both hospitals throughout much of the century. Even taking into account the periods when the hospitals seriously cut back on provision for foul patients, they still represented 18 percent of all patients treated at Bart's for the entire period 1622–1696. Despite an historiography that has stressed the refusal of such patients, the evidence is impressive that both hospitals saw fit to extend charity to such patients on a major scale. Yet despite whatever pity or sympathy hospital administrators may have felt towards these patients, each set of governors, independently, made cuts to the foul wards when facing constraints on resources. There was a clear hierarchy in the royal hospitals, and foul patients found themselves at the very bottom.

What of these patients? It is difficult to know who they were. Even determining something as basic as the gender breakdown of the foul wards proves difficult. Had Bart's listed monies paid to the male Lock and female Kingsland separately, we would quickly know the proportion of foul men to foul women. The spotty evidence we do have suggests that Bart's treated more men than women in its foul wards. For example, at points the outhouse surgeons' salaries were scaled according to quantity of work, as evidenced by a 1651 petition, in which John Kent, surgeon at the female Kingsland sought a raise based on the claim that he now treated as many patients as Richard Eden at the all-male Lock.[60] Yet only four years later Eden's replacement at the Lock, John Hasselock, successfully petitioned for a raise of his own based on a series of demands, including the argument that "the Lock hath recepcon for a third part more than Kingsland."[61] Later in 1690 the board ordered that while the Kingsland would continue to operate at a capacity of twenty patients, the Lock could receive thirty.[62] So the evidence that is available suggests that during the seventeenth century St. Bartholomew's tended to appropriate more resources to the care of foul men than of foul women. Unfortunately, the records do not say why. One might speculate that the rates of venereal infection in men outstretched the rates in women, and that the greater number of men cured represents a response to the actual number of foul men and women who sought relief. This remains a viable hypothesis, but one that warrants a reminder that such "rates of infection" are extremely elusive given the difficulty in identifying early modern diagnoses and the early modern tendency to confound several conditions that we now separate. If more men "had" the pox than women, it may only mean they were more likely to be so diagnosed.

Of course, the possibility also exists that the governors steered greater resources towards the treatment of men because of contemporary gender prejudices. Hospital authorities may have assigned greater value to men's health, or have been less forgiving of women who became infected. Considering the ideological and medical assumptions that portrayed women as the cause of the disease, this must remain a consideration. But the issue of gender and hospital provision is further complicated when we consider where poor patients sought care. Some of the data about the capacity of the Lock and Kingsland comes from the period following the fire when Bart's had charged fees. This is a key development, because we cannot assume that men's and women's abilities to pay such fees were equal. To the contrary, the gendered economic realities meant that women had more difficulty accessing hospital care once fees entered the equation. We will see that admission fees made it more difficult for women to access hospital care, and forced many to seek care elsewhere. Economics has to be taken into account when considering the gender breakdown at the royal hospitals once fees became permanent.

What did patients face when they entered a foul ward? To answer that question we should begin at admission. Both hospitals had weekly committees to consider the prospective patients who gathered at the hospital gates each Monday morning and to choose the lucky few who gained admission. Even though they would send them on to an outhouse, Bart's demanded that foul applicants first apply to this committee at the main hospital. Awarding a place in the hospital was one aspect of the operation that the governors were not willing to subcontract to the outhouse surgeons. At both hospitals the "takers in" committee was comprised of a handful of governors and some members of the medical staff. They were instructed to admit only "the most necessitous persons then present and curable . . . and none others."[63] Patients, therefore, had to prove that they were proper objects for the hospitals' charity on two counts, economically and medically.

The first point meant that applicants were likely interviewed about the gravity of their need—did they have resources, employment, any kin that might be able to take care of them? The second factor meant that applicants had to disclose to the committee the nature of their medical condition. Here the unique nature of the foul disease came into play. Evidence from both hospitals makes it clear that foul patients frequently resented, indeed resisted, having to reveal their condition. In 1642 St. Thomas's governors complained that patients consistently tried to avoid entering the foul wards by denying their infection. Attempting to curb such behavior, the hospital threatened to refuse treatment to any such patients.

> It is ordered that any patient come to this house for cure of the Pox and there be then noe roome in the sweatwards, yf they shall thereupon feyne [feign] some other disease and after it be found to be the Pox, the said patients are then to be held incapable of cure within this house.[64]

Bart's fared no better. There, too, patients lined up on Monday mornings and promptly lied about what ailed them. Bart's governors similarly tried to police what they viewed as a nagging problem throughout the century. And they, too, issued orders to cast out such patients upon discovery. For example, in June 1687 the governors passed the following resolution.

Patients that have the French Pox to discover it: Ordered that all patients that have the French Pox and do not discover the same att their admittance into this hosp[ital] & obtaine admittance upon the p[re]tence of some other distemper shall be forthwith discharged and shall not be sent from this hospitall to either of the outhouses.⁶⁵

The identical tendency for foul patients to conceal their conditions at both hospitals indicates a general aversion on the part of such patients to public pronouncements about their infection, not a particular aversion to entering the former leper houses. Put another way, it seems to have made little difference whether foul patients were to be segregated *within* the hospital at St. Thomas's or *outside* the hospital in Bart's outhouses. A similar stain seems to have accompanied either option, and many infected Londoners attempted to negotiate the admission process so as to avoid segregated and stigmatized care. ⁶⁶

Thus public admission of their condition constituted part of the price plebeian foul patients had to pay for treatment at early hospitals. This issue, it will be seen in coming chapters, emerges at institutions that treated the pox throughout two centuries. One has to handle the data carefully. On the one hand there seems clear evidence of deception. The governors certainly perceived it that way; their complaints indicate that they felt they were being duped by sly patients who lied their way into clean wards. Early modern patients certainly had several good reasons to try to avoid being labeled "foul." As we saw, the damage that public exposure of infection could do to plebeians' reputation could have serious economic ramifications. They may have had other reasons, too. In addition to the clear stigma that accompanied the disease, patients also may have denied the diagnosis in order to access a hospital bed when the foul wards were full, to avoid paying fees once they were required, or, at Bart's, for fear that they may be housed with patients considered too contagious for the main hospital.

However, while we can imagine that the governors' descriptions of these events were often accurate, and that patients were simply lying, we must also imagine that frequently other patients merely refused to believe that their condition was the French pox. Here the muddy nature of this diagnosis is relevant. When declared "foul" by the surgeon on the admissions committee some patients may well have honestly disagreed with the diagnosis. Given a medical culture in which doctor and patient frequently negotiated a diagnosis, with patients coming to the doctor with their own sense of the nature of their sickness,[67] patients may have felt legitimate in debating the conclusions reached by the admissions committee. Here governors' assumptions about these patients become crucial. Accusations that patients were "feigning" another disease indicate significant tension, stemming from underlying assumptions that hospital authorities made about such patients' honesty. Because they came to see patient-deception as such an ongoing dilemma governors may have come to doubt the integrity of any patient whose own diagnosis differed from that of their surgeons. In such cases, these patients may well have been labeled not just "foul," but liars as well.

Speculation about foul patients' characters held other important ramifications. At both hospitals governors also reached conclusions about those patients who applied for admission a second time. Since contemporaries believed strongly in a cure, hospital authorities assumed that any foul patient who returned to be readmitted must have reinfected themselves by returning to a life of vice. Both hospitals made it a standing rule not to readmit foul patients a second time; they were given one chance, a rule that did not apply to clean patients.[68] At Bart's some such returning patients tried to avoid detection by circumventing the admissions committee and applying for readmission directly to the outhouses. But in 1684, the hospital ordered the outhouse guides to report all such returning patients. To prevent what they considered recidivism the governors reserved the right to examine returning foul patients prior to admission.[69] They threatened to fine the guides a guinea "ev[er]y tyme that they shall take any p[er]son into either of the said houses that have been there before."[70]

But while some patients must have honestly denied that they had the pox, others openly lied. It is just as instructive to consider those patients who believed themselves to be poxed, but who subsequently tried to negotiate their diagnosis so as to avoid the label "foul." Such folk essentially sought to access hospital resources without a public admission that their case was venereal. Evidence that impoverished patients maneuvered to guard their medical and sexual history suggest that elements of the modesty discussed in the preceding chapter, that have been explored most thoroughly for the elite classes, were not a patrician monopoly. At hospitals, patients confronting the foul wards tried to guard their dignity just like their wealthier neighbors. Plebeian Londoners sought the same medical privacy that their social betters demanded. They just could not always afford to buy it.

The admissions process may have been even more involved for patients at St. Thomas's than at Bart's. Beginning in 1640 the Grand Committee considered requiring all patients to produce a certificate verifying their parish of settlement.[71] They seem not to have acted on this initial suggestion. However, in 1670, the same year that the hospital closed two venereal wards due to the financial crisis caused by the fire, the governors revisited the idea. This time, they targeted foul patients for the new policy: "It is ordered that from henceforth no pson shalbe received into this Hospitall for cure of the foul disease w.thout a Certificate from the Parson and Churchwardens of the parish where such pson then inhabites or last inhabited. . . ." The purpose of the certificate was to bind the parish to take the patient away if he or she proved incurable, or pay for their burial if they died, freeing the hospital of the financial burden. It is interesting that almost as an afterthought the hospital seems to apply this resolution generally. But the wording of the order itself as well as the marginalia indicates that this was first and foremost a stipulation for foul patients.[72]

This meant that prospective patients first had to apply to their parochial officials before they could apply for a bed in the venereal wards. Of course, this meant another level of bureaucracy to confront, and another face-to-face exchange with one's social superiors at which paupers had to declare their condition publicly. Chapter 4 will discuss the poor law system and parochial medical provision in greater detail. What is important here is that at St. Thomas's the admissions process for venereal patients was made even more extensive and more intrusive after 1670. Patients who leave clear evidence that they wished to keep their medical condition a private affair now had to discuss it with twice as many public authorities in order to access a hospital bed. So after 1670 patients at St. Thomas's may have had a new reason not to declare themselves foul at the hospital gates. Some may not have wished to involve their parish officers, while others may not have been able to produce the certificate, having migrated to London and left their parish of settlement far behind.

Once they got in, patients generally faced comparable care at the two hospitals. For example, surgeons administered the majority of treatment to foul patients in both hospitals. Of course, the outhouses only had a tiny medical staff of a lone surgeon and a single nurse, while St. Thomas's had a staff of multiple physicians and surgeons. But this seems to have made rather little difference. The pox fell more into the domain of surgery,[73] and it seems that in the main it was the surgeons that tended to the foul wards at St. Thomas's, too.[74] Because Bart's placed the entire operation of the outhouses in the hands of a single surgeon—leaving him to run all aspects of the outhouses, from medical treatment, to upkeep, to cooking and cleaning—patients' experiences and the quality of their care was entirely dependent on each surgeon's skill and integrity. Occasionally there were problems, as in 1649 when surgeon John Topliffe was fired for neglecting the patients and running an alehouse within the Kingsland outhouse.[75] Patients at the outhouses also complained that the guides did not grant them their proper food allowance.[76] However, at St. Thomas's the traditional medical division of labor forced its surgeons to consult with physicians before they administered internal medicine; the care they could provide was thus sometimes delayed while they waited to clear their decisions with physicians.[77] The traditional institutional structure of St. Thomas's Hospital imposed the traditional medical roles on its staff. By contrast the unique set-up at Bart's outhouses eliminated this kind of interprofessional friction, and may have actually promoted better, or at least speedier, care.

As for therapeutics, salivation dominated, but there is evidence of some alternative therapies, notably the use of guaiac. At Bart's, for example, outhouse guides received annual supplementary pay "for severall cures done by them by the dyett for the said yeare."[78] This refers to the use of so-called diet drinks, herbal concoctions brewed to treat the pox.[79] Although we know that such diet drinks contained a range of herbal substances in the period, guaiac was far and away the most common.[80] Like many alternatives to mercury throughout the centuries, doctors lauded the drug as much milder. In 1693 surgeon Charles Peter described the advantages of such treatment thus: "Diet . . . is a safe way, working moderately, and yet surely." [81] Peter's treatise demonstrates that diet drinks might contain herbs like china root and sarsaparilla, but St. Thomas's records indicate that doctors there relied on guaiac.[82]

In 1662 the board of governors granted a budget of £20 to the house apothecary for guaiac on the grounds that "a dyet drinke p[re]pared and given to the patients in this house of the foul disease after their unction, salivation, or sweating, would much conduce to the perfecting of their cures."[83] They doubled the expenditure for the drug to £40 two years later, marking a twenty percent increase in the apothecary's budget.[84]

At Bart's, the governors never explain why they were willing pay extra for those patients cured by guaiac as opposed to those salivated. But the ledgers indicate that they remained committed to providing the treatment (to at least some patients) throughout most of the century. An optimistic hypothesis might suggest that governors recognized the debilitating nature of salivation, and felt it worth the expense to spare those patients who might be treated without resorting to the operation. This is worth considering because through the middle of the century the numbers of guaiac "dyett" cures were not insignificant. These extra payments seem to begin in 1628, and they are a feature of the annual account for more than fifty years. At times the guides each received what seems like a set gratuity or these treatments every year. But in other years, they received payment on a per-patient basis at the impressive rate of twenty shillings (or one pound) per cure.[85] These reached a peak in 1682 when the two guides earned an extra £118 for completing 118 of these "cures."[86] Gender differences as to who received the mild but costly treatment are difficult to demonstrate. At certain points, as in the 1630s, surgeons at the male Lock recorded more treatments. But at other points, as in the 1670s, the surgeon at all-female Kingsland regularly administered more of the drug. The four years in which the Lock recorded significantly more "dyett" cures (1638–42) stand as anomalous to the larger trend of rather equal distribution throughout much of the century.

Perhaps unsurprisingly, such expensive treatment came to be seen as a luxury. Indeed, it went on the chopping block when resources became scarce in the period after the Great Fire. In 1669, just seven years after the committee had expanded the expenditure for guaiac and lauded the drug as quite useful, St. Thomas's treasurer now attacked its utility, claiming to be "credibly informed that the diett drinke . . . is not so effectual for . . . speedy curing of patients of the foul disease." He invited the "Committee on Burnt Houses" to consider the necessity of the expense.[87]

By June 1670, the same committee that closed half of the hospital's foul wards, now concurred that the drug was no longer necessary.[88] The treatment survived a bit longer at Bart's, but there, too, economics seem to have determined therapeutics. After the unprecedented expenditures on guaiac treatments in 1682, such payments come to an abrupt halt. From 1683 onwards the hospital stopped recording payments for the treatment altogether.

This means that Wilson's claim that hospital patients almost exclusively faced the torments of mercury in the eighteenth century may also be true from roughly the last third of the seventeenth century forward.[89] By 1710 one finds open criticism of hospitals for relying almost universally on salivation, to the detriment of patients' health. "Hence it is their Church-Yards are so well fill'd with dead Bodies, for they use it [salivation] . . . promiscuously without distinction, or consideration had of their peculiar and different cases."[90] There can be no doubt that the experiences of hospital patients were largely determined by their encounters with quicksilver, for even when guaiac was in use salivation still constituted the main therapy at both hospitals. For example, the figures support that only a relatively small portion of Bart's outhouse patients received guaiac. Taking a single year, in 1634/35 Bart's recorded 33 guaiac cures, in a year when the outhouses admitted, by rough estimate, more than 700 patients.[91]

The increasing restriction of treatment to salivation begins to demonstrate how the therapeutic opportunities for hospital patients were much more limited than for contemporary private patients. In an important early article on the history of syphilis Owsei Temkin compared mercury treatment with guaiac. He suggested that the gentler herbal treatment was essentially monopolized by the rich, and priced out of reach for the poor, who were left to suffer through quicksilver. The hospital records suggest that his overall impressions are generally correct. For while some poor patients may have had some access to guaiac in hospitals, they were exceptions, growing rarer as the seventeenth century wore on and as mercury increasingly became the universal therapy. Temkin also suggested that the harsher mercury treatment served as a kind of social discipline, a rough therapy that doubled as moral punishment.[92] The hospital records do not indicate that this was an openly acknowledged motive. Fiscal constraint seems to have played a primary role driving policy. But it hardly matters. For cold economics determined that Temkin's picture is basically correct. Those forced by need to seek hospital care could not negotiate with their doctor for a milder treatment regimen, as could their wealthier neighbors. Milder care existed, and for a time both hospitals tried to provide it. But as resources became scarcer each hospital grew less inclined to provide such expensive treatments to paupers in the foul wards.

That said, later attempts by hospital authorities to provide the drug make a simple verdict again difficult. In 1684, records indicate that St. Thomas's tried to renew the use of guaiac, although they now acted to limit overproscription, noting that none should be treated with it "without the advice of ye Physicians." Two years later they reiterated the order, again relying on physicians to police its use very closely.[93] So while it is clear that the treatment options for venereal hospital patients narrowed, and that most would have had no choice but to undergo salivation, governors did try to steer resources to the foul wards when they could. Regardless, as F. G. Parsons described, by roughly 1670 "systematic treatment by salivation was now becoming established."[94]

The question of whether or not salivation may have doubled as a form of hospital discipline brings us to the issue of hospital discipline itself. Curfews and other house rules meant that all patients, regardless of diagnosis, confronted attempts to discipline them when they entered a seventeenth-century hospital. What of our patients? While Bart's and St. Thomas's seem to parallel one another on issues like therapeutics, their approaches to discipline varied significantly. Overall, it seems that patients in St. Thomas's faced a much more stringent regime than their counterparts at either of Bart's outhouses.

In fact, discipline at Bart's outhouses seems quite lax. The "Lock" hospital was not in fact locked, as many have assumed; the name probably derived from the French "loques," which referred to the rags used to dress lepers.[95] Patients at the outhouses were free to leave the premises, and it seems that many did so daily. Often, the surgeons chose not to cook for patients and instead distributed their daily four-pence meal allowance so that they could purchase food in the Southwark streets.[96] This mobility of patients inevitably caused problems. Governors complained that patients drank away their food allowance in a local alehouse.[97] The outhouses must have had a curfew, and attempts were made to keep track of the patients' whereabouts.[98] But the records do not leave evidence of very strict control over patients' mobility. The debilitating regimen of mercurial salivation may have sufficiently limited patients' mobility to make such disciplinary measures unnecessary.

The extent to which outhouse patients had to work is also unclear. Early modern hospitals commonly expected ambulatory patients to help out with basic chores while in the house. While there is evidence that the Kingsland's women were made to cook, clean, spin, and sometimes nurse other patients, the Lock's men seem not to have done any of this, a point about which the Lock surgeon complained, asking for a raise to compensate him for the tasks which his counterpart could make his female patients perform, but which he could not make his men do.[99]

St. Bartholomew's tried to make sure that outhouse patients received some religious instruction while in the hospital; however, its efforts do not constitute a particularly strong campaign to reform patients. Beginning in the 1630s Bart's budgeted to pay a minister to read services at the outhouses. By the middle of the century, they contracted with a separate minister for each one. By doing so, they merely arranged for basic services to be read on a weekly or bi-weekly basis.[100] It would only be after the turn of the eighteenth century that the religious presence increased in the outhouses. It would not be until 1706 that Robert Hawkins and James Bisse became the first ministers who actually visited with patients in the wards, twice a week.[101] We must thus remain somewhat reserved about the intensity of Bart's moral mission to impress religion on foul patients in the seventeenth century. It is most telling that throughout the seventeenth century there seems little difference between the religious instruction of Bart's clean patients in the main hospital and its foul patients in its outhouses. Because it situated venereal patients at a such distance from the hospital, they could not attend the religious services in the main hospital. Therefore the governors simply provided for equivalent services at the outhouses, but no more.

St. Bartholomew's clearly did try to instill in foul patients a healthy sense of gratitude for the charity that they had received. Starting in the 1680s, patients cured in the outhouses were made to appear before the board of governors to give thanks for their care. The board threatened to withhold the surgeons' salaries unless they brought patients "to return thanks to the Governors for their said cures and likewise that the Governors may be satisfied that the said patients are perfectly cured."[102] The journals are silent on the nature of this exchange between wealthy benefactors and poor charity patients. We can assume that governors took the occasion to remind patients of the errors of their ways, of the dangers of repeated transgressions, and that they would not be readmitted if they returned infected. The ritual was clearly designed to humble supplicants and instill in them a sense of gratitude, cementing the "gift relation" that Porter described so well for the eighteenth century.[103] This ceremony mirrored the one that accompanied admission; once again foul patients had to discuss their illness publicly and assume a deferential role before their social betters. This by itself would have had a disciplinary impact on patients.

But it is worth pointing out that, while this rule targets foul patients, the governors again may have just been trying to get foul patients to do what was already standard policy in the main hospital. As early as 1652 patients in the main hospital were ordered to memorize a prayer of thanks, which they had to recite publicly before discharge. Before they could leave the hospital, patients had to recite the prayer thanking both God and governors "upon their knees in the Hall before the Hospitaler, and two Masters of this House at the least."[104] So again, the attempt to discipline foul patients at Bart's does not stand out as entirely unique from general hospital policy in the period. Moreover, the order to force surgeons to bring outhouse patients before the governors may have been as much a way of disciplining the guides as the patients. The surgeons now had to bring evidence of their work (in the form of healthy patients) to their employers for inspection in order to secure their pay.

Although scholars of early modern charities have stressed the roles of strict disciplinary regimens and religion when describing such places as reforming institutions, we are forced to admit that in the case of Bart's outhouses, attempts at reform do not impress.[105] Those expecting to find evidence of concerted efforts to develop a unique course of moral therapy specific to the foul wards do not find them at Bart's outhouses. Outhouse patients were disciplined, to be sure, but hardly more so than any other patients in the main hospital. In fact, one can even argue that because they were situated so far away from the main hospital they may have actually faced less overall scrutiny than so-called clean patients. By and large, Bart's sent foul patients to the outhouses and do not seem to have paid very much attention to them after that.

A very different picture emerges at St. Thomas's. Whereas the movements of Bart's outhouse patients were not stringently controlled, St. Thomas's foul patients found themselves strictly confined to the foul wards. As early as the mid-sixteenth century governors stipulated that the sisters of the foul ward "shall not suffer the poor to go a brode in the streets" at penalty of expulsion.[106] Almost a century later in 1647 the governors still promised to expel any foul patient found out of their ward,[107] and in 1699 they reiterated the order as a principle component of a new set of rules drawn up that year. Hereafter, St. Thomas's foul patients "shall not goe out of their ward nor come into the house to fetch any thing nor within the chappell nor sit upon the seats in the court yards upon pain of expulsion."[108] On this issue, the two hospitals could hardly have been more different.

The differences likely stem, at least in part, from the different institutional structures of Bart's distant outhouses and St. Thomas's centralized foul wards. The sheer proximity of "clean" patients to "foul" ones made necessary a much stricter set of regulations at St. Thomas's. At first these regulations represented a biological quarantine to prevent the spread of the disease. However, the belief in casual contagion—that simply being too close to someone with the pox posed a health risk—was waning by the middle of the seventeenth century.[109] After that the continued strict segregation and monitoring of patients' movements must be viewed as a kind of moral quarantine, which sought to prevent any dangerous contact with those designated foul.

But the differences between the two hospitals were far greater than just their surveillance of patients' movements. Claude Quétel has suggested that the seventeenth century brought "a change in the moral landscape," with the pox coming to be viewed with increasing contempt.[110] When one looks at St. Thomas's disciplinary policy it seems difficult to disagree. There, governors mandated corporeal punishment for foul patients. The use of corporeal punishment was not entirely restricted to the foul wards. There is evidence that the governors ordered lashes for various patients who broke house rules. For example, as early as 1567 the board ordered that patient John Marten be "whipped at the crosse and have XXV stripes" for "misusing his bodye with a poor Innocent of [the] House called Thomasyn Holden."[111] But such instances were rare, especially as the seventeenth century progressed. Moreover, they were penalties for specific breaches of hospital regulations. But in 1599 the governors put "the crosse" to new use, and initiated a radical new approach to the care and prevention of venereal disease. It was ordered in July that patients cured of venereal disease be whipped at the cross before being discharged. The governors stated:

> forasmuch as it is credibly informed to this court by some of the almoners that very many of the poor people that are taken in to be cured in this hospitall are visited with the French Disease and that it is very likely that many of them do get the same by lewd and incontinent life. Therefore it is ordered that from henceforth such of the same people as shall be noted and known as notorious and lewde livers shall, after they be cured, have some punishment inflicted upon them at the cross before they be suffered to be pass away. That by the terror thereof others may be admonished from falling into the like vice.[112]

Whereas Bart's foul patients might have to give thanks, St. Thomas's foul patients faced the lash. This order gave foul patients frightening new incentive to conceal the nature of their condition at the time of admission. From this point forward we should hardly be surprised to find patients contesting their diagnosis at St. Thomas's gates when the admitting surgeon declared them poxed. The order does not stipulate that each and every foul patient must be flogged, only those deemed to be "lewde livvers." But there is no evidence to indicate how or by whom that decision was made. Presumably, married people and known wet nurses stood a chance of avoiding the punishment, because they could blame their infection on either an unfaithful spouse or an infected nurseling. But one has to assume that the majority of those not fortunate enough to be armed with such an excuse suffered the punishment. The readiness of the hospital administration to impose physical punishment also makes one reconsider the decision to limit therapeutics to salivation only. Temkin's suggestion, that the physical torture brought by salivation may have served a purposeful disciplinary role, again resonates. It is hard to believe that a governing board that thought foul patients deserved flogging did not similarly believe that they also deserved the grief that accompanied large doses of mercury. Moreover, if administrators were rendering judgments on whether or not patients were "lewde," perhaps such judgments also determined who received the gentler guaiac treatment and who the rougher salivation.

It is also clear that the punishment was not intended just to make patients see the errors of their ways. The governors clearly meant the punishment to prevent *others* from sinning; the whipping was a public affair. We can surmise therefore that this flogging served the same purpose as other public punishments and shaming rituals that were central to social policing efforts in the period. Historians have made the sixteenth and seventeenth centuries renowned for rough music, the ducking stool, the stocks, and the pillory.[113] St. Thomas's public whipping is right in line with other contemporary forms of ritual shaming and physical punishment that aimed to preserve the social order by tarnishing transgressors in a public forum, while teaching them a painful physical lesson.[114] When exploring the social meanings of violence Susan Dwyer Amussen has stressed that "spectacle was central" to corporeal punishment in early modern England, noting that the rituals that combined violence and shaming, such as public whipping or the pillory, were most often connected with moral offenses.[115] St. Thomas's punishment of "lewde" foul patients falls right in line with that description.

Another reference to corporeal punishment indicates that the practice continued at least until the middle of the century. In 1649, the hospital reiterated that foul patients, once cured, were not to be readmitted to the hospital. In January of that year governors amended the rule, probably due to the high rates of remission. Instead of refusing to readmit venereal patients altogether, the governors ordered that they be whipped: "that henceforward none of the Fowle Disease once p[er]fectly cured and put out be after received in for the same disease except hee or shee first receive the punishment of this house for the same if by the opinion of the surgeons or credible report the same be received by ill courses."[116] Again, we see the importance of the assumptions that governors might make about the character of those who came to the hospital with this disease.

This order also put the surgeons in a rather powerful, yet possibly compromising, position. On the one hand, they could vouch for any returning patient and save them the pain and embarrassment of public whipping. However, to do so was to make a public claim (to their employers) that their original work was faulty. In other words, for a surgeon to save a patient from being whipped he had to claim that his original treatment had failed, and therefore the patient's symptoms represented an old pox, not one newly acquired through "ill courses." Considering the contemporary confidence that the disease could be cured, the importance of doctors' reputations to their livelihoods, and the assumptions made about foul patients' integrity and propensity to deceive, it becomes difficult to imagine that this happened in the majority of cases.

Needless to say, the disparity between St. Thomas's draconian approach to discipline and Bart's rather lax regimen could hardly be greater. It is difficult to pin down the reason for this difference. I can only offer an hypothesis with a suggestion for further research. There can be no doubt that early modern hospital policy reflected the values of the governors who controlled the administrative board. One possibility is that these different strategies resulted from fundamental differences in the values of those in charge of the two hospitals. Craig Rose contrasted the political and religious makeup of the two sets of governors in the later seventeenth century.[117] Rose showed that the governing board of St. Bartholomew's tended to be Tories and High Anglicans. The governors of St. Thomas's, by contrast, were Whigs, many of whom were nonconformists who used hospital patronage to foster dissent. In light of Rose's study it becomes possible to consider the effects of a religious divide on hospital policy. Certainly, the public punishment of "lewde livers" seems quite consistent with other Puritan attempts to police immoral behavior. Moreover, the chronology of this disciplinary regimen (ca.1599–ca.1649) is perfectly suited for it to have been the work of Puritans, who succeeded in the Interegnum to affect such infamous moral policing measures as the Adultery Act of 1650.[118]

However, there is a danger in pressing that argument too far. Rose's research picks up after the solid evidence of corporeal punishment, but references to whipping from the period that Rose studies do not survive. For this suggestion to hold true, then, the patterns that he describes for the late seventeenth century would have to be continuations of earlier patterns of hospital governorship, which we simply do not know. Moreover, it is questionable whether one can equate late seventeenth-century

Whiggery with early seventeenth-century Puritanism, though there is certainly some connection between the two. Muddying the waters even further, the one figure actively involved in the royal hospitals' medical operations in the early seventeenth century whose Puritanism is well documented, surgeon William Clowes, worked at St. Bartholomew's, not St. Thomas's.[119] So it is entirely possible that something similar may have taken place at Bart's, but simply evaded recording. Here is one point where further research is needed. Still, the contrast between St. Bartholomew's and St. Thomas's on the issue of discipline seems striking, and it does not seem that the different institutional structure of the two hospitals can explain that difference by itself. The differences in institutional structure can explain why one hospital enforced greater restrictions on the movements of venereal patients, but not why one hospital publicly whipped them and the other did not.

Throughout the seventeenth century, then, London's royal hospitals treated venereal patients on a large scale as part of their general day-to-day operations. Because they survived on rental incomes, and did not, like the voluntary hospitals of the following century, rely heavily on subscriptions, they did not have to advertise what they were doing in fundraising pamphlets. So while they did not prominently broadcast that throughout the century almost one out of every five patients entering their hospital went into the foul wards, the records clearly indicate that such was the case. The moral climate of the seventeenth century produced many commentators who saw the disease as part of God's design, as a just punishment for sinners who brought the disease on themselves. But despite the fact that English writers seemed especially prone to take a hard line on the moral transgressions of venereal patients, and in spite of the gruesome evidence of corporeal punishment, we are still left with the fact that those in charge of both institutions saw fit to devote significant hospital resources to the treatment of venereal disease throughout the century. Some, like those at St. Thomas's, may have thought it a useful way to impose discipline on the bodies of the unruly poor who contracted this disease. But others seem driven by different, albeit elusive, motivations. One begins to get the sense that that the problem of venereal disease among the poor was simply too big and too threatening to be handled through moral chastisement alone. The sheer numbers of paupers flooding the foul wards throughout the century suggest that twin problems of poverty and the pox proved massive and recalcitrant. Whether motivated by a sense of civic obligation, a sense of pity, a sense of Christian duty, or a deep fear of an unchecked epidemic raging through the city, authorities at both hospitals turned significant and sustained energies to the campaign against the foul disease.

These hospitals provided medical resources to those economically excluded from the medical market. They served these patients, but the foul assumed a clearly defined position at the very bottom of each hospital's hierarchy. They were set off in quarantined wards that held the cultural memory of leprosy, where authorities might also place patients deemed too contagious for the general population. They were barred from readmission, if they became reinfected. And most importantly, when times got tough and resources grew scarce, they could find themselves on the outside looking in. Authorities seemed unable to consider providing care for a foul patient if it meant withholding care from someone else, and thus they privileged the care of clean patients over foul ones as a kind of unwritten rule. Moreover, the assumptions that underlay foul patients' special status led at the end of the century to new policies that demanded fees from them not required of other patients. Those who believed that foul patients were responsible for their own illness were likely happy to see them become at least partially responsible for the cost of their care. These fees would become a standard feature of both hospitals' policies throughout the following century, creating a new obstacle for the very poor who contracted the foul disease. As the seventeenth century wore on, hospital patients also faced treatment that became increasingly limited to mercurial salivation alone. Governors' attempts to provide expensive but milder therapeutics to some patients mark a notable, perhaps even a noble, effort, which further undermines the notion that hospital authorities were without pity for those entering the foul wards in the seventeenth century. But again, as finances tightened such luxuries were among the first to go. The differences in the illness-experiences between the poxed wealthy and the poxed poor grew starker.

Precisely when salivation was becoming less common for private patients, it was becoming increasingly uniform in hospitals. And therapeutics marks just one aspect of the widening gap between private patients and hospital patients. Paupers quickly found that the medical privacy easily obtained by their social betters was nearly impossible for them, as they lined up at the hospital gates hoping for a bed. If entering a hospital in this period was generally something of a last medical resort, and if entering the public eye with the pox was similarly something that most people sought badly to avoid, we have to consider the people who took the steps to enter seventeenth-century foul wards as being driven there by desperation. They were driven by economic need, to be sure; they could not afford private care. But they were also driven by a kind of corporeal need. For the people who risked not just public exposure, but even physical punishment, to access medical care must have been those enduring the very worst of what the foul disease had to offer. If there was a great range in illness experiences of the seventeenth-century "foul," from minor urinary tract infections to raging cases of secondary syphilis, we are probably safe in assuming that the people who undertook to enter a hospital in this period were those suffering from the worst of physical symptoms.

In the end, the ambiguous characterization of the foul wards that began this chapter seems well placed. Understanding VD in the royal hospitals means understanding how the foul wards could be simultaneously merciful and despised. They were certainly far more merciful than most scholars have assumed, devoting considerable resources to the care of people with the most odious disease then known. However, when we consider the individual men and women who found themselves humiliated before admissions committees, forced into stigmatized wards, who struggled to endure the tortures of mercury, who humbled themselves before the governors to give thanks, and who even found themselves strapped to the cross and publicly beaten, we should hardly be surprised to find that they absolutely despised the foul wards, with every ounce of strength.

3

THE FOUL DISEASE IN THE ROYAL HOSPITALS: THE EIGHTEENTH CENTURY

As the eighteenth century dawned, beds in royal hospital foul wards were becoming harder to get. The financial effects of fire and war in the 1690s severely depleted the coffers as the seventeenth century drew to a close. This fiscal pressure forced St. Bartholomew's to stop paying to support foul patients in the outhouses after 1696. So as the new century began, venereal patients now had to come up with the four pence per day in order to stay in the outhouses, even though the hospital continued to pay to support the hundreds of patients treated each month in the clean wards of the main hospital. This two-tiered fee structure would last in one form or another throughout the century.

Bart's did resume paying to support some venereal patients in 1703. However, the figures betray a clear shift in policy. In 1703 the governors spent just £93 to support outhouse patients while they spent in excess of £1,900 to feed clean patients. When translated into fiscal terms this means that that the hospital paid to support about 314 clean patients at a time, while they supported on average only about fifteen foul patients throughout the course of the year. In stark contrast to the pattern in the seventeenth century, when venereal patients represented such a significant portion of St. Bartholomew's charity cases—well over 20 percent in some years—in 1703 they represented less than 5 percent of the patients supported by the governors.

However, it is important to bear in mind that these figures do not represent all the patients treated at the two outhouses, but only those receiving full hospital charity. The hospital now classified foul patients as either needy of charity or capable of paying their own way. The hospital consented to support only those who were "entirely destitute of Mony, or friends, & parish settlements."[1] So in late 1702 or early 1703 the governors resumed their charitable support for poor foul patients, but for a much more limited group. This renewal shows that the hospital had not entirely abandoned the people struck by the dual dilemmas of poverty and the pox. There remained some commitment to helping impoverished foul patients. However, governors now drastically limited the amount of money devoted to the cause, and began to call on parishes to contribute towards the support of their own venereal paupers. In plainest terms, a hospital that had at one time supported a significant body of foul patients, now supported a select few.

An examination of the annual diet payments provides evidence about this change. In the eighteenth century the sums spent to feed venereal patients dropped off substantially from those spent in the seventeenth century. Even when one includes the period of fiscal crisis that followed the Great Fire of 1666, Bart's governors spent an average of over £365 each year to feed the patients in its outhouses during the period 1622 to 1696. However, from 1703 until the closure of the outhouses in 1760 these payments fell 72 percent; on average the hospital paid less than £100 pounds per year to support venereal patients. When translated into actual numbers of patients the diet payments show that throughout the seventeenth century Bart's charity supported over sixty venereal patients at a time, while in the eighteenth century the average number of foul patients who received full hospital charity was merely sixteen. Figure 6 illustrates the drastic reduction in the overall number of charity foul patients in the eighteenth century. Hardly a period of progress, from the point of view of the venereal poor this was a step in

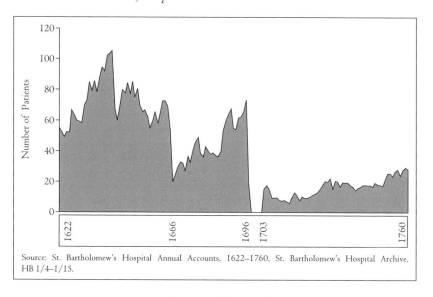

Fig. 6: Annual Average Number of Venereal Patients Supported by St. Bartholomew's Hospital, 1622–1760.

the wrong direction. It is further useful to contrast these figures with the data for clean patients. Figure 7 charts the average number of clean patients supported over the same period, 1622–1760. Generally speaking, the figures show that the hospital continued to expand, supporting an increasing number of clean patients as time went on. Therefore, not only did the absolute number of venereal patients receiving charitable care drop off in the eighteenth century, but these patients also came to represent a smaller fraction of Bart's total charity recipients, as figure 8 illustrates. In contrast to the seemingly generous seventeenth century, in no year between 1703 and 1760 did venereal patients who received the support of the house ever represent more than 6 percent of all charity patients. In many years that figure was below 2 percent, and for the entire fifty-seven-year period to 1760 they represented merely 3.5 percent of all the patients supported by St. Bartholomew's Hospital.

However, the outhouses also received fee-paying patients that the financial records simply do not register, making it difficult to gauge the overall scope of the operations in this period. Paying patients are simply invisible in the records, and so the numbers received at the outhouses

The Foul Disease in the Royal Hospitals: The Eighteenth Century 99

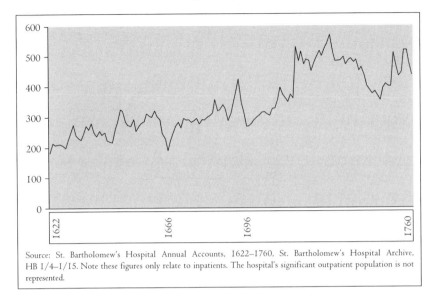

Source: St. Bartholomew's Hospital Annual Accounts, 1622–1760, St. Bartholomew's Hospital Archive, HB 1/4–1/15. Note these figures only relate to inpatients. The hospital's significant outpatient population is not represented.

Fig. 7: Annual Average Number of "Clean" Patients Supported by St. Bartholomew's Hospital, 1622–1760.

appear overly anemic if one examines just the diet payments. One surviving account that indicates the capacity of the outhouses in the eighteenth century might help to forge a rough estimate of the balance between paying patients and charitable patients in Bart's outhouses. In 1724 the surgeon at the Lock outhouse, Samuel Palmer, wrote a letter in support of salivation during the debate occasioned by the challenge from Montpellier doctors, which surgeon Daniel Turner reprinted in two of his treatises.[2] Palmer gave an account of all the patients received at the Lock outhouse from January 1, 1719, to January 1, 1720. According to Palmer, he treated 115 patients during that twelve-month period; 108 were cured and discharged, and seven died. We must double that figure to incorporate the approximate number of female patients also treated at the Kingsland that year (with the caveat that this is a very rough estimate). Since treatment lasted approximately 40.5 days,[3] the cost to feed all 230 venereal patients would have been approximately £155.5.0. Yet the ledgers show that the hospital paid just £58.7.0 to feed outhouse patients that year. This means that the governors supported only (approximately) 37.5 percent of the foul patients treated in the

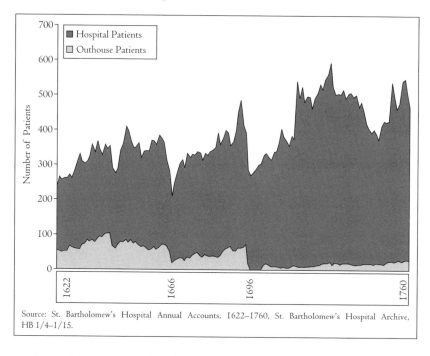

Fig. 8: Annual Average Number of "Clean" Patients and Venereal Patients Supported by St. Bartholomew's Hospital, 1622–1760.

outhouses that year. Almost two-thirds were not deemed destitute of friends, money, or a parish settlement, and therefore had to support themselves. However, the figures that arise from Palmer's account also indicate that the overall capacity of the outhouses remained low in this period. Regardless of who paid for their care, only around 230 foul patients received care from Bart's in the entire year of 1719. Yet in a typical year in the seventeenth century the outhouses combined to treat around 481 patients.[4] Moreover, when related to the sums spent to feed clean patients in 1719 it emerges that the 230 venereal patients would have still represented just over 5 percent of all of Bart's patients that year.[5] Of course, these estimates are rough, and it would be dangerous to extrapolate too much from Palmer's account of one outhouse during one year. But it seems clear that Bart's accepted fewer venereal patients in the early eighteenth century than it had during much of the previous century.

These numbers seems to have rebounded by the middle of the century, something which, again, the graphs above do not reflect. A 1754 report reveals the capacity of the outhouses, thirty-three beds in each house.[6] Although Bart's provided sixty-six beds for venereal patients by 1754, the diet figures for that year show that the outhouses paid to support an average of about twenty-three charity patients. In other words it continued to support only about one-third of their total possible capacity, still leaving roughly two-thirds of the beds for paying patients, whom the records do not register. Still, knowing the outhouses' capacity is helpful. If the outhouses operated at full capacity all year long they would have cared for over five hundred patients. Anything even approaching this would represent a dramatic rebound from the first decades of the century.[7] In fact, the City of London's annual reports on city hospitals confirm that by the 1750s the outhouses indeed did operate quite near capacity. Each Easter the city published the *True Report of the great number of Poor Children, and other Poor People maintained in the several Hospitals under the pious care of the Lord Mayor*, which recorded the number of patients treated in hospitals, and those for 1753–55 and 1757–60 record the number of outhouse patients. In most years all sixty-six beds were full, and they never recorded fewer than fifty-eight patients in the outhouses.[8]

So indeed by the 1750s Bart's was treating about as many foul patients as it had through much of the previous century. Sixty-odd patients could be found in the outhouses throughout many years of the seventeenth century. Therefore, the diet figures taken alone slightly overstate the extent to which Bart's cut back on VD care in the eighteenth century. The raw numbers of patients treated was somewhat more constant over the long period, with two main dips following the fires of 1666 and 1696, the effects of the latter lasting at least until 1720. However, expansion seems not to have happened. Rather it seems that the same facilities accommodating roughly the same number of beds continued to serve Bart's from 1622 right through to 1760. Of course, over that time the main wards had expanded, rendering these sixty-six foul patients a smaller portion of Bart's total clientele. When related to the number of patients reported in the main wards in the Easter reports of the 1750s, outhouse patients now represented between 16.1 percent and 19.3 percent each year.[9] Ultimately, then, the figures we have suggest that Bart's VD operations had diminished significantly at the turn of the century, but had rebounded somewhat by 1750. The outhouses operated at full capacity again, but this was the same capacity as in 1622, and London had grown significantly in the interim. Finally, we cannot forget that by 1750 roughly two out of three patients who occupied one of these increasingly difficult-to-obtain outhouse beds had to pay fees.

So did patients at St. Thomas's, which also initiated fees for foul patients in the early eighteenth century. The first clear description of patient fees emerges in 1715.[10] At St. Thomas's all patients, regardless of diagnosis, had to pay fees, but foul patients had to pay quite a bit more. All patients (or their parishes) had to pay a per diem subsistence fee and post a bond of one guinea to cover burial costs in case of death. In addition to these charges, clean patients had to pay a six-pence fee to process their petition, one shilling, two pence to the sister of their ward for "cleane sheets, earthen ware, vialls, cork and other necessaryes," and six pence to the nurse in their ward. Their fees totaled two shillings, two pence. By contrast, foul patients also had to pay the six-pence petition fee, but they had to pay their sister two shillings, six pence, and their nurse one full shilling. Their fees totaled four shillings. This unequal fee structure became more pronounced as the century progressed. By 1730/31 admission fees had risen.[11] Clean patients' fees had increased four pence, while those of foul patients had risen a full shilling: clean patients now paid two shillings, six pence, while foul patients paid five shillings, or fully double.[12]

It is tempting to conclude that the higher fees demanded of foul patients represented a manifestation of discrimination. There is likely much truth to that hypothesis. But evidence also suggests that salivation itself was not cheap, and that the hospital may have charged higher fees in an attempt to put some of that cost onto the foul patients themselves. Consider that the sisters and nurses of the foul wards consistently commanded higher salaries than the nurses of other wards.[13] Salivation was labor intensive. Bringing patients through the dangerous procedure likely required greater skill. It is also possible that the foul wards were simply so gruesome that the hospital had to pay the nursing staff higher salaries to persuade them to take the jobs. Whatever the case, in terms of labor, the foul wards were the most costly in the hospital. Salivation also required an extra quantity flannels in which to dress patients.

Discussions of salivation constantly comment that the operation required many flannels, and parish records show that churchwardens often had to pay extra to cover the cost the flannels used by their parishioners while salivating in hospitals.[14] When reviewing what was covered by foul patients' higher fees, the board stipulated that over and above the standard fees, foul patients were charged for "what is further necessary or usuall upon salivations."[15] Taking these costs into consideration, it becomes difficult to explain the hierarchy of fees merely in terms of discrimination against unpopular venereal patients.

But we cannot entirely rule out the influence of such attitudes when examining St. Thomas's fee structure in the eighteenth century. Evidence shows that the disproportionate fee schedule became even more radically pronounced as the century wore on. Interestingly, the hospital experimented with eliminating all fees for all patients in 1758.[16] Unfortunately, there survives no record of the debate that led up to this rather progressive step. The moratorium on fees lasted for a decade. But in 1768 the Grand Committee, complaining of a budget shortfall, reversed itself, and advised the General Court to reinstate admissions fees. Clean patients' fees remained at their original level of two shillings, six pence. However, the Grand Committee suggested, and the General Court ordered, to raise foul patients' fees more than 300 percent, to seventeen shillings, six pence.[17] As figure 9 shows, from 1715 to 1768 the fees of clean patients rose barely 15 percent, while the fees demanded of foul patients skyrocketed over 337 percent. Such a disparity in the fee schedule not only calls into question the underlying attitudes of the hospital administrators who set it, but also raises wider issues touching the availability of hospital care and the social class of those who could afford to access beds in the foul wards.

Such issues were not lost on the board of governors. Three years later, St. Thomas's administrators acknowledged that the fees demanded of foul patients were excessive. They admitted that "many Foul people are greatly distressed in not being able to raise 17/6."[18] The governors offered relief by decreasing the admission fees for foul patients to ten shillings, six pence. To make up for the loss of income, they simultaneously agreed to increase the fees of clean patients—which had only risen four pence in close to sixty years—by one shilling, to three shillings, six pence.[19] Despite this move to ease the disparity in the fee schedule, foul

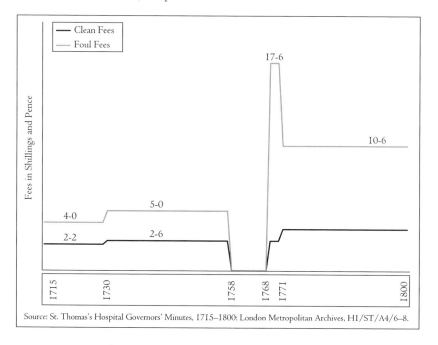

Fig. 9: Admissions Fees at St. Thomas's Hospital, 1715–1800.

patients continued to pay three times more than clean patients for admission to St. Thomas's. This fee structure lasted for the rest of the century. Moreover, the situation was even worse at St. Bartholomew's. By the end of the eighteenth century Bart's was actually a cheaper option for clean patients, but a much more expensive choice for the foul. Clean patients there were charged just two shillings, while the foul had to pay £1.5.8, or almost thirteen times more.[20]

Of course this gave foul patients yet one more reason to conceal or challenge their diagnosis during the admission process. While some patients surely avoided the foul wards because of the stigma, others did so out of sheer economic need. But there can hardly be doubt that the fee structure was, to an extent, punitive. Even if they did not originate for that purpose, surviving discussions show that by the mid-eighteenth century fees had come to be perceived as a just punishment for the foul. Trustees of the Westminster Infirmary, one of the voluntary hospitals that refused venereal patients on moral grounds, as much as said so in January 1738/39. When considering whom to accept into their hospital, a committee of governors recommended refusing venereal patients, noting that "persons infected with the Venereal Disease do generally bring it upon themselves by their own lewd and vicious Habits." Tellingly they cast the admissions fees charged at the royal hospitals as a kind of punishment for such patients: "That upon Enquiry, your Committee are informed that . . . the Great Hospitals in the City and Suburbs of London . . . do not Admit Venereal patients without subjecting [them] to an Extraordinary Contribution for their own support."[21]

The move to make foul patients bear a greater cost of their care at the royal hospitals undoubtedly affected access to these medical resources. By and large hospitals in this period remained for those at the bottom of the social ladder. People of means continued to rely on private physicians. But the "poor" treated in hospitals was a stratified group. For example, at any given time the royal hospital foul wards would have housed patients who paid for their own care, patients supported by the resources of friends and family, patients with access to parochial support, and other patients in abject poverty without recourse even to the parish, whom the hospital agreed to support. According to a mid-century report on the outhouses:

> [G]reat numbers of patients were now admitted therein of a better rank in life than formerly such as tradesmen & manufacturers being housekeepers & gentlemens servants, who can very well afford so small a fee on their Admission, that there were many poor who have parish settlements, for whom the parish officers paid the fees and others who belong to clubs of tradesmen & merchants who were paid for by the stewards thereof and all soldiers by the agents of their regiments. And that all those persons who were destitute of money friends & parish settlements were admitted on the house account free from fees of any kind . . .[22]

Therefore, when we consider the overall picture of institutional venereal care for the poor, it is clear that the royal hospitals in the eighteenth century came to care for those at the top end of that demographic group. Increasingly Bart's and St. Thomas's foul wards came to care for the working poor, reserving fewer and fewer beds for the absolutely destitute. On average the majority of the clientele at Bart's and St. Thomas's in the eighteenth century were simply not as poor as those they treated earlier. Of course the very poor continued to require help when they became infected, but as the eighteenth century wore on, they increasingly had to look elsewhere.

One of the more important policy decisions affecting venereal care at Bart's was undoubtedly the decision to close the outhouses after roughly two centuries of service. This resulted from yet another financial crisis. The idea to close the outhouses first emerged in the mid 1750s when the hospital found its finances tight thanks to a rebuilding project that proved more expensive than the board had envisioned. In 1754 Treasurer John Tuff prepared a report on finances that warned of fiscal hardships and which specifically targeted the outhouses as a way to cut expenses.[23] The report estimated the total cost of the outhouses (including staff salaries, medicines, and upkeep) at over seven hundred pounds per annum. Moreover, it estimated that the hospital could generate one hundred pounds per year if it rented out the buildings. So for the third time in its history, when faced with a financial crisis, Bart's governors hit upon the idea to close the outhouses.

Tuff's report relates the history of Bart's lazar houses and the origin of its policies towards venereal patients, most of which has been already covered in chapter 2. But it also includes some invaluable new information. For example, it demonstrates beyond doubt that by the mid-eighteenth century hospitals segregated foul patients for reasons other than a fear of biological contamination. The report acknowledged that the hospital had originally sent venereal patients to former leper houses because of the fear that the disease might spread in the wards. But Treasurer Tuff now acknowledged that people no longer believed this. He portrayed the outhouses as obsolete, claiming that the original reason for segregating patients "is now exploded, there being no such danger of infection as was formerly apprehended."[24] The continued segregation of foul patients well after the decline of that belief could only stem from the contemporary stigma of the disease itself.

In fact, that policy of segregation was routinely compromised. Tuff also acknowledged that venereal patients regularly entered clean wards, and everyone knew it. Since that was the case, he asked, why keep the outhouses? He argued:

That it doth also appear to us that notwithstanding the Orders, that no Foul Patients be admitted or continue in this hospital, there hath been & now are great numbers of patients with Venereal Complaints admitted into and salivated in several of the wards of this hospital, especially in the two fluxing wards, the one for men & the other for women, which do contain twenty beds each . . .[25]

This surprising admission raises additional questions. Certainly it confirms that creative patients commonly tried to access hospital care while avoiding the foul wards well into the eighteenth century. Moreover, the frankness with which Tuff suggests that everyone knows this, suggests that some members of the medical staff and the admissions committee may have helped them do so. Doctors and possibly some governors may have played along with patients' fictitious claims at the time of admission, agreeing to salivate them under the guise of a different ailment. Perhaps some aspiring patients came to the hospital armed with the knowledge of another condition that they could claim which also demanded "fluxing." We can imagine patients and surgeons engaging in a kind of agreed fiction at the hospital gates, as some patients quietly slipped in. Given the excessive admission fees, it is possible that some patients' claims of poverty may have convinced hospital staff to admit them to a clean ward as an act of charity. Conversely, since the hospital came to treat patients of a "better rank," we cannot rule out the possibility that some used old-fashioned graft to secure a bed in the main hospital rather than in an outhouse.

However they did it, some venereal patients successfully negotiated their way around the official admissions policy and avoided entering the outhouses. Furthermore, this also meant that the hospital treated even more venereal patients than the financial figures suggest. To the numbers treated in the outhouses, we would have to add an unknown portion of men and women who occupied the forty beds of the fluxing wards. Finally, none of this seemed a great secret. Since everyone knew this to be commonplace, Tuff argued, they might as well make it official, close the outhouses, and consolidate operations in the main hospital.

However, at a meeting of nineteen governors the board decided against Tuff's proposition. For the moment the outhouses survived.[26] If we assume that the four authors of the report endorsed for the plan, at least ten of the remaining fifteen governors present at the meeting voted to keep the outhouses open. Someone must have made a strong case against Tuff's proposition, though the grounds for that opposition are unclear. Two quite different possibilities exist. First, it is possible that certain governors argued to support Bart's current venereal operations. Despite the cost, they may have felt that it was money well spent to provide extra resources specifically devoted to the treatment of foul patients. In other words, they may have voted to protect the outhouses from closure out of sympathy for the plight of plebeian venereal patients. But it is equally possible that some governors may have opposed the closure because of prejudices *against* foul patients. Regardless of medical theories about contagion, it is clear that many people in the eighteenth century still felt strongly about segregating the foul. Thus it is possible that some governors may have opposed the plan to close the outhouses, because they still believed in their original purpose: namely, to keep venereal patients as far away from clean patients as possible.

Closing the outhouses remained a dead issue until January 1757, when Tuff and two of the other three governors originally committed to their closure (John Wallington and Robert Pyecroft) reintroduced the proposal. The board began discussing the possibility of building a new structure on the hospital grounds to accommodate foul patients if they closed the outhouses.[27] As they considered the move an unexpected financial blow devastated Bart's, and sealed the fates of the Lock and Kingsland. John Tuff, the treasurer, absconded with an estimated £4,100 of hospital funds. Details of the incident are sketchy.[28] It seems that Tuff lived well beyond his means and embezzled money. The board called an emergency session on June 16, 1760, because the hospital could not meet its financial obligations. They drew a Commission of Bankruptcy on June 20, and printed a notice in the *London Gazette* announcing the state of the hospital's finances.[29]

It was in the wake of this fiasco that the outhouses finally closed. It is unclear what happened to the opposition to Tuff's scheme to close the outhouses; it is likely that dire financial straits now silenced anyone who held hopes of saving them. On October 23, 1760, a committee of eleven governors reread the 1754 report and the 1757 proposal and issued a final resolution to close the outhouses forever. In the resolution, they again admitted that "a great number" of venereal patients "have been admitted into and now are under the care of this hospital, under pretence of other distempers, and are and have been salivated in all the wards of this hospital, contrary to the rules. . . ."[30] They complained that such patients proved an unnecessary cost because they were provided with their diet by the hospital, when they should be made to pay for their own subsistence.[31] As had happened twice in the seventeenth century, financial crisis spelled bad news for the outhouses. But this time it was permanent. The outhouses closed in 1760, and foul patients now took up beds in the newly rebuilt main hospital. Men would enter Job and Lazarus wards; women would enter Hope and Charity.[32]

It becomes more difficult to ascertain what happened to venereal patients at Bart's after the closure of the outhouses.[33] There may have finally been some increase in the number of venereal patients accommodated, though this seems to have largely benefited men. When debating the closure of the outhouses the board indicated that they originally envisioned men's and women's foul wards of forty beds each.[34] However, when they moved the patients into the main house their four wards actually totaled just sixty-eight beds, a net gain of two beds. However, these were not evenly distributed; men gained three beds, but women actually lost a bed.[35] There remained need for more space and the board quickly debated adding another ward for foul women. In 1761 they seemed to commit to increasing the capacity of the women's wards from thirty-two to forty beds, but they subsequently postponed the decision and seem not to have enacted the order.[36] Instead they seem to have increased the capacity of the men's ward by as many as twenty beds by 1762.[37] That trend continued, and the gender gap became more pronounced later in the century. By 1788 Bart's devoted twice as much space to foul men, reserving four wards for foul men and just two for women.[38] Yet one finds the board still discussing the need for more women's ward space as late as 1793.[39]

That is a brief overview of the scope of hospital provision in the eighteenth century. But what of the patients themselves? St. Thomas's admissions registers offer some of the best surviving data about the population of the royal hospitals' foul wards. An impressive set of weekly admission registers survive covering much of the period from 1773 to 1800.[40] No similar records survive from St. Bartholomew's for the period. At first glance, the registers appear to offer rather little information: patient's name, ward, date of admission, fee, and the name of their surety (i.e., the person, parish or agent that posted the bond to cover burial costs in case of death). Though this may not seem to offer much, the registers prove to be immensely helpful for a study of the foul disease. Because by this point venereal patients had to pay ten shillings, six pence rather than the standard three shillings, six pence, it is immediately apparent which patients were designated foul. Moreover, since St. Thomas's continued segregation, the name of the ward indicates whether patients were in clean wards or foul wards. Throughout this period foul men went to Job and Naples wards, foul women to Magdalen and Susannah wards.[41] If these two indications were not already enough to identify the venereal patients, the records themselves are divided so as to list the foul patients separately. The male and female patients were indexed on separate pages, and foul patients were listed separately at the bottom of each page, their names physically segregated on the page. Therefore, the admissions registers offer the most concrete data on the breakdown of the patient population. They also shed light on the strategies that the poor employed to access the care they needed. Finally, the registers prove beyond any doubt that VD care remained a major component of the royal hospitals' services in this period.

The first three registers comprise the surviving records from forty weeks falling between June 1773 and August 1776. During these forty weeks St. Thomas admitted a total of 2,432 patients. Of those, 684 were venereal patients. In other words, over 28 percent of St. Thomas's clientele occupied the foul wards. Although Bart's seems to have committed only about 16 to 19 percent of its ward space to VD by mid century, at St. Thomas's well over a quarter of all patients went to the foul wards. An even larger and more complete sample corroborates this figure. The hospital compiled an index of all its patients for the years 1768–71.[42] Out of the 5,329 men who entered the hospital between 1768 and 1771, 27.7 percent (1,481) entered the foul wards. Regardless of how they gained admission, or who paid their fees, these patients prove beyond doubt that VD care remained one of the primary functions the royal hospitals served in the eighteenth century. Each year one single diagnosis accounted for more than a quarter of all of St. Thomas's patients and approached one-fifth of Bart's patients. That fact alone requires historians of London medicine and London charity to reevaluate the role played by these institutions in this period.

Who were these patients? The admissions registers show that more of them were men than women. The figures drawn from the 1768–71 index for the entire hospital indicate a 59/41 percent split: out of 9,107 total patients listed in the index, 3,778 (41 percent) were women.[43] The administrative records offer no indication as to why St. Thomas's executive reserved more ward space for men. One deduces that plebeian women who fell ill must have faced a much more competitive and difficult admissions process as they scrambled to occupy fewer available beds. Women represented no less a portion of the total London population than men, and they were no less likely to fall ill. In fact, sheer economics determined that women were more likely than men to need charitable care. Yet the admissions fees demanded at the royal hospitals may have priced them out of reach for many women. Women would have had greater difficulty paying fees generally, but the exaggerated fees demanded of foul patients resulted in an even greater gender split in the foul wards. Of the 684 foul patients that can be identified from the surviving 1773–76 registers, 63.3 percent (433) were men and 36.6 percent (251) were women. Thus, the gender breakdown in the foul wards was slightly more pronounced than that of the main hospital. The divide would become wider as time went on.

First, foul women did not have two full wards, as did men. Almost two-thirds of the beds in Susannah Ward actually went to clean women, suggesting that the "strict" policy of segregation was compromised at St. Thomas's, too. Women listed among the clean patients paying the lower fees accounted for 63 percent of all the patients who entered the ward in the early 1770s.[44] The evidence perplexes, for the board never discussed a change to the segregation policy, and a similar phenomenon of clean and foul men sharing a ward did not happen. One can only hypothesize that the hospital appropriated ward space for clean women, leaving fewer beds for foul women. Foul women did not have two wards; they had one and one-third.

A series of questions immediately rises. Was there any sort of moral hierarchy that existed between Susannah and Magdalen wards? Did members of the admissions committee steer "innocent" women to Susannah (e.g., married women infected by their husbands, or wet nurses infected by nurslings) and "corrupt" women (e.g., single women suspected of promiscuity or prostitution) to Magdalen? If there was a stigma attached to the foul wards, was the stigma of Susannah ward mitigated since almost two-thirds of its clientele were clean patients? One must also look at it from the opposite perspective, that of the clean women who entered Susannah. If it may have been advantageous for some foul women to enter a less dubious ward along with clean patients, was it some kind of punishment for clean women to have to share a ward with poxed women? In other words, did women with questionable reputations have to enter Susannah ward when they merely broke a leg? Of course, patients' diagnoses may have determined this, with the ward accommodating patients requiring fluxing for a variety of ailments. But at least one piece of evidence suggests that some clean women resisted entering Susannah. On February 22, 1770, Susan Bird was admitted into Susannah, but she refused to enter. In fact, she discharged herself the same day and chose not to enter the hospital at all if it meant entering Susannah ward. The note pertaining to her discharge reads: "went away because warded in Sus[annah]."[45] Regardless, by mixing foul and clean patients in one ward, the hospital may have also unknowingly afforded some women a better chance of attaining treatment for their pox under the guise of a less dubious diagnosis at a cheaper cost. But while this may have benefited some women, it certainly hurt others, for simply fewer beds were made available for plebeian women with the pox.

The situation grew worse. The board decided in 1781 to turn Susannah into a men's ward for clean patients to be called "Henry." No other changes compensated for the loss of this highly valued space for female venereal patients. So after 1781, foul women faced even stiffer competition for a hospital bed; the hospital still reserved two wards for its venereal men, but now just the Magdalen ward for women. We can chart the impact of that loss. During the period 1773 to 1776 St. Thomas's admitted an average of 6.4 foul women each week. Contrast that to the period after Susannah closed. By 1787, a year with particularly good surviving registers, the average number of foul women admitted had fallen to 4.7 women per week, a decrease of over 26 percent.[46] The gender disparity in ward space would become even more pronounced in the early nineteenth century, when one finds the gap growing from two-to-one to three-to-one, thus mirroring the situation at Bart's.[47]

The different strategies employed by men and women also become clearer when one views the gender breakdown in combination with the data on patients sent from parishes. Parish patients—identified in the registers with "PP" next to their name—accounted for just over 9 percent of the total patient population.[48] Foul patients were slightly more likely than clean patients to have to rely on their parish to get in; almost 11 percent of all foul patients came from parishes.[49] This again highlights the effect of the unequal fee structure at this hospital; since a bed in the foul wards simply cost more, more people had to rely on parish support when applying. Women made up the clear majority of these patients: over 56 percent of all patients sent from a parish were women.[50] But the effect of VD is striking. The gender breakdown of parish patients in the clean wards is a virtual dead heat. Among clean patients sent from parishes we see a 51/49 split between men and women, with women representing the slight minority. The picture for foul patients is radically different. Women comprised 73 percent of all patients sent by parishes into the foul wards.[51] In other words, of all the venereal patients who had to rely on their parish in order to get into the hospital, women outnumbered men by almost three to one.

These data begin to show how crucial parish relief was to the poor women who contracted this disease in eighteenth-century London. Women were simply much more dependent on parish resources when they fell ill. Consider that 21.5 percent (54 of 251) of all the women who entered St. Thomas's foul wards in the early 1770s could only do so with the help of their parish. Contrast that figure to the men's wards, where just 4.6 percent (20 of 433) of foul men had to negotiate their admission through a parish. We should consider the implications of this for privacy. Patients who had to arrange their petition through their parish first had to confront the parochial bureaucracy where they almost certainly would have had to disclose their infection. Such patients would have already gone through one public declaration by the time they arrived to present their petition at the hospital gates. In simple terms, patients who entered the foul wards by way of their parish faced an admissions process that was twice as public an affair as that of other patients. Moreover, patients who had to apply first to a parish might also face other risks, such as having to enter the workhouse as their only medical option, a development that we will explore further in the next chapter.

More than one out of every five women who entered the foul wards had to endure that process, while less than one in twenty men did.

As we have seen, getting a bed was not easy. One way that patients could improve their chances at the admissions committee was to obtain a governor's nomination. We will see that such a nomination was required by voluntary hospitals like the Lock Hospital. They were not, however, required at St. Thomas's or St. Bartholomew's. It is interesting, therefore, that the admissions registers show that many patients were admitted with the help of a governor. For some patients, a particular governor's name appeared alongside the name of the parish or person who posted their bond. Certainly those paupers who managed to arrange for a governor to vouch for them stood a better chance of getting a bed than those who did not. Precisely one-third of all the known patients admitted between 1773 and 1776 attained their admission with the aid of a nomination.[52] Patients sent from parishes were slightly less likely to be nominated, but it was not uncommon. Over one-fifth of all patients with a parish petition also had a governor's nomination.[53] As we will see in the discussion of the Lock Hospital, where such nominations were required, these exchanges between supplicant and benefactor likely saw paupers assume a deferential role, trying to emulate the necessary qualities that made up contemporary notions of what constituted a "worthy object of charity." We must also assume that similar dramas played themselves out between St. Thomas's governors and the paupers who aspired to be patients.

It is telling to analyze who was successful. Among clean patients, there is virtually no difference between the number of men and women who received nominations: governors nominated 34.6 percent of all clean men and 34.8 percent of all clean women.[54] Gender appears to have played no significant role in determining which nonvenereal patients received nominations. Because foul patients tended not fit the eighteenth-century ideal of worthy charitable objects, it is not surprising to learn that governors nominated them less often; only 29 percent of all foul patients won admission with the help of a governor.[55] However, when that group is further broken down by sex it becomes overwhelmingly clear that gender significantly affected the charitable exchange between hospital governors and paupers with the foul disease. Foul men were actually *more* successful than clean men when seeking a nomination; governors nominated almost 36 percent of all men admitted to the foul wards.[56] The smaller percentage of nominated foul patients owes itself completely to the infrequency of governors nominating foul women—only 18.7 percent of foul women were nominated by a governor.[57] The significantly smaller portion of nominated foul women is not skewed downward as a result of the large number of foul women who entered the hospital on a parish petition.[58] In other words, men with the pox were almost twice as likely as women to have the advantage of a governor's nomination when they maneuvered to get into St. Thomas's Hospital.

It is likely that two equally important factors contributed to this phenomenon. First, and perhaps most obviously, this data seems to speak to governors' differing attitudes towards men and women with this disease. In the eyes of many of the elite men who came to be hospital governors, poor women who contracted the foul disease did not fit their conception of a worthy object of charity. They clearly did not hold prospective male patients to the same moral standard as that which they used to judge the women who came before them suffering from the pox. However, one must also consider the evidence from chapter I showing that infected women were extremely selective when choosing to whom they disclosed their infection. In addition to being less successful when petitioning for nominations, it is also probable that poor women were less likely to approach socially superior male governors in the first place. Commentators on the pox noted as much when they discussed hospitals, speculating that "those of the female sex . . . would be perhaps detained by shame from applying to such a place."[59] As public as the admissions process already was, it is likely that many women may have simply opted not to disclose their condition to yet another man. Such figures as these indicate the extent to which contemporary cultural attitudes informed institutional healthcare in early modern London. By shaping the worldviews of those who ran early modern institutions—as well as those who utilized them—such attitudes influenced most every aspect of early modern institutional medicine.

Perhaps even more important than governors' nominations were the people who stood as patients' sureties. With the exception of a few rare cases when the hospital took pity on destitute paupers—and it is worth pointing out here that unlike Bart's, which reserved some of its foul ward space for charity patients, St. Thomas's very rarely waived foul patients' fees—no patient entered St. Thomas's in this period without having someone post a bond of one guinea.[60] The purpose of the bond was to cover the costs of burial in case the patient died. Churchwardens posted the bond for parish patients. Regimental agents posted the bond for soldiers, a group which made up a notable portion of men in the foul wards.[61] It was incumbent on everyone else either to post the bond themselves, which very few patients could afford to do, or ask a friend or family member to post it on their behalf. In the majority of cases it is impossible to determine the nature of the relationship between these guarantors and the patients. For example, surprisingly few patients listed sureties with the same last name and who could thus be assumed to be relatives. However, no hard and fast assertion can be drawn from here, since any number of kin might not have shared a surname. It is probably safe to assume that some patients continued to rely on the age-old support network provided by the family.

Whoever these various sureties were they played a crucial role in the admissions process, which has not been fully appreciated. Since patients had to arrange their bond before they came to seek admission, this represented the first stage in the admissions process. Someone hoping for a bed in St. Thomas's would have first needed to seek out someone to stand as their surety. One can only speculate on the nature of the exchange when poor sick people had to ask others to put up the bond to help them into the hospital. Presumably, it was not unlike the negotiations that took place between hospital governors and the patients who sought nominations. Of course, when thinking about medical privacy, here was yet another way in which the admissions process for VD patients became that much more public. It is hard to imagine that when an aspiring patient sought out a friend, relative, neighbor, or employer to ask them to put up the money to get them into a hospital, that the person being approached did not reply with the natural question: "Why do you need to go into the hospital?"

Yet again, gender proves to have been a crucial factor in the patient-experience of this disease. Considering who controlled wealth in the eighteenth century we should hardly be surprised to find that the majority of those who posted bonds were men. When one eliminates all those who relied on their churchwarden or soldiers for whom a regimental agent posted the bond, over 92 percent of all the individuals listed as patient sureties were men.[62] Here again was another situation in which women might find themselves pressed by need to reveal their illness and sexual history to a socially superior man.

Many women responded by relying on another woman. During the forty documented weeks between 1773 and 1776, 148 of all patients secured their bond from a woman; this represented 7 percent of the total population. Among clean patients, women were more than twice as likely as men to secure a bond from a woman; London women posted the bond for 9.6 percent of clean women and less than 4 percent of clean men.[63] This by itself says something about the importance of female support networks, exclusive of the influence of a sexual disease and its stigma. But the figures for venereal patients show that this divide was far more pronounced in the foul wards. Foul men were even less likely than clean men to rely on a woman for their bond, showing that the gender-dynamic cut both ways. Just as women in the period seemed reluctant to approach men when they were infected, so, too were men reluctant to approach women; only 2.5 percent of foul men listed a woman guarantor.[64] However, a far greater portion of foul women—23.3 percent—who arranged their own bond chose to confide in another woman.[65]

Such evidence offers substantial support to the claims made thus far about foul women's tendencies to rely on other women when facing sexual infection. Foul women were almost ten times more likely than foul men to impart the nature of their disease to another woman, and well more than twice as likely as clean women. Just as foul women of means sought out the services of women doctors, so, too, did poorer women with the pox seek to confide in other women when negotiating to access institutional health care. It remains true that the majority of women who entered the foul wards had to arrange their bond through either a male churchwarden or a male surety. Contemporary political and economic realities determined that that this was so. But of those who arranged a bond for themselves, nearly one out of every four women sought out another woman.

The information on patients' sureties illuminates another revealing aspect of patients' strategies to negotiate the admissions process. While the relationship between most patients and their guarantors remains obscure, it seems that some small-time London businessmen acted as professional bondsmen, furnishing patients' admissions security for a small fee. This hypothesis stems from the curiously frequent recurrence of the names of particular men over the course of several years. Certain sureties stand out in the registers because the name of their business was listed alongside their name. Many of these common bondsmen were alehouse keepers, or victuallers of some sort. Some of the names that appear frequently as the guarantors for multiple patients during the 1770s and early 1780s were Thomas Swaine at the "Ship and Shovel," Lawrence Carlson and Thomas Warden at the "Bird in Hand," John Liberty and Thomas Scott at the "Cooper's Arms," and William Ellis at the "Spur Inn." All of these men posted the bonds for both clean patients and foul patients alike. None of them, however, were as active as Nicholas Harris of the "Guy's Head Tavern" in Southwark. Harris posted the one guinea bond for each of twenty-five foul patients—to say nothing of all the clean patients he secured—during the forty weeks covered by the 1773–76 registers. In other words, Harris put up the bond for more than one foul patient every other week throughout the period. None of these patients appear to have any familial relation to Harris, and it is hard to imagine that he posted these bonds out of the goodness of his heart.

Instead, it is seems that these victuallers engaged in a side-business of providing security bonds for poorer people for a small fee. The demand for this service appears to have been high enough to support many different professional bondsmen. In other words, a small industry developed to provide temporary credit to those people who generally had none. It is notable that most of the identifiable alehouse-bondsmen lived and worked near the hospital. The Guy's Head Tavern, Spur Inn, Ship and Shovel, and Cooper's Arms were all in Southwark. Over the years they became well known to the staff at the hospital, and presumably their reputation, literally their "credit," grew stronger with the hospital governors. So established was the professional relationship between the hospital and Nicholas Harris, that frequently his name merely appeared without reference to his address or the Guy's Head Tavern. The takers-in often noted that he provided the bond merely by jotting down his name, "Nich. Harris, Southwark."[66]

Some women seem to have worked in this industry, too. The hospital matron and several of the sisters of the various wards all posted bonds for patients. The regularity of the practice suggests that it may have been one way that hospital staff made extra money on the side. But women unassociated with the hospital also provided these services. Guy's Hospital's admission records from the same period show that widows like Elizabeth Morton, Jane Fortman, Mary Burroughs, and Elizabeth Horton all posted bonds for multiple patients over the course of 1770. None were as active as widow Sarah Harrison. Sarah stood as the surety for ten different patients, both men and women, clean and foul between January and June 1770. In fact, Sarah worked at the Guy's Head Tavern, just like Nicholas Harris. It may even be that Nicholas took over from Sarah that year, because Nicholas begins appearing in June, posting the bond for seventeen patients over the next six months, right when Sarah seems to disappear from the records.[67]

That these men and women were victuallers is supported by a letter contained in the Lock Hospital records. That hospital, we shall see, had its own admissions policy. There were certain occasions when patients could not gain admission without first acquiring a bond similar to the one demanded at St. Thomas's. In November 1766 Mary Negus tried to convince the Lock Hospital to admit her seven-year old daughter, Elizabeth, who had caught the disease as the victim of a rape. The governors were sympathetic to the child's case, but they would not admit her unless Mary could first secure a bond.[68] Mary was told to return the following week. She did return, accompanied by Charles Atkins and John Howles of the Bird in Hand tavern. They signed the following promissory note, entered into the Lock Hospital's administrative minutes:

> We Charles Atkins at the Bird in Hand the corner of Chapel Street Oxford in the Parish of St. Ann Westm[inster] in the county of Middlesex Victualler, & Howles of the same place Chairman, Do hereby joyntly and severally promise and undertake to and with Jonathan Durden Esq. as Chairman of the Weekly Committee at the Lock Hospital near Hyde Park Corner of Thursday the 13th Instant, and on behalf of the Gov[ernor]s of the said charity, That we or one of us, or that Francis Negus of Tottenham Court Road in the parish of St. Giles in the Fields in the said County and Mary his wife or one of them shall and will receive again with their or our custody Eliz[abe]th. Negus and infant daughter of the said Francis and Mary Negus brot into the said hospital and admitted as a patient by the above committee for the cure of the Venereal

Disorder, as soon as she shall be cur'd thereof and we or they shall be thereto requir'd by the Governors or the Weekly Committee of the said hospital; And that the said Elizabeth Negus the Inf[an]t shall not be left as a burthen upon, or be any incumbrance to the said Charity, after she shall be cur'd as afores[ai]d. And we do further promise to save the said charity harmless and keep them indemnified therefrom & from any expence that shall attend the same, as witness our hands this 15 day of [November].

Charles Atkins
John Howles[69]

This bond is more general than St. Thomas's bonds, protecting the charity from any unexpected expense, not just burial costs. But it makes little difference. It took Mary Negus just a week to find Atkins and Howles at the Bird in Hand and reach an agreement to attain the credit she needed to get her daughter into the hospital. The evidence of the activities of victuallers at the Bird in Hand is particularly notable. This tavern was not in Southwark, but in Westminster. So clearly Atkins and Howles provided their service for patients at different hospitals. While some of the Southwark victuallers may have dealt only with nearby St. Thomas's Hospital, these entrepreneurs seem to have had a wider business network. Presumably, they provided similar bonds for patients at a host of hospitals and institutions around London. Who knows what other kinds of credit-related financial services they may have also provided to the poor?

This low-level credit industry for prospective hospital patients had special significance for people hoping to get into the foul wards. It is impossible to calculate what portion of foul patients used their services. Only a few have been identified thanks to their sheer ubiquity in the registers. But while such exact data is elusive, one can assert with confidence that foul patients represented a significant portion of their clientele. First, we should consider the perspective of the bondsmen. In a sense they provided a sort of early modern life insurance. They agreed to pay certain expenses in the case of death. Therefore it was in their interest to secure patients unlikely to die. In that sense, foul patients represented good prospective clients. Public perception continued to hold that the foul disease was treatable. One cannot know the conditions of the clean patients secured by these bondsmen. But it is likely that such businessmen considered prospective clients' diagnoses when deciding whether or not to provide a bond or what they should charge. It would not be surprising to find that many clean patients who secured a bond this way suffered from broken limbs or other non-life-threatening conditions.

It is also worth considering the services of these alehouse-bondsmen from the patients' perspective. They provided a service that may have met the needs of various sorts of people. First, there were those people without a parish settlement in London who fell ill but could not rely on a parish to get them into a hospital. However, such folk could get a bed if they could scratch together the admission fee plus whatever fee the bondsmen charged. More importantly, these bondsmen may have provided some amount of anonymity for those foul patients who sought it. Men and women who contracted the pox and needed hospitalization, but who did not want to confide in a churchwarden, relative, or friend to finance their hospitalization, had another option. Widows providing this service may have been especially attractive to infected women.

It was not completely confidential, to be sure. Such patients had to approach a stranger and discuss the nature of their illness. But it may have been more desirable to discuss such details with a stranger, someone less likely to pry into the intimate details about how one acquired the disease, than with one's relatives, neighbors, or employers. Again, this option was only available to those who could afford it. Just as medical confidentiality was increasingly provided to London's patrician classes by private practitioners, so also was a sort of partial confidentiality being made available to those further down the social ladder by these alehouse-bondsmen. Their services may have at least mitigated some of the problems posed to foul patients by the public nature of the hospital admissions process. That the demand for this service could sustain such a wide range of businesses indicates in a different way the powerful effects that the force of patient-demand and the desire for privacy could have on the medical marketplace. But one still needed some amount of disposable cash to take advantage of this service.

Beyond the admission process, surviving documentation also offers insight into the experiences of patients once they entered the foul wards. Even if one could arrange their financing discretely, there were no guarantees that the details of one's case would be confined to doctor and patient. St. Thomas's surgeon William Beckett published the details of several "most extraordinary" cases from the hospital wards in 1740, which included cases from the foul wards. Here we learn that a range of people might be invited to comment on a patient's case. A young man of eighteen suffering from symptoms initially thought to indicate a bladder stone found that his surgeon interviewed his master and several other people about his case. Beckett suspected his symptoms might be venereal. To confirm this he discussed the case with several of the young man's acquaintances, including his employer. Beckett reported that his suspicion that the man was foul "seemed to be fortified by the young Man's Master, and some other persons asserting he was venereal, he himself not positively affirming the contrary." Here it was *others'* speculations about the man's health and habits that carried the day. Beckett ordered the man under a salivation, during which he died. A post mortem revealed that indeed the man had a bladder stone.[70]

One of the clear differences between Bart's and St. Thomas's witnessed in the seventeenth century concerned discipline. That difference remained, although it was hardly as stark it had been in the seventeenth century. It is impossible to know for certain whether St. Thomas's continued to inflict public corporal punishment on foul patients. It is unlikely that this rather brutal policy continued. No mention of corporal punishment survives in the eighteenth-century administrative records. Indeed no mention of it survives after 1649.

From this silence optimists can infer, or at least hope, that the practice probably died out by the dawn of the Enlightenment. This belief finds some support in a surviving mid-century account of hospital rules.[71] The lengthy account lists all the duties of the various employees and the rules for the patients. The only specific punishments mentioned are the withholding of patients' food rations for minor offenses, and immediate discharge for more serious transgressions. Moreover, no officer of the house was charged with the duty of inflicting punishment. However, because of certain purposefully vague references to punishment it remains impossible to discount completely that corporal punishment may have continued in some form. For example, when highlighting that patients must attend religious services, the rules stipulate that patients who resisted were to be "punished at [the] discretion of Treasurer or Steward," whatever that meant.[72]

That mid-century list of rules also illustrates that many common regulations of the seventeenth century carried over into the following century. St. Thomas's continued to control patients' eating and sleeping schedules, prevented patients from leaving the premises, and imposed mandatory religious instruction. Typical of hospitals at this time, the governors also tried to impose some sense of moral order by outlawing things like swearing, theft, fighting, and drinking.[73] The rules specific to foul patients illustrate that the two most common stipulations governing foul patients in the seventeenth century remained on the books in the eighteenth century. Venereal patients were forbidden to enter any but their own wards. And the governors remained convinced that many foul patients were lying their way into the clean wards.

> That patients admitted into this house who have the Foul Disease shall not go into any of the clean wards, officers houses or chapel. And if any patient that have the Foul Disease, shall knowingly, or willingly conceal the same at the time of taking in and shall be placed in a clean ward, every such patient when discovered shall be immediately discharged the house.[74]

The issue of the chapel is important because it raises the question of what form of religious instruction St. Thomas's provided for foul patients. Hospital regulations had specifically forbidden venereal patients from entering the chapel in 1699 and now again in 1752.[75] The reiterated stipulation that clean patients must attend religious service or face (vague) punishment indicates that St. Thomas's governors remained committed to the general cause of moral improvement. Yet their own rules seem to exclude foul patients from the chapel, presumably those most in need of reformation. But while foul patients were forbidden to receive religious instruction with the general population, we should not think that the nonconformist Whigs who ran St. Thomas's neglected the chance to impart a good dose of moral education. A 1709 decision illustrates how the hospital moved to merge spiritual reformation with medical therapy. The governors ruled that foul patients were to be brought for religious instruction immediately before and immediately after they entered the fluxing ward to receive their mercury treatment. On December 14, 1709, the board ordered,

That the patients in the Foule Wards should be brought by the sisters before and after salivations into the Cutting Roome to the Minister at such time as he shall appoint, to be instructed; And that if any patients in those wards . . . shall refuse to be instructed by the minister, the respective sisters should give notice to the Steward or to Mr. Trea[sure]r . . . to the end that they may be dismissed from the house or otherwise dealt with as Mr. Trea[sure]r. shall think fitt. Was approved and agreed to by this Committee and fair coppies ordered to be hung up in the Foul Wards and other wards of this hospitall.[76]

This rule is fascinating because it illustrates how intricately St. Thomas's moral mission became bound with its medical one. They bound salvation and salivation into a single ritual. Religious instruction became an integral part of the medical operation itself, as patients were reminded of their moral transgressions immediately before and immediately after their painful experience of mercury. By 1709 St. Thomas's may have moved beyond the gross physical punishment of the whip. But there was certainly plenty of precedent for viewing the torture of mercury as a form of punishment in and of itself.[77] As in the previous century, it seems that the dissenting Whigs at St. Thomas's continued to make a much clearer and more pronounced effort to save foul patients' souls than their Anglican Tory counterparts at Bart's, where no mention is made of any similar practice.

In the same period Bart's efforts to save venereal patients amounted to purchasing copies of Lewis Bayley's *The Practice of Piety* to give to outhouse patients as they left the hospital. However, it is not clear that they were made to read it while in the house, and more importantly this seems to have been a one-off experiment. They bought one hundred copies in 1708, and did not order subsequent batches of the book to replenish their stock.[78] This hardly compares with St. Thomas's, where they literally transformed the operation of mercurial salivation, where what had previously served as a physical purge now also served as a spiritual purge. It is important that this religious instruction took place in the surgical ward, "the cutting room." Fear of moral contamination prohibited foul patients from entering the sacred space of the chapel. This not only contributed to the general stigma of the pox itself, it drove the governors to contain venereal patients' religious instruction within the confines of a medical space. Situating the meeting between foul patient and minister in a medical ward rather than the chapel helped to further fuse this spiritual and medical healing into a single ceremony. The governors' threats to reprimand those who refused to comply underscore the seriousness of their purpose. Moreover, the vague phrase that patients who refused could be "dealt with as the Treasurer shall think fitt" again opens the possibility that the hospital may have inflicted corporal punishment on the bodies of foul patients who resisted. Whether or not beatings occurred, the evidence indicates that St. Thomas's continued to employ a stricter disciplinary and reforming regimen for its venereal patients than did St. Bartholomew's, which leaves no evidence from over two centuries that it disciplined its foul patients much at all.

The development of hospital education also had an impact on the experiences of patients once they entered the foul wards. In several European cities hospitals became incorporated with universities that offered medical degrees.[79] In the case of London, however, Susan Lawrence has shown how the major hospitals themselves became teaching institutions independent of any university affiliation.[80] In all these places walking the wards became a crucial eighteenth-century component of medical students' education. All the major London hospitals, including Bart's and St. Thomas's, allowed their staff to take on students who followed them as they made their rounds. The result for patients was that they now became teaching tools. Not only did patients have to pay admissions fees, by the second half of the eighteenth century they also had to agree to give up their bodies to the clinical gaze of scores of medical students in order to obtain the care that they needed.

The exposure of their bodies to medical students was troubling to many patients, foul and clean alike, who did not always appreciate being the object of display while suffering. However, the unique sexual and stigmatized nature of the foul disease made such public scrutiny that much more invasive for venereal patients. The list of people to whom one had to reveal him or herself if they sought care at a hospital grew longer all the time. A patient in one of the royal hospitals' foul wards may have already had to discuss her condition with her churchwarden, friends, family members, or semi-professional bondsmen to secure a bond, a hospital governor for a nomination, and the various governors and staff members at the time of admission, not to mention the physician or surgeon, nurses, chaplain, nonmedical staff, and fellow patients once she entered the ward. Now, in the midst of debilitating mercury treatment, patients had to agree to lie obediently while doctors discussed their conditions and treatment to whole classes of medical students, who, it must be remembered, were all men.

Occasionally, members of the medical staff were sympathetic to foul patients' desires for privacy in the face of medical students. The notebook of an early eighteenth-century student from St. Thomas's survives.[81] He described the case of a female foul patient who asked not to be examined in front of the class. The woman claimed to be a wet nurse who had caught the disease from one child and passed it to another. The student describes his first encounter with her:

> The nurse who I believe was innocent of the matter came in great concern to Mr. Oxley and myself & desird we would look upon her breasts, each of w[hi]ch. had a little round ulcer near the nipple with a hard basis & a redness round both very nearly resembling a shanker, she press'd us to give our opinions, & accordingly we pronounc'd it Venereall, she was vastly uneasy because of her husband, and knowing that I attended ye H[ospital], apply'd her self to us upon that Account, & begg'd I woud assist her, & meeting me at ye H[ospital] next morning, I introduced her to Dr. Coatsw[orth]. & Mr. Fern when all the pupils were in another ward, because she was a shy modest woman & desird it, They talked with her & look'd upon her breasts and confirm'd our opinions, they encourag'd her & civilly advis'd her to come into the House & be salivated, w[hi]ch by my perswasion she consented too.[82]

It is encouraging to learn that some doctors were willing on occasion to take such patients' concerns to heart and save them the humiliation of exposure before the class.

But this case was clearly an occasion when special arrangements had been made. The patient clearly took the initiative and sought out the student outside the hospital and made her case in an attempt to arrange preferable circumstances for herself. Moreover, we get the sense that this special treatment hinged on the medical men's evaluation of the woman's character. Because they perceived her to be an innocent wet nurse, they agreed to treat her differently. She was likely a fortunate exception. The pox remained a standard subject of hospital teaching curriculum throughout the century.[83] Therefore exposing venereal patients to students remained common practice. Even in the mid-nineteenth century William Acton could still complain that women in the foul wards continued to face exposure before classes of male medical students, describing the procedure as "painful and demoralising to the unhappy patients."[84]

The genesis of medical education in eighteenth-century hospitals touches on another important development affecting our story. The late eighteenth century began a process that would lead to a change in medical perception, what Foucault called the birth of the clinical gaze, and which others have termed the rise of hospital medicine.[85] As the eighteenth century winds down we begin to see a new corporeal paradigm emerging as a result of hospital teaching and new medical relationships. This new vision saw disease increasingly in terms of physical lesions located in various bodily tissues, and it finally (definitively) replaced the humoral explanation of disease. Hospitals, where doctors and medical students performed an increasing number of dissections and saw an increasing number of cases, provided the forum for the birth of this new medicine. Over time this produced a more objectified image of the body and a more statistical approach to medicine, which became increasingly anonymous.

In addition to changing the view of the body and disease, the new hospital medicine had transforming effects on the doctor-patient relationship. Whereas patients had formerly been active participants in a negotiated medical exchange, offering their own interpretations on their symptoms, they now became silent bodies to be read by their doctors, teaching tools on display before classes, both when alive and later as corpses. Whereas their diseases had formerly been individual to them, resulting from their unique humoral make up, they now became owners of uniform bodies which (doctors presumed) functioned in identical ways to all other bodies. Scholars contend that this cost patients much of their voice within the doctor-patient exchange.[86] No longer necessary were the long interviews between doctor and patient, as practitioners sought to learn who their patients were. No longer important were details about lifestyle, constitution, personal history, or patients' own descriptions of symptoms. Now the physical exam, increasingly devoid of such an exchange, came to replace the patient interview, and thus patients lost much of their voice. Hospital patients were the first to experience this new medical role of silent patient, and as Mary Fissell has shown, doctors further distanced themselves from their hospital patients by relying increasingly on Latin and medical jargon, which the poor simply did not understand.[87]

What of this new medical relationship for venereal patients? Scholars like Fissell have been right to cite this new silent role for patients as gravely disempowering. The early modern medical exchange had offered great latitude for patients who had a chance to cast their own illnesses in their own terms. For early modern patients facing the foul wards this chance to interpret their own symptoms was particularly vital. The premodern exchange, flexible, open, based on two voices, gave patients a chance to participate in the construction of their own diagnosis, and we have already seen just how empowering this could be for patients. Here was a chance to contest the diagnosis "foul." Important ramifications and real issues of power followed that labeling, but the traditional medical exchange allowed patients some room for maneuver. We have seen that the ability to diagnose oneself as not "foul" was an important tool in the repertoire of survival strategies employed by plebeian patients facing hospitalization, and the evidence from hospital records shows that patients contested that diagnosis throughout at least two centuries. Hospital governors might accuse some of these folks of lying and even expel them when they entered "clean" wards under what they considered to be false pretences. But the evidence shows that early modern hospital patients used the latitude afforded them within the traditional medical exchange creatively, in attempts to affect at least some amount of control over their hospitalization.[88]

In addition to challenging the diagnosis altogether and casting their symptoms as something other than the pox, the conversational give-and-take built into the traditional medical exchange also afforded patients a chance to save face by spinning their stories about how they acquired their infection. Patient agency was not exerted just by those who refused the diagnosis altogether, but also by those who took the opportunity to cast the circumstances of their infection in their own light. The patient's "history"—the patient's own story of how and when they became sick, along with their early symptoms—formed a crucial component of early modern diagnosis. The taking of that history afforded venereal patients (whether inside or outside hospitals) the chance to construct their own tale of how they came to be infected. Here was the chance to explain away their infection, for an unmarried woman to claim she had nursed an infected infant, or for a prostitute to claim she had been betrayed by a rogue promising marriage who had been her first and only lover.

The example of the woman examined privately by the medical student a moment ago is a prime example. In that case, the patient was able to make her case to the doctor and arrange slightly favorable conditions for her treatment. It is notable that this account comes from so early in the century, before the birth of the clinic. Moreover, it is clear that the story of *how* she contracted the disease made all the difference. By convincing the doctors that she caught the disease through wet-nursing, she convinced them that she was worthy of special care. The doctors believed she was "innocent of the matter" and treated her accordingly. The woman's voice is not recorded, but there was clearly a negotiation of a plurality of voices, doctors' and the patient's.

Such chances were lost with the birth of the clinic, the new medical exchange that considered what patients had to say unimportant. Doctors now afforded patients less discursive space, less opportunity to speak. That loss of voice was disempowering for all patients, but it may have been a loss that venereal patients felt particularly acutely. Furthermore, a class dynamic again shows itself. As Fissell and Lawrence have capably shown, the birth of this new medical relationship affected plebeian (hospital) patients first. In the second half of the eighteenth century patients who availed themselves of doctors' private services would continue to benefit from the traditional medical exchange, which continued to allow them a voice, while their poorer counterparts in hospitals were being increasingly silenced. The class chasm dividing the experiences of venereal patients grew wider.

Of course, how one experienced the medical exchange would have varied from patient to patient. Given the sensitive and personal nature of this particular diagnosis we cannot rule out the possibility that some patients may have preferred the new silent role assigned to them by modern clinical medicine. The birth of the clinic may have cut both ways in the foul wards. While it may have stolen the voice of some patients who wished to try to steer the medical exchange to their advantage, it may have saved other patients from having to endure an unwanted interview. For some, allowing a doctor to inspect their bodies and reach their own conclusions while they remained silent may have been preferable to a medical exchange that required them to reveal personal details in lengthy conversation. In other words, some patients with the pox may have chosen to retreat into this descending silence, and welcomed the new option. But while this may be another way in which a sexual disease had a unique trajectory within the history of the doctor-patient relationship, one must still see the disappearance of the patient-narrative in the foul wards as further evidence of the general disempowerment of patients in this period.

Although they may have been losing their voices, patients found other ways to express their dissatisfaction in the foul wards. Such dissatisfaction could be expressed by disobeying hospital rules. St. Thomas's 1768–72 patient index includes comments on patients' behavior, usually as justification for premature discharge. An analysis of such comments reveals that foul patients tended to disobey house rules in specific and telling ways. Of course patients of all kinds might act up, disrupting the wards, harassing nurses or other patients, committing acts of petty theft, or lying out of the ward at night without permission. Foul patients were guilty of all these transgressions. But foul patients did not exhibit these kinds of generally disruptive behavior with any significant frequency. Foul patients accounted for only around one-fifth of all those discharged for being disorderly or lying out of the ward without permission. If anything they were slightly less likely to exhibit such behavior since they represented close to 28 percent of the total population.

However, foul patients were much more likely than other patients to run away from the hospital prematurely. Fully two-thirds of all the patients that quit the hospital early left from the foul wards. Fifty-five percent of all patients who left the very same day that they arrived also ran away from the foul wards. And finally, foul patients represented almost 80 percent of all those patients that the index indicates arranged to be admitted but then never even showed up. Put simply, when they faced the reality of the life in the foul wards, when they came face to face with the horrors of salivation, many patients voted with their feet.[89]

One cannot underestimate the effects of mercury treatment on patients' experiences in and opinions of these wards. Several entries in the index indicate that patients decided to leave while in the midst of their salivation. It did not take some patients very long to decide against enduring their mercury regimen. St. Thomas's patient William Grocer entered Naples ward on October 3, 1771. Five days later he "pulled off his flannels & went away." Both James Steers and James Mills stayed in the foul ward for about a month before they, too, pulled off their flannels and escaped the hospital. Some patients, like John Swallow, did not even bother to remove their dressings before leaving, simply walking out when the opportunity presented itself. Mary Wooding may not have wanted to quit the hospital permanently when she "went to the alehouse in her flannels," but she was discharged nonetheless.[90] There is a clear indication that some of these patients made their decisions based upon their own ideas about what level of treatment they needed or could endure. Such seems to have been the case with Elizabeth Stocker, who the hospital described as "very disorderly" because she "threw off her flan[nels] at least a week before the time."[91]

These comments indicate that foul hospital patients, like many patients who encountered institutional medicine, tried to keep some amount of control over their own treatment, even if they may have been losing their voice within the emerging medical exchange. Moreover, these comments indicate how distasteful life in the foul wards was in the opinion of many eighteenth-century Londoners. Hospital foul patients tried to use hospital resources creatively. Many patients did what they had to do in order to access the care that hospitals like St. Thomas's provided, but then only stayed on their own terms. Once they felt that they had received a sufficient dose of quicksilver, or they decided they could not endure the pain of salivation or the strain of being displayed before a class of students, many foul patients took matters into their own hands and made for the exit. However, one must remember that getting into the hospital in the first place was not easy. A hospital bed was a hard-won resource, not easily squandered. So it is not surprising that most foul patients did not run away prematurely, and did not disrupt the wards. With so much at stake they dared not risk it. The vast majority of patients obediently followed doctors' orders in the (vain) hope that they would one day get better.

St. Bartholomew's and St. Thomas's remained at the heart of London's medical world, and they continued to treat large numbers of foul patients throughout the eighteenth century. So prevalent were foul patients in these hospitals that one is forced to reconsider the civic missions of these institutions. By and large, their care for venereal patients has flown beneath the historical radar, receiving only brief attention. Yet year in and year out St. Thomas's saw well over a quarter of its patients enter the foul wards. Bart's seemed to offer fewer beds in the early part of the century, but by mid-century it, too, treated hundreds of patients a year. Responding to the twin problems of poverty and the pox represented an essential service that both hospitals provided to the city, and a major component of their overall operations commanding a significant portion of hospital resources. That one diagnosis could account for such a massive portion of hospital clientele speaks volumes about the omnipresence of the foul disease in eighteenth-century London. Regardless of what it "was" in microbiological terms, eighteenth-century Londoners saw the foul disease constantly around them. How could Bart's or St. Thomas's have excluded such a clearly huge category of patients? Certainly, the argument persisted that these patients brought on their own misery and were unworthy of care. Voluntary charities like the Westminster Infirmary made that argument in those terms, declaring foul patients unworthy of their charity and excluding them. It is telling that over more than two centuries, the royal hospitals never did.

The pox among the poor was one of London's most pressing medical problems, and its hospitals had to respond. Perhaps because they could largely subsist on rents rather than by voluntary donations, the royal hospitals did not need to justify what they were doing, as did voluntary hospitals, which were forever fundraising and were highly sensitive about their public appearance. By contrast Bart's and St. Thomas's with their rather steady income simply treated the patients who applied—and it just happens that an overwhelming number who applied had the pox. But a sense of civic obligation radiates from both hospitals, especially when compared to the voluntary hospitals that excluded the foul. The Crown had granted these hospitals to the City, and endowed them with property; the evidence of their commitment to treating patients that other institutions refused may suggest that they were further endowed with certain expectations to serve the medical needs of London. And treating the poxed poor happened to be one of the most pressing of those needs. Segregating them, charging them, at times even punishing them, was acceptable. Excluding them altogether, it seems, was not.

So the royal hospitals continued to make VD care a primary component of their operations, and they devoted significant resources to that end. However, the picture is not entirely rosy for foul patients facing these hospitals in the eighteenth century. As the century dawned administrations at both hospitals increasingly shifted the cost of care onto the patients themselves. The disparate fee structure that emerged highlights the hierarchy of patients within these hospitals. It officially solidified the reality of two classes of patients, clean and foul. Moreover, those fees had real effects on who could access care. The very poor increasingly found themselves shut out as beds in the foul wards became more and more expensive. Economics determined that women felt the crunch more acutely than men. By the end of the century both hospitals reserved much more ward space for men, perhaps because they privileged men's health, perhaps because more men could afford the fees. The competition for fewer beds and the increasing cost of care meant that very poor women would have to seek care elsewhere. Many women who did succeed in getting a bed were able to do so only by first going to their parish, which meant first negotiating an entirely different level of bureaucracy. To gauge how important parish resources were for poor women struggling with this disease consider that women were four times more likely than men to have rely on their parish to get a bed in the foul wards. The discriminating fee structure that made foul beds more costly than clean ones made a bad situation worse: foul women were more than twice as likely as clean women to have go through their parish to access the hospital, pointing towards the particularly impoverishing effects of this disease on London's plebeian women.

So as the very poor became increasingly shut out of the royal hospitals in the eighteenth century, they fell back more and more on parish relief. Some would be passed onto the royal hospitals, their fees paid. But these were a minority. Many more who applied to their parish when they were sick were treated locally. If one hopes to find what happened to a great many of the people struggling against the specters of poverty and the pox, they need to look beyond the royal hospitals, to the parish, which is what the following chapter will do.

4

THE FOUL DISEASE AND THE POOR LAW: WORKHOUSE MEDICINE IN THE EIGHTEENTH CENTURY

On August 22, 1728, Flora Price applied to her churchwardens in the parish of St. Margaret's Westminster. When questioned by the overseer's of the poor she admitted that she was poxed and sought their help. The clerk recorded that she "be admitted into ye House till such time she can be got into ye Hospital for cure of the foul distemper." However, she never entered a hospital. Instead, it seems she entered the workhouse and underwent mercury treatment there. Following her salivation she was discharged on October 21. The workhouse committee ordered "That Flora Price be discharged ye house & to have some old cloaths & to be sent to Bridewell upon any new Application." Such stories are ubiquitous in eighteenth-century parochial records. Workhouse admission records register the admission of sick paupers week in and week out throughout the entire eighteenth century. Given the great prevalence of the pox one should not be surprised to learn that foul patients like Flora Price were ever-present in these institutions.

Yet the medical role of the eighteenth-century workhouse has received little attention. Studies of early modern English medical institutions have generally focused on the large hospitals like St. Bartholomew's and St. Thomas's or on the growing number of private specialist charities like the Lock Hospital. Yet, in parochial workhouse infirmaries there existed an important level of institutional health care for the very poor. Many paupers like Flora Price did not run immediately to a hospital when they became ill. Often their first stop (or their last resting place) was the workhouse.

Over the past decades social historians have tirelessly explored the massive landscape of English parochial records, which has yielded a wealth of rich data on the English poor. However, too few early modern medical historians have mined this body of material, much of which concerns issues related to health and healing.[1] Focusing on a single disease in these institutions allows access to the much larger issue of eighteenth-century workhouse medicine, which still awaits proper investigation.[2] Overall, the assumption continues that the medicalization of workhouses was a nineteenth-century phenomenon and a product of the New Poor Law. Just one example is the recent reaction to evidence of medical care in London workhouses from 1837, which, we read, "was important from an early date."[3] In fact it was important from more than a century earlier. I suggest that workhouse infirmaries became the primary medical institutions for the very poor in London from early in the eighteenth century, long before the well-known reforms of 1834. Situated in parishes throughout the city, they collectively represented an important level of institutional health care. Though smaller and less renowned than institutions like Bart's or St. Thomas's, they played a crucial role in the overall network of London medical charity. From the particularly good medical records that survive for workhouses in parishes such as St. Sepulchre (London Division), St. Margaret's (Westminster), and St. Luke's Chelsea (Middlesex), we can begin to gauge the importance of these institutions to wider medical charity. And specific to the goals of this book, we can chart the strategies and illness-experiences of the very poor in London who came to be diagnosed with the foul disease. As the evidence from the royal hospitals began to suggest, parish medical authorities came to cater to a particular segments of the London underclass, especially poor women. Workhouse infirmaries were London's other hospitals, the primary healthcare institutions for poor urban women in the eighteenth century.

Early modern England created a system of poor relief based on the parish as its key organizational unit. Under the old Poor Law paupers could seek assistance from the officers of the parish in which they had a "settlement." In most cases this meant their parish of birth, although there were a host of other ways to establish a settlement, including through marriage, by working, or by serving an apprenticeship in a parish.[4] In each parish, officers collected poor rates from wealthier members of the parish and used these funds to relieve the poor who could prove that they were "settled" therein. Those who sought relief at parishes other than their own were usually refused, as churchwardens tried to limit relief to their own paupers. Throughout the sixteenth and seventeenth centuries the most common form of relief was the dole. Each parish differed, but in the majority of cases parish officers simply gave cash payments to those paupers who qualified, either as pensions or as one-time payments.[5] The local government apparati that emerged to collect the poor rates and administer relief have recently begun to receive the attention they deserve as crucial sites of local politics, and historians have recently placed the parish at the center of debates on the history of English governance and state formation.[6]

In the early eighteenth century poor relief changed dramatically. In parishes throughout England social reformers became excited by the numerous advantages that seemed to be offered by the establishment of workhouses. Tim Hitchcock has given the most thorough account of this development.[7] Originally, the driving force behind the workhouse movement was the Society for the Promotion of Christian Knowledge, the SPCK. This organization helped establish several early workhouses, published pamphlet literature to promote others to do likewise, and, Hitchcock suggests, pushed the 1723 Workhouse Test Act through Parliament.[8] Advocates of workhouses offered these institutions as a panacea for several perceived contemporary social problems, both economic and moral. First, reformers promised that workhouses would tap into an important national resource that lay stagnant, namely, the labor of the unemployed. Proponents of workhouses felt certain that once this labor was organized and set on task they would turn a profit and ease the burden on parish rate payers. No longer would the poor receive charitable handouts; they would earn their own keep. In addition to these economic benefits, workhouses would also provide a forum in which the bad habits of the English poor would be reformed. The SPCK promised that workhouses would eradicate idleness, laziness, and apathy towards religion through toil and spiritual instruction. In addition to these stated goals, Hitchcock has persuasively argued that workhouses also served a somewhat darker purpose. Workhouses were constructed so as to frighten the poor and thus act as a deterrent from seeking relief in the first place. Hitchcock supports this assertion by showing that the number of applicants for relief dropped by over 50 percent in parishes that established workhouses.[9] While in practice English workhouses almost unanimously failed to turn a profit, the concept of workhouses largely succeeded in capturing the imaginations of early eighteenth-century reformers, and by mid-century over six hundred workhouses were in operation in England.[10]

The SPCK's promotional literature and Hitchcock's analysis clearly indicate that the original plans for workhouses revolved around the issues of work, profit, and moral reform, not health care. Yet the provision of care for the sick and infirm quickly became a crucial function of most London workhouses. Once in place, these institutions originally erected for very different purposes evolved into important local medical institutions. This evolution of workhouses was a product of the gap between the designs of reformers and the actual needs of the poor. In simple terms, reformers sought to fill workhouses with able-bodied paupers who needed work, but parish officers throughout London quickly found themselves overwhelmed by the throngs of poor, sick people who sought relief and who, by virtue of their settlement, were entitled to it.

At first most parishes simply planned to handle medical needs as they had done for much of the seventeenth century. They continued to contract a surgeon or physician to care for the parish poor who fell ill.[11] Apothecaries and surgeons almost always drew salaries for their services, while physicians often worked for free.[12] For example, the officers of St. George Hanover Square, London, paid their apothecary, Samuel

Despaignol, the handsome salary of £60 per annum, while their physician, Dr. Hardy, received no money for his services.[13] That physicians worked for free indicates that the post itself was valuable in its own way. The position as a parish physician likely lent significant weight to one's reputation in the community. Paying customers could be assured that a parish physician was experienced, having had ample opportunity to practice his art on the sizeable body of paupers. His interaction with members of the vestry provided an opportunity to network with important members of the community, and could well build his (paying) customer base. Moreover, the fact that a physician donated his services helped cast him as a charitable man and a good Christian, qualities that appealed to many prospective wealthy clients. Just as in the case of contemporary hospital physicians, the benefit arising from the position of parish physician could not be measured in pounds and pence, but rather in terms of local repute and connections.[14] Such positions may have played an important role in physicians' career structures, just as it did for their hospital counterparts.[15]

Apothecaries also sought these coveted positions, and on several occasions they tried to work for free as well. But drugs do not come cheaply, and frequently it did not take long for such apothecaries to find that they could not afford to supply an entire parish with medicines out of their own pockets. Thus they, too, demanded salaries.[16] In addition to standard medical figures like physicians, surgeons, and apothecaries, most parishes also kept midwives on the payroll,[17] and continued the early modern tradition of utilizing female nurses, usually local women on relief and widows, to perform a host of medical tasks.[18] In St. Sepulchre the parish continued to pay women termed nurses to care for sick people in their own homes throughout the eighteenth century.[19] Court testimony illustrates that such nurses did indeed treat the foul disease.[20] Nurses were often sent to examine paupers who were too sick to apply for relief in person; and, in some cases, they examined sick applicants on those occasions when a doctor could not be present.[21]

However, it became quickly apparent to parochial officers that their new workhouses were ill equipped to handle the parish's sizeable medical responsibilities. The speed with which parish officers had to build infirmaries in their workhouses indicates how original planners likely did not anticipate the sheer number of sick people who would turn to the workhouse for relief. Consider that barely a year had passed following the opening of the workhouse in St. George Hanover Square in 1726[22] when the workhouse committee ordered that an infirmary be built in January, 1727/28.[23] In some instances parish officials did consider the provision of medical care when originally designing the workhouse, but in these cases they often grossly underestimated the amount of medical resources they would need. For example, St. Sepulchre established its workhouse in 1727, and in May of that year parish officers agreed to set aside space for the ill.[24] Whatever this space consisted of, it was insufficient. By November the house surgeon requested that a proper infirmary be established with at least a room each for male and female patients, and a room for lying-in women. Only two weeks later, the officers also reserved additional space for "mad people."[25] The officers of St. Margaret's Westminster also allotted space for an infirmary at the outset of its plans to erect a workhouse in 1726. They set aside several rooms for the sick, as well as wards for "lunaticks" and patients with the "Itch."[26] Yet a year later St. Margaret's medical staff declared the need for more medical space. In October, 1727 house physician Dr. Harvey and surgeon John Westbrook appeared before the committee and "represented ye Great want of a ward being particularly Assigned for the reception of the sick."[27] The committee considered their advice and on November 16 ordered that six more rooms be converted into an infirmary."[28] Within a year even this infirmary proved too small, and Westbrook had to apply for further rooms for sick boys and sick girls and three separate rooms for "lunaticks." Westbrook's comments indicate that the medical space in this workhouse also included a "Medicine Room" and a "Salivating Ward."[29]

An examination of the reasons given by paupers who sought relief immediately demonstrates the pressure put upon workhouse committees to step up their medical operations and establish infirmaries. Moreover, that data clearly show why the meager space originally devoted to healthcare frequently proved insufficient. In his original study, Hitchcock made extensive use of the unusually rich records of St. Luke's Chelsea. He tabulated the reasons given for admission to St. Luke's workhouse for a group taken from the years 1743–50.[30] Of the 446 inmates for whom complete information is available fully 55 percent were not the able-bodied workers that the SPCK had hoped for, but instead listed as "sick," "infirm/lame," "injured," or "foul." Even if we discard the 15.5 percent of inmates who were "infirm/lame" (and therefore probably not candidates for immediate medical care) roughly 40 percent of all workhouse inmates were admitted specifically to receive medical treatment. More recently, data has been made available for an even larger sample group of workhouse inmates in St. Luke's.[31] These figures show that the general picture Hitchcock described for the 1740s holds true throughout the second half of the eighteenth century as well. The combined inmates classified as "Ill/Lunatic/Injured/Foul," "Infirm/Lame," and "Returned from Hospital" make up 1,807 of 4,352 total inmates, or 41.5 percent.[32]

The records of St. Luke's workhouse are particularly rich. However, further research into other parishes indicates that it was not an anomaly. Applicants for relief who cited illness as their reason for claims represented significant portions of the total applicants for relief at the workhouses in both St. Sepulchre and St. Margaret's Westminster. Consider the data from both workhouses for the twelve-month period January 1 to December 31, 1735.[33] Of all 659 applicants for relief at St. Sepulchre, 177 (26.9 percent) cited some form of illness or injury as their reason for applying. An even greater portion of St. Margaret's parishioners turned up at the workhouse that year due to illness. That parish's admission records resemble those for St. Luke's. Of 733 total applicants, 276 (37.6 percent) were driven to the parish in search of medical care because of illness or injury.[34]

The impact of this significant plebeian medical demand is crucial for understanding the medicalization of workhouses. Recent historians have tended to agree that English paupers indeed came to view parochial relief as something of a customary entitlement over the course of the long eighteenth century.[35] While saying the poor had a "right" to relief might be slightly too strong, it seems clear that it was widely agreed that parishes had a strong obligation to take care of settled parishioners. Just as the royal hospitals may not have been able to refuse all venereal patients out of hand, neither could parishes simply ignore the sick who applied. Moreover, paupers clearly recognized this, and those who fell ill without other options applied to the parish demanding aid on the basis of sickness. We should remember that as the workhouse movement was just underway, St. Bartholomew's and St. Thomas's were initiating admission fees, pricing hospital care further out of reach for the very poor and likely forcing an increasing number of people to look to the parishes when they fell ill. Parish officials frequently decided that they could best meet their mounting medical obligations by establishing an infirmary within their new workhouse. But the role of the poor in driving this medicalization is worth considering. While certainly not orchestrated in any sense, the London poor simply kept turning up at the parish door, exercising what they perceived to be their entitlement, and making what they believed were legitimate demands for medical care. Parish officers were reactive, not proactive when it came to setting up workhouse infirmaries. In a very real sense it was the poor who medicalized London's workhouses. Workhouse proponents leave no evidence that they initially intended workhouses to become infirmaries. The poor, on the other hand, had other ideas for these new institutions.[36]

Parishes each handled their medical obligations differently, making generalizations difficult. A comparison of single year in two parishes shows this. Consider St. Margaret's and St. Sepulchre. St. Margaret's workhouse infirmary appears to have been a good deal larger than St. Sepulchre's. During the twelve months of 1735 churchwardens in the former parish chose to admit 177 of the 276 sick or injured paupers who applied for relief into the workhouse. In other words, they admitted roughly two out of every three sick applicants. By contrast, St. Sepulchre admitted only 29 percent of its sick applicants into the workhouse. Instead of an infirmary bed, St. Sepulchre's officers gave cash doles to over one-fifth (38 of 177) of sick and injured applicants, while another 5 percent were given medical supplies to treat themselves. In addition to these handouts, St. Sepulchre's officers sent another 22.6 percent of ill applicants to other hospitals. This marks a contrast to St. Margaret's, which sent only 8.7 percent of its sick and injured parishioners to outside medical institutions. Perhaps most importantly, St. Sepulchre denied relief to over 19 percent of the sick people who applied, while St. Margaret's denied relief to only two ill people all year.

An inverse picture of the two parishes emerges quite quickly. St. Margaret's granted relief to virtually every sick pauper who applied, and treated the majority of them within its own infirmary. This indicates that the officers of that parish were willing and able to devote significant resources and space to its medical operations. By contrast, the figures for St. Sepulchre suggest that that parish could not (or would not) devote quite as much in terms of resources to treating the sick of their parish. Its infirmary cannot have been as large as St. Margaret's; it admitted an average of 4.5 patients per month, while St. Margaret's averaged over 14 patients per month. As a result, St. Sepulchre had to rely on other institutions to treat a much greater number of its patients. As we learned, this was becoming an increasingly expensive option. That St. Sepulchre denied relief to so many could indicate many things. It could reflect less sympathetic attitudes by members of the workhouse committee towards those who claimed to be ill; it could reflect limited space and resources; or it could indicate that the demographic make-up of that parish assured that it received a higher percentage of non-settled claimants. Unfortunately, the workhouse committee minutes themselves do not answer this question. Regardless, the differences between the two parishes are striking.

But so are the similarities. Though they differ, the two parishes share important features common to many London parishes of the day. First, although St. Sepulchre sent a large number of parishioners to hospitals, these still represented a minority of the patients. This was true of St. Margaret's as well. The fees demanded by hospitals convinced most vestries that it would be cheaper to deal with sick applicants locally whenever they could. Although they differed in certain respects, both St. Margaret's and St. Sepulchre—and most London parishes for that matter—took this approach. Lists of patients cured in St. Margaret's workhouse survive from 1733 and 1734, which indicate what a broad range of care was available by then. The surgeon registered 152 cures in 1733, and 192 in 1734. A wide range of conditions was represented: fevers, rheumatism, dropsy, measles, broken limbs, "cholick," asthma, itch, and of course the pox. These in addition to more vague descriptions of patients described as "vomiting," "looseness," or just suffering from "pains."[37] These figures did not include outpatients. Clearly this was a significant medical operation.

Still, neither parish could completely meet its medical obligations without relying on existing hospitals. No matter how large they may have been, workhouse infirmaries could not provide the same range of care as such large institutions as Bart's and St. Thomas's. Despite building workhouses, London parishes continued to rely on other medical institutions, much like parishes in other English towns.[38] London overseers had a long tradition of relying on hospitals prior to the era of workhouses. A collection of receipts survives for parishioners of St. James Westminster, that provides the details of patients sent from that parish to various London institutions throughout the 1690s, including several who were sent to the Lock and Kingsland outhouses.[39] That this continued as standard practice is clear from the fact that parishes kept on hand pre-printed hospital petitions in which they merely filled in the name and affliction of each sick parishioner whom they sent for admission.[40]

Churchwardens also made sure to grease the palms of hospital staff to ensure that their parishioners got in and were well looked after. The workhouse committees of St. Margaret's, St. Sepulchre, St. George Hanover Square, and the combined workhouse of St. Andrew's and St. George the Martyr all contributed money during the 1720s and 1730s to the "Christmas Boxes" of the beadles of important London hospitals, which included those of St. Thomas's, St. Bartholomew's, Guy's, and Bethlem Hospitals.[41] Such payments were not limited to the beadles. In March 1729 the workhouse master in St. Sepulchre was ordered by the committee "to give nurse Graham of Kingsland Hospital half a crown in order that she may take better care of our poor."[42] In addition to the money that parishes had to lay out in admissions fees and "gifts," workhouse committees also had to supply paupers with clothing and/or linen for their stay in a hospital.[43] Parishes also relied on hospitals like Bethlem and other asylums to care for some mentally ill parishioners,[44] though we have reason to believe that families also received parish pensions to care for such patients in the home, making the household an important loci of care deserving greater attention.[45] The workhouse also had to receive paupers after they were discharged from hospitals if they were still too weak or infirm to survive on their own. So while parishes tried not to resort to the use of hospitals to treat their sick poor, arranging for the admission of patients into hospitals that they could not otherwise afford was a key medical service that workhouse committees provided throughout the eighteenth century. Clearly, paupers understood that, and they frequently came to workhouses seeking a hospital petition.[46]

Another commonality illuminated by the records of St. Sepulchre and St. Margaret's was the provision of assistance for the relatives of ill paupers. Sickness could have devastating economic effects on poor families in eighteenth-century London, especially when it was the principle wage earner who fell ill.[47] A mid-century letter by a gentleman recommending a poor man infected with the pox to an apothecary highlights the impoverishing nature of the disease to the man's entire family, and speculates on their possible reliance on parish support.

> 8 May 1742
> Dear Sir, Here is Poor Tho. S . . . . . . s (who you know) I fear in a bad way as the Apothecary, who has him in hand, tells me; he takes it to be venereal, and that he thinks he can't be cured without a Salivation. . . . Now I desire the Favour of you to be so good as to call on Mr. Watson . . . and shew him this case, and desire his opinion whether he thinks it would cure him, and let him work at his Trade all the Time. If he gives you any encouragement, I would not stick at three or four guineas if I could get the poor Fellow cured; for if he is forced to go into the Hospital, his Family must all go to the Parish, at least during that Time; and if he should lose his Customers by his long Absence, he must follow em to.[48]

The dependents of the sick represented another burden for the parish. Thus, throughout 1735 we find both parishes relieving the families of those who became too ill to work. For example, on July 17 the churchwarden of St. Sepulchre gave the wife of Jonathan Smith five shillings to support her family, because Jonathan suffered from "rheumatism." On the same day the churchwarden also granted two shillings, six pence to Elizabeth Rumball, whose husband also lay ill.[49] Similarly, on June 16 St. Margaret's parishioner Sarah Earl successfully applied for relief for her children, citing her sick husband as her cause of need; the churchwarden granted her two shillings, six pence.[50] Applicants who cited sick relatives when seeking relief were common and frequently successful throughout the eighteenth century. Thus, the parochial response to sickness was often aimed not only at the immediate ill parishioner, but also his or her dependants.

Of course, highlighting the poor health of a family member—or completely fabricating such a story—might well have increased an applicant's chances of success, and should also be seen as part of the larger set of strategies employed by the London poor who sought to tap into parochial resources, and which in turn needs to be set against a much wider set of plebeian survival strategies about which we still know too little, but which certainly went far beyond parochial aid. In the case of street begging, for example, we witness similar tendencies of supplicants to cite poor health or injury, again demonstrating the impoverishing effects of illness on the one hand, and on the other the agency of those paupers who sought to utilize a tale of malady to vie for resources.[51]

One final instructive similarity between St. Sepulchre and St. Margaret's involves the percentage of sick applicants that each parish relieved. Although St. Sepulchre denied almost one-fifth of all sick applicants, that workhouse committee still relieved, in one way or another, four out of every five applicants who cited illness. The fact that such a large portion of St. Margaret's and St. Sepulchre's applicants cite ill health as their reason for need illustrates how contemporary parishioners perceived parish workhouses and sought to use them. That the vast majority of sick paupers were successful indicates that by 1735 these workhouses were already established as important local medical institutions. Within a short time after their foundation London workhouses had evolved into institutions that were part hospital, part dispensary, and part outpatient clinic for the poor people of their communities. While those who originally planned workhouses may not have intended them to provide these services, the real demands of parishioners rapidly forced parish officials to recast the form and function of their young institutions.

By the second half of the eighteenth century it was an accepted reality that workhouses were not primarily for the able-bodied, as intended, but rather for the sick, weak, old, and infirm. The allotment of space to these sorts of inmates indicates how prevalent medicine was in late-eighteenth-century workhouses. A 1791 description of the various wards that comprised the workhouse in the parish of St. Andrew's Holborn, shows that this parish devoted fully fifteen of just twenty-two total wards to inmates described as "lying-in," "infirm," "sick," "foul," or "insane."[52] Moreover, late-century treatises on workhouse management clearly indicate that the provision of care for the sick and weak had long since surpassed work and profit as the primary function of workhouses. In their advice manual on workhouse administration originally published in 1781, members of the vestries of St. Giles in the Field and St. George

Bloomsbury (who founded a single workhouse to serve both parishes) clearly show how their vision of the workhouse had radically altered from the one put forth by the SPCK in the 1720s. They instructed prospective churchwardens "to consider the workhouse an Asylum for the aged, for orphans in an infant state, for idiots and the lame, blind, sick, or otherwise infirm and diseased persons; and that no persons who are able to earn a livelihood . . . should be admitted to remain therein."[53] This statement from the end of the century merely shows the acceptance of a situation that the records indicate was prevalent at a much earlier date.

Although paupers undoubtedly considered the workhouse infirmary an important medical resource, it is equally clear that entering a workhouse was a distasteful option. As Hitchcock has shown, paupers found many aspects of workhouse life unpleasant, to say the least. For instance, the living conditions varied considerably from one workhouse to another. Many vestries contracted an individual (the master) to handle the day-to-day administrative duties, so the integrity and talent of this person affected greatly the experiences of those living under his care. An ill-chosen master could translate into horrible workhouse conditions, as the vestry of St. James Westminster, discovered in 1741. John Tucker seems to have run that parish's workhouse quite poorly, allowing the house to remain in a filthy and unhygienic state. The conditions of the house shocked officers who inspected the premises, finding the inmates in a deplorable state and a "Scorbutick distemper" ravaging the house.[54] In some extreme cases workhouses suffered extremely high rates of infant mortality, insufficient diet, infestation of vermin, and periodic outbreaks of disease.[55]

Paupers also avoided workhouses because they were, by design, disciplinary institutions. When parishioners became inmates they had to submit to the rules of the house that governed almost every aspect of life. Inmates rose, ate, worked, and slept when they were told to do so. Moreover, they were not free to come and go as they pleased, but had to remain within the confines of the workhouse unless they had arranged special permission in advance. In a word, people lost a great deal of personal freedom when they entered a workhouse. Poor parishioners usually had to give up their personal clothing and don a parish uniform that immediately marked them as inmates.[56] Furthermore, parish uniforms often included numbered badges that parishioners detested because of their accompanying stigma. In certain instances uniforms were even color-coded to classify inmates, as in the case of St. Martin's in the Fields, which made mothers of bastards wear blue or yellow.[57] In parishes like St. Giles in the Fields, which numbered their badges according to an inmate's ward, inmates of the salivating wards were immediately marked off as foul by their clothing.[58] Donna Andrew has provided strong evidence that a stigma accompanied accepting parish relief in eighteenth-century London, especially entering a workhouse. In begging letters published in newspapers supplicants often highlighted that they sought alms specifically to stay out of the workhouse and avoid the shame that it incurred, a point made especially emphatically by formerly middle-class Londoners who had recently fallen into destitution because of an unlucky turn in trade, illness, or other misfortune.[59]

Workhouse masters went to great lengths to enforce house rules; often they resorted to corporeal punishment. Many workhouses utilized some form of solitary confinement to punish those who disobeyed. St. George Hanover Square had a chamber called "the Dark Room" and St. Sepulchre had "the Hole" in which punished inmates were confined for extensive periods.[60] The masters of the former parish also flogged their inmates on many occasions. The officers in St. George's clearly endorsed this practice, ordering that "some Cats with Nine Tails be provided by the Master to correct Patients adjudged guilty of offences in the House."[61] It is unclear how committees adjudicated inmates' guilt, but some workhouse committees did not hesitate to send offenders to public houses of correction like Bridewell.[62] Many supporters of workhouse schemes did not shirk from proclaiming that workhouses served as harsh disciplinary institutions. The SPCK's 1725 *Account* stated bluntly: "A *Work-House* is a name that carries along with it an idea of correction and punishment; many of our poor have taken such an aversion to living in it upon that account. . . ."[63]

That claim was accurate, because many plebeians actively resisted entering the workhouse, and others ran away shortly after admission. For the impoverished who fell sick, workhouses were a mixed blessing. They continued the long tradition of parochial responsibility for the ill, and in some ways may have even improved upon it. Rather than receiving a mere handout, which may or may not have been enough to purchase competent medical care, the poor of many parishes now had free access to rather sophisticated infirmaries staffed by apothecaries, surgeons, and in some cases even physicians.

But to access that care, paupers had to give up quite a lot. They had to submit to the work and disciplinary regimens of the house, and risked stringent punishment if they failed to do so. In some cases they entered houses with terrible, even dangerous, living conditions. In light of these factors, we must consider workhouse infirmaries the last resort for many sick paupers who probably first attempted to obtain some form of medical care on their own—likely through self-medication with whatever cheap medicines they may have been able to procure—and probably employed a range of survival strategies to sustain themselves outside the workhouse, strategies about which we are just beginning to learn.[64] The patients of workhouse infirmaries, therefore, represent those at the very bottom of the social ladder with few, if any, other medical options available to them.

Turning to venereal disease, we find the pox prevalent in parish after parish.[65] Like the royal hospitals, a model of hard discrimination against venereal patients fails for workhouses. None of the London parishes examined here made it a policy to exclude venereal patients from its infirmary. Nor, given their obligation under the Poor Law, could they, a point that Dorothy George casually observed as long ago as 1925: "Even in the eighteenth century workhouse infirmaries in London were beginning to count as supplementary hospitals; they could not, like other hospitals refuse patients suffering from certain diseases. . . ."[66]

It is difficult to determine just how prevalent venereal disease was in workhouse infirmaries. The available figures vary. For example, less than 5 percent of sick applicants for relief in St. Sepulchre in 1735 were diagnosed "foul," while in the same year over 11 percent of the sick applicants at St. Margaret's were so noted. Even in a single parish, the figures could fluctuate from year to year, as evidenced by the lists of patients "cured" in St. Margaret's for 1733 and 1734 . Whereas those listed "Pox'd" make up just 3.9 percent of all infirmary patients in 1733, the twenty-five foul patients treated in 1734 represent more than 12 percent of all infirmary patients.[67] The figures that emerge from the records of the Shoe Lane Workhouse in the parish of St. Andrew's

Holborn offer a striking comparison. Foul patients represent fully 70 percent of all sick or injured inmates during the five-year period for which good records survive, 1776–81.[68]

A host of problems makes it impossible to generalize about the prevalence of venereal disease in these infirmaries. First, as these figures indicate, situations seem to have varied over time and from one parish to another. Presumably, certain parishes encountered the disease more frequently. Perhaps those parishes serving the neighborhoods renowned for prostitution, such as those around Covent Garden or in Southwark, had higher rates of infection among their applicants; the pox was, after all, nicknamed the "Covent Garden ague."[69] The nature of such variables cannot be specified without a complete study of all parochial records for the century, and the records are far from complete.

Another factor that thwarts the attempt to fix firm estimates is the tendency towards vague descriptions of illness on the part of the clerks who recorded information of admissions. If "foul" were not shadowy enough already, researches are further frustrated to find the vast majority of infirmary inmates simply noted "sick" when admitted to the workhouse. For example, the 1735 records of St. Margaret's show that 199 of 276 applicants for medical care (72 percent) were merely listed as "sick" or "ill" with no further description of their condition. The pox was one of the few medical conditions that many workhouse committees even bothered to mention by name, indicating that it was indeed a unique condition in contemporaries' eyes. Yet further inspection discloses that many patients merely termed "sick" when they arrived later turned out to be "poxed," suggesting that the patients we can identify represent only a portion of their actual numbers.[70] Thus, the figures that do exist can indicate the presence of venereal disease only in a broad sense. Too many factors obstruct more precise quantification.

Regardless of the difficulties involved in calculating exact numbers, the records of workhouse infirmaries clearly show that these institutions treated the pox regularly. Workhouse infirmaries followed the lead of contemporary hospitals and segregated foul patients in separate wards. For example, St. Margaret's had separate chambers for foul patients by the late 1720s and its surgeon mentions the "Salivating Ward" as early as 1728.[71] Thereafter foul patients admitted to the infirmary were frequently said to be "in a salivation,"[72] leaving little question as to the form of therapy that dominated. Similarly, the workhouse that served the parishes of St. Andrew Holborn and St. George the Martyr also had specific foul wards set aside. The master devoted three separate rooms to foul patients, and evidence indicates that salivation occurred there as well.[73]

Even though salivation remained the dominant treatment in most infirmaries, several parishes experimented with alternative treatment strategies. On December 5, 1731, the workhouse committee of St. Sepulchre invited Mr. Lewis LeBarr to appear before them.[74] Presumably the parish officers had heard about LeBarr from officers of other parishes, because he appeared the following week with a certificate signed by the churchwardens of St. Martin's-in-the-Fields and St. James Westminster, praising his acumen in curing venereal patients in their infirmaries. It is quite possible that the French LeBarr employed the Montpellier method outlined in chapter 2, because the certificate claimed that LeBarr "hath an excellent new way of curing the Venereal Disease without salivation or sickness or alteration of diet or hinderance of business. . . ."[75] One cannot discount the possibility that this method attracted parish officers because they felt sympathy for the pain venereal patients suffered while undergoing salivation. But it is much more likely that churchwardens saw the financial advantages offered them by a practitioner who promised to cure all their venereal patients at the rate of one guinea per head. Moreover, the promise of a cure that did not "hinder business" must have also held the added appeal of promising to make venereal patients available to perform house labor while undergoing treatment.

Whatever the reason, St. Sepulchre followed in the footsteps of St. Martin's and St. James and contracted LeBarr to act as their official parish VD specialist. They ordered LeBarr to begin by treating Mary Matthews, a woman they had treated for the venereal disease in the infirmary on several occasions and whom they had sent to St. Thomas's Hospital the previous year.[76] Matthews seems to have resisted LeBarr's treatment, because the master had to threaten her with expulsion from the house if she did not "submit" to Lebarr's treatment.[77] The churchwardens eventually put three separate patients under LeBarr's care during the spring of 1731/32, but his tenure was short lived. No patients were directed to undergo his treatment after June 1732. This is not surprising, considering that all three of LeBarr's patients returned to the workhouse complaining of symptoms shortly after finishing his treatment. The parish had to pay to send all three to city hospitals. By September Mary Matthews reapplied to the workhouse "to be gott into the Hospital being much afflicted with a pain in her bones the relicts of the foul disease," while in August the parish paid to get Susannah Pain into Guy's Hospital just two months after LeBarr began treating her.[78] By all accounts the experiment had failed.

But the episode indicates that venereal patients represented a significant problem for parish workhouses. The disease was extremely prevalent and very difficult to treat. Foul patients were likely too ill from their disease and the effects of mercury to contribute any significant labor to workhouse production. Moreover, their hospital fees were three times higher than those for general patients. The comments of members of the vestries of St. Giles in the Fields and St. George Bloomsbury indicate that parish officials recognized the problems, especially the economic burden, posed by large numbers of venereal patients in the workhouse. The very same vestrymen who advised their churchwardens to view the workhouse as an "asylum" for the sick and infirm, simultaneously cautioned parish officers to limit the number of venereal patients in their infirmaries. They warned,

> too indiscriminate an admission of persons infected with the venereal distemper, would not only encourage such persons to persevere in a wicked and abandoned course of life, but be a temptation to those who contract that distemper in a state of profligacy, to flock from all quarters and procure a temporary residence in these parishes, with a view of being admitted into the workhouse and cured—by which means the parishes would be continually infested with great numbers of the most profligate of casual paupers, and put to much charge in maintaining and curing them. . . .

To combat this problem they advised parish officials to use "the greatest caution" when confronted with foul applicants, and make it a rule to admit only "cases of the greatest necessity."[79]

The records show that churchwardens took this advice and frequently resisted admitting paupers diagnosed with the pox. All three of the parishes examined in detail for this chapter, St. Luke's Chelsea, St. Sepulchre (London), and St. Margaret's Westminster, refused applicants infected with venereal disease at one time or another. On occasion, these decisions seem to have been motivated by moral indignation toward the applicants and their lifestyles. For example, the churchwardens of St. Sepulchre refused the application of Susannah Montague in July 1761, claiming that she "had ye foul disease twice & also [had] ye character of a very slothful person. . . . [She] was denied it being suspected that she only wanted an opportunity to practice lewdness."[80] While examples of this kind do exist, they are rarities. It was not common for workhouse officers to turn away foul applicants simply because of their dubious moral character. Like it or not, churchwardens had some obligation to those parishioners with legitimate settlements.

For that reason, it was much more common for parish officers to use settlement enquiries to limit the number of foul patients in their infirmaries, a tactic that they used to rid themselves of many different burdens.[81] Officers could legitimately refuse those applicants who could not sufficiently prove that they were legally settled in the parish. For example, in 1758 the officers of St. Luke's Chelsea turned out foul patient Ann Harris after seventy-one days in their workhouse infirmary; the comment in the register reads "proved no parishioner."[82] Often, parish officers would "pass" such paupers to their legitimate parish of settlement, where they could expect relief. "Passing" a pauper essentially meant that the parish would not just pay the carriage to return someone to his own parish, but also process the paperwork necessary to legally remove him and force his parish to accept him. Although it involved some cost to do this, it was well worth the expense to rid one's parish of infected paupers who would otherwise pose a significant burden. As a result, parishes commonly passed any foul paupers that they could. Evidence shows that foul paupers both entered and left workhouse infirmaries because of this ongoing attempt by churchwardens to avoid accepting responsibility for venereal patients. For example, in November 1744, Thomas Morris stayed just five days in St. Luke's infirmary before they passed him across the river to Battersea. Martha Jackson arrived at St. Luke's in April 1749, having been passed from St. George Hanover Square. It took just one month for the churchwardens to maneuver to pass her on to Battersea as well. Similarly, Samuel Russell spent twenty-five days in St. Luke's foul ward before the officers arranged to pass him all the way to Cork, Ireland.[83]

Technically, all applicants for relief were supposed to prove their settlement. In practice, the minuscule parochial bureaucracy could not investigate the details of very many of the scores of applicants who sought relief each week. The records show that parish officers tended to reserve such investigations for those applicants who represented significant burdens to the parish coffers. Just as the vestrymen of St. Giles-in-the-Fields and St. George Bloomsbury had warned, venereal patients were one such group. Churchwardens often demanded that foul applicants explain their settlement claims before admitting them to the workhouse. Such explanations usually involved details of births, marriages, apprenticeships, or occupations in the parish that might entitle them to relief. Typical is the following statement made by Susannah Pain, one of the three venereal patients put under the care of Lewis LeBarr.

> Susannah Pain wife of john pain passed from St. Giles in the Fields to Mr. Churchwarden Ward she was ordered to be admitted into the House for the present, confess[in]g she is foul, giving the following acct. of her settlement Viz. that she was married to John Pain (of 18 years of Age) on a Tuesday a fortnight before Jonathan Wild Sufferd, he was the son of Pain a watchmaker in Meers Court in warders Street near old Soho in St. James parish who made oath of the settlement Susannah Pain says there were present at the marriage Mrs Saunders who kept a chandlers Shop in Ducking Pond Alley in New Bond Street as also one Symonds a washerwoman and Norton a mantua maker who lodged in Saunder's house, the master is ordered to enquire into the above particulars &c and Susannah Pain is ordered under the care of Mr. Lewis Lebarr in order for cure.[84]

One of the workhouse master's tasks was to conduct the necessary interviews either to corroborate or disprove such claims. In some instances, foul applicants were instructed to return with known parishioners who could vouch for them. Sarah Jones admitted that she was foul and applied for a hospital petition from the officers of St. Sepulchre in 1728, claiming to be the daughter of a local innkeeper. The churchwardens delayed their decision and told her to "come back on Sunday and bring somebody that knows her."[85] In the meantime, an overseer looked into Jones's claim and discovered that she had obtained a subsequent settlement in another parish, freeing St. Sepulchre of any responsibility.[86]

At times the arguments concerning which workhouse was responsible for particular foul patients lasted quite a long time, as parishes met claims with counter claims in attempts to pass sick paupers on to one another. The story of Ann Cox suggests the lengths to which churchwardens might go to fix the responsibility for a venereal patient on another parish. Ann applied for relief in St. Sepulchre in November 1730, but the committee denied her, claiming that she did not live in the parish.[87] She returned in January "in a very poor condition and almost naked" claiming relief on the grounds that her husband was settled in St. Sepulchre, but the committee again tried to refuse her. In a novel turn of events, they accused the master of the workhouse in the parish of St. James Clerkenwell of arranging her marriage to a St. Sepulchre parishioner just to get rid of her:

> Ann Cox her husband having been gone some time since applied for relief being in a very poor condition and almost naked it appearing she had been married by contrivance of the Mas[ter] of the workhouse of St. James Clerkenwell and also of one Gill the beadle of the same parish who paid John Cox 40s to mary her his settlement being in this parish. . . .[88]

Despite their obvious anger at this scheme, the committee had to relieve her, and she was ordered into the house for treatment. If the St. James workhouse master did indeed arrange her marriage for forty shillings, it may have been money well spent. St. Sepulchre had to pay for Ann's treatment in the foul wards of city hospitals on two separate occasions over the next twelve months.[89] However, when she again applied for a hospital petition in October 1732 they again denied her, showing what a game of cat and mouse this could be. Stubbornly, she reapplied a week later, and the committee again took up the issue of her settlement, resolving to demand that the churchwardens of St. James Clerkenwell officially pass Ann to St. Sepulchre.[90] The officers of St. James happily complied. St. Sepulchre's committee had to pay for another hospital petition for Cox in January.[91] She continues to appear in the records through 1737, going to the hospital on at least one more occasion.[92]

Some were not so lucky. Ann Edmunds could not prove her settlement, and like many such people she was declared a vagrant and incarcerated in Bridewell while city authorities arranged to pass her to a parish. She must have been in clearly poor health. The beadle's wife testified that Ann appeared "very ill" upon entering the house, and that she "understood from the Deced. that she had a veneral [sic] complaint." Ann continued in the house for a few days before being taken to appear before the Lord Mayor a second time. The men deciding Ann's fate could not have been ignorant of her waning health. Witnesses testified that when she returned from that hearing she "appeared [so] ill that she took to her Bed and hardly ever got out of it." She died two days later.[93]

It is worth considering what stood behind the tendency to enquire so strictly into venereal patients' settlements. Considering the handful of examples of parishes citing poor moral character when refusing applicants, one must ask whether settlement enquiries of the foul represented churchwardens' attempts to discriminate against sinners. To answer that question it is instructive to compare parishes' policies toward venereal patients with their handling of bastard bearers. In the eighteenth century both unmarried pregnancy and venereal infection were considered proof of moral, sexual transgression. It was assumed that neither would have existed were it not for extramarital sex. With the exception of those who caught the foul disease by nursing an infected child or thanks to an unfaithful spouse, contemporaries could (and did) assume that the vast majority of foul patients became infected through promiscuity. Therefore, single mothers and foul patients were guilty, in the eyes of most eighteenth-century folk, of the same moral transgression.

It is telling, then, that churchwardens were much more stringent in their settlement enquiries of single mothers than of foul patients. Officers looked more frequently and much more keenly into the details of bastardy enquiries than settlement enquiries of venereal patients.[94] For example, bastardy enquiries always involved an attempt to discover the bastard's father and his settlement details. The testimony of the women, of witnesses and of acquaintances were all taken down in the official record. Moreover, as Laura Gowing has described, the procedures could also hinge on invasive examinations of humble women's bodies for various signs of pregnancy, often performed by bourgeois women or parish midwives.[95]

By contrast, the settlement enquiries of foul patients were not nearly as thorough. Rarely were the details considered important enough to copy down, and in no instances do the records include any discussion of how a foul patient became infected to parallel the enquiry of how single mothers became pregnant. Moreover, evidence shows that parish officers went to much greater lengths to punish single mothers. Consider that the churchwardens of St. Sepulchre passed the resolution to punish bastard bearers and published that resolution in the newspaper.[96] Although rather few women were actually punished, these policies were put into practice frequently enough. For example, in 1729 the officers of St. Margaret's ordered that Sara Pagett be sent to "ye House of Correction" after giving birth to a bastard in the workhouse.[97] The following year they similarly ordered that single mother Jane Herbert be sent to Bridewell after her period of lying-in.[98] However, even the threat of imprisonment could work as a powerful tool for workhouse committees trying to limit applications for relief. Workhouse committees commonly threatened women with imprisonment if they returned seeking relief.[99]

The stricter process of settlement enquiry of single mothers and the greater propensity to punish these women suggests strongly that economics, not morality, lay behind the actions of eighteenth-century workhouse committees. Venereal patients were costly to workhouses, and officers clearly tried to limit the number of patients they accepted into their foul wards. They could face punishment, too, as is amply demonstrated by the threat to send Flora Price to Bridewell (in the example that opened this chapter).[100] But the economic burden presented by a bastard—a poor infant for whom the parish might be financially responsible at least until his or her apprenticeship[101]—dwarfed the economic drain of a foul patient. As a result of this fiscal reality, churchwardens regulated bastardy much more stringently than venereal disease, even though the two were believed to stem from essentially the same deviant behavior.

While churchwardens clearly tried to limit the number of venereal patients that they relieved, the records also show that officers were sometimes reluctant to turn obviously ill people into the streets. In fact, churchwardens often agreed to treat foul patients who may not have had legitimate settlements, at least temporarily, while they conducted their enquiries. The aforementioned Susannah Pain, for example, was ordered under the care of Lewis LeBarr before the workhouse committee had actually enquired into the legitimacy of her settlement claim.[102] Likewise, in 1730 the churchwardens of St. Margaret's ordered their surgeon to begin treating Mary Lloyd for the pox while her settlement enquiry proceeded.[103] The health of some patients was sometimes so poor that churchwardens had to halt settlement enquiries altogether. Such was the case in November 1728 when St. Margaret's officers ordered "That an enquiry into Mary Meldans settlm[ent]t be suspended till she is out of her salivation."[104] Records also show that city officials sometimes intervened on behalf of infected paupers who were too ill to be passed. Even though Sarah Law was proven not to have a settlement in St. Sepulchre, the Lord Mayor refused to grant the churchwardens a pass to remove her. The parish was forced to get her a bed in a hospital foul ward.[105] Margaret Coleman must have been similarly debilitated by her bout with the pox. The committee at St. Margaret's ordered that she and her two children be treated for the disease before being removed to her husband's settlement in Suffolk.[106]

These instances are instructive because they begin to illustrate how serious contemporaries often considered the pox to be. Contemporaries did not always view infection with flip sarcasm or light-hearted satire. Nor did moral indignation always dominate contemporaries' responses to real people in need. In many instances throughout the century workhouse officers faced truly ill people whom they could have legitimately refused, but chose to help. While every parish had official policies governing infirmary admissions, in practice the face-to-face nature of the charitable exchange left significant room for flexibility.

Further evidence that parish officials viewed venereal infection as a serious condition that demanded attention are the numerous cases of individuals who were originally denied relief but later granted aid when they confessed their illness. The parish of St. Sepulchre, which (it will be remembered) frequently denied relief to sick applicants, provides some of the most illuminating cases. Typical was the story of Martha Pye, also known as Mercy Pye, who applied for relief on July 25, 1736, but did not cite any illness when applying. The churchwardens decided against granting her relief, on the grounds that she was a "young person." Yet less than two weeks later Martha reapplied, this time citing her infection. She based her settlement claim on having served an apprenticeship in the parish. Typically, the parish looked into her settlement details.

Finding her story to be true, they admitted her into the workhouse and ordered a hospital petition on her behalf.[107] The churchwardens also denied relief to Elizabeth Hodge on several occasions. They turned her away in May 1749 and again on June 17, 1750. On neither occasion did Hodge claim illness. But four days after her second refusal Elizabeth returned and "applied for relief being afflicted with the Foul Disease." This time her application succeeded; they ordered her a hospital petition and admitted her into the workhouse until she could obtain a hospital bed.[108]

Even when paupers were well known and disliked by the churchwardens, their applications for venereal care frequently succeeded. Such was the case of Teresa Tiplady, who applied to be admitted to St. Sepulchre's workhouse on October 4, 1750. The churchwardens refused her application, because they had previously relieved her only to watch her pawn the clothing that they had given her. However, one week later Teresa again applied and confessed her venereal infection. The parish agreed to arrange her hospital admission, and did so again almost four years later when she returned seeking help in February 1754.[109] These cases show that churchwardens' attitudes often changed once they discovered that an applicant suffered from the foul disease. People to whom they had just refused help now became worthy objects of charity. Surely, the pox did not suddenly endear these people to the members of workhouse committees. But it demanded their attention. Whether churchwardens responded out of pity for foul patients' plight, or out of the fear of what might happen if the pox was allowed to go unchecked among the poor population of their communities, the evidence is clear that the foul disease posed a problem that churchwardens felt they could not ignore.

Parish officers' seemingly contradictory responses to the foul disease—at times reluctant to grant treatment, at other times sympathetic—make generalizations impossible. But these varied responses are nonetheless illuminating of the ethical and economic struggle that churchwardens faced when confronted by foul paupers who sought help. It was not uncommon for churchwardens to contradict themselves, reversing former decisions either to grant aid or to deny aid to foul patients. The case of the aforementioned Sarah Howard helps demonstrate the tension. Sarah applied for relief on May 21, 1732, and was ordered under the care of Lewis Lebarr on June 1.[110] Two years later, pregnant and unmarried, Sarah applied for admission to the workhouse to give birth. On December 8, 1734, they officially denied her any relief. Yet at some later date they admitted her, because she gave birth to her child in the house and was ordered discharged "without the punishment [that] the law inflicts" on March 9, 1735.[111] By December she returned to the workhouse complaining of symptoms that the infirmary staff diagnosed as venereal, and she requested a petition for St. Bartholomew's. Even though the workhouse had already seen fit to aid her in the past, and admitted her to the house to give birth, the churchwardens now refused her application and told Howard that that "she must prove her settlement to their satisfaction in a better manner."[112] Two weeks later Sarah reapplied; again the churchwardens denied her. Somehow Sarah gained admission to a hospital on her own, and upon discharge she applied to convalesce in the workhouse. Her condition was described as "very much weak." Inexplicably, the churchwardens now admitted her, letting her rest in the workhouse infirmary for a fortnight. Less than a month later a woman applied on Sarah's behalf, describing Sarah's condition as "very ill and helpless." Again the churchwardens granted her a space in the house.[113]

What is one to make of these whiplash decisions? The churchwardens seemed willing to help Sarah up to a point. But either the accumulated cost of her claims or the growing disgust for an applicant that officers judged to be sexually and morally loose (or a combination of these factors) pushed the churchwardens to use a settlement enquiry and then straight denial to refuse Sarah's applications. Yet after her trip to the hospital Sarah again became a worthy object of charity. It is difficult to know exactly why. The decisions could have turned on economics; perhaps the officers were no longer willing to pay the steep price for another hospital petition, but were willing to grant Sarah a bed in the workhouse. Or perhaps Sarah's declining health now evoked pity, which overcame their negative feelings.

Whatever the reason, Sarah's story illustrates how workhouse committees' responses to foul patients were not marked by rigid policies, but often by inconsistency. Moreover, her case shows that the unique exigencies of foul patients challenged the old Poor Law system and the very nature of what constituted a worthy object of charity. Finally,

Sarah's case also begins to illustrate the strategies of poor foul patients. Sarah did not give up after her first rejection. She was stubborn. Despite repeated rejections, she returned and made her case for the medical care that she (and others like her) viewed as rightfully hers. The parish had an obligation to help someone in her condition and situation, and Sarah knew it. She did not hold the upper hand in this exchange, but neither did she go quietly away after the first rejection. The actions of the recipients of workhouse medical charity like Sarah Howard are just as illuminating as the actions of churchwardens. It is to their stories that we must now turn.

Who used parish workhouse infirmaries? In some ways foul patients generally mirror the overall population of parish workhouses, as described by Hitchcock. His work has shown that workhouse inmates were disproportionately female, disproportionately old, and disproportionately young. In other words, women, the elderly, and children were much more likely to find themselves in a position desperate enough to force them to accept admission to a workhouse than were adult men. In Hitchcock's sample group from St. Luke's, over 50 percent of all inmates were adult women, 34 percent were children under the age of sixteen, and only 15 percent were adult men. Moreover, roughly half of the adult male and adult female populations were over 55 years of age. Women, children and the elderly were more likely to enter the house when healthy and they stayed for longer periods than did the small population (approximately 7.5 percent) of adult men below the age of 55. The gendered disparity of workhouse inmates surely stemmed from the corresponding economic disparity between men and women in eighteenth-century London. Put simply, women were more likely than men to be poor—so poor, that is, that they had to rely on their parish workhouse, distasteful as it was. Working-aged men in the metropolis stood in a stronger economic position than their female peers. As a result, it was relatively rare for them to fall on parish relief; and it is furthermore quite informative when they did. According to Hitchcock, the small group of men aged 16 to 55 entered workhouses "almost invariably in response to serious physical illness."[114]

The evidence of several parish workhouses in various periods shows that the sub-group of foul patients follows, to some degree, the pattern described above: many more foul women applied seeking care

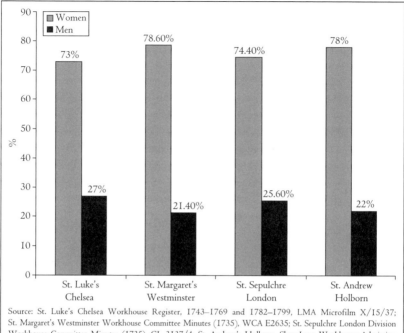

Fig. 10: Gender Breakdown of Venereal Patients in Four London Parishes.

than did foul men. As the hospital registers indicate, women relied much more heavily than men on their parish when they fell ill. And as we now know, parishes with workhouses treated most of their cases locally. Hence workhouse infirmaries were filled with women. The data indicates that the gendered disparity of foul patients was even greater than that of the general population; women made up virtually three quarters of all workhouse venereal patients in every parish studied. As the graph in figure 10 shows, women comprised over 70 percent of all identifiable foul patients in four separate parishes.[115] In St. Luke's Chelsea, for example, over 73 percent of all foul inmates identified throughout the century were women. The populations of the foul wards in St. Margaret's Westminster and St. Sepulchre bear great similarity. In those parishes women represented 78.6 percent and 74.4 percent of all identifiable venereal patients respectively. Looking later in the century

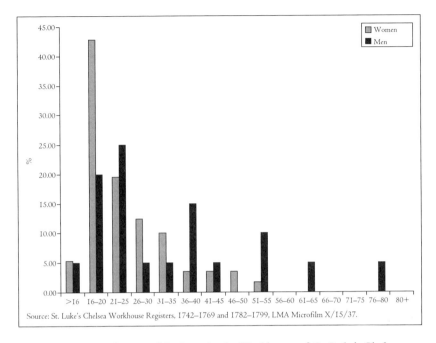

Fig. 11: Ages of Venereal Patients in the Workhouse of St. Luke's Chelsea, 1742–1769 and 1782–1799.

and in yet another parish one finds more consistency. At the Shoe Lane Workhouse in St. Andrew's Holborn women represented 78 percent of all identifiable venereal patients during the five-year period for which good records survive, 1776–81.[116]

The ages of these women are even more revealing, though only the records of St. Luke's workhouse and St. Andrew's indicate the ages of workhouse inmates. Figure 11 charts the ages of all known foul patients admitted into St. Luke's workhouse. Female foul patients deviate from the general pattern of workhouse inmates in several ways. First, the vast majority of female foul patients—over 90 percent—were 35 years of age or younger. And this figure is not skewed downward because of an abundance of children. Children (younger than sixteen) diagnosed as foul represented just slightly more than 5 percent of all female foul patients and only 5 percent of all male foul patients. Fully 84.9 percent of all female foul patients were between the ages of 16 and 35. Perhaps more striking still is the fact that 42.8 percent of foul women had already caught the disease and been admitted to the workhouse between the tender ages of 16 and 20, and almost two out of three (62 percent) during the ten year span from 16 to 25. While Hitchcock's data shows that over half of the general adult workhouse population were over 55, very few (less than 5 percent) of women in the foul wards were even over the age of 40, and none exceeded 50. The ages of St. Luke's foul women are grouped rather closely together in the twenty-year range from 16 to 35, and the average age of female foul patients is just under 24 years old. St. Luke's men, by contrast, were not nearly as young. While less than 10 percent of foul women exceeded 35 years of age, 40 percent of St. Luke's foul men did. And while many men also fell between the ages of 16 and 25, the ages of male foul patients spread out over a much greater range, including men in their forties, fifties, sixties, and seventies. Finally, the average of these diverse ages, just over 33, is almost ten years higher than the average for foul women.

Although it does not cover the same long period, the evidence of female venereal patients in the Shoe Lane workhouse corroborates the situation described above (see fig.12). In that infirmary, 96 percent of all foul women were younger than 35; 60 percent were younger than 25. None exceeded 37. The average age of all St. Andrew's female foul patients is quite consistent with that of St. Luke's at just over 25 years old. It is difficult to draw conclusions from so small a sample group as the seven identifiable male venereal patients in the Shoe Lane workhouse infirmary. However, it is still informative that the average age of these men is over four years higher than the female average, and that the ages of male patients is also spread out over a greater range, including the oldest two patients recorded.[117]

The figures for the two simple variables of gender and age already begin to elucidate key elements of eighteenth-century workhouse foul wards. In terms of gender, foul patients present an exaggerated form of Hitchcock's general pattern. Women outnumbered men by a sizeable margin. Yet in terms of age, foul patients represent an inverse of Hitchcock's general model. The very age group that he shows to be the least likely to rely on the workhouse, people between the ages of 16 and 55, comprise virtually all foul patients; the bulk of whom fell into the first half of that category. In light of this, Hitchcock's claim about the role played by "serious physical illness" in forcing working-aged people

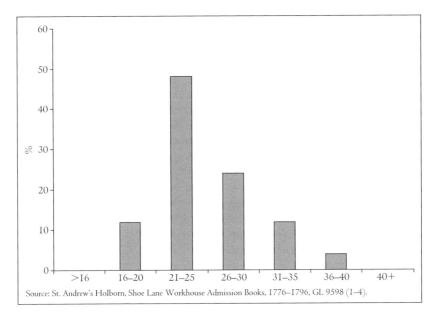

Fig. 12: Ages of Female Venereal Patients in the Shoe Lane Workhouse, St. Andrew's Holborn, 1776–1781.

into the workhouse seems particularly insightful. Perhaps unsurprisingly, venereal infection seems to have plagued the (sexually active) young. Little surprise that when placed alongside the data on the age of prostitutes in London, or the age of single mothers, the data on women in workhouse foul wards is strikingly similar.[118] Of course, in the era before penicillin the disease could remain to plague people throughout their years, hence cases like David Davis who in 1789 was admitted into St. Luke's foul ward at the ripe age of 76.[119] However, such cases were anomalies. Far more common in workhouse infirmaries were young women in their late teens and early twenties.

This is of paramount importance when considering the role of these institutions. All other evidence would have one believe that men had higher rates of infection than women. Certainly one might assume that by looking at the royal hospitals, where more men received care. Yet in every workhouse studied foul women far outnumber foul men, and upon closer inspection, these women prove to be young. When one expands this picture to the more than one hundred London parishes, it begins to appear that the incidence of the pox among men and women—taking into account the shadowy nature of the diagnosis and the impossibility of ever determining rates of infection with complete accuracy—may have been much closer than we have tended to assume.

One must also conclude from the demographics of workhouse foul wards that women who contracted venereal disease, especially young women, had far fewer medical options available to them than did men. Private care of all sorts abounded. The royal hospitals treated the disease for set fees. But these avenues seem to have been closed to many young women who could not afford them. Workhouses provided care for those at the very bottom of the social ladder, and as we have seen this group was disproportionately female. The figures for workhouse foul wards suggest that the economic effects of venereal disease exacerbated this already existent gender gap among the very poor. The symptoms frequently complained of by foul patients could prevent one from working and earning. Exposure of infection could also cost plebeian folk their employment. Furthermore, attempts to purchase medicines for self-medication could quickly sap what little money one might have. In short, the foul disease (like disease in general) was itself impoverishing. Among the London poor who lived on the margins it seems that men who caught the disease were more successful than women at dealing with their illness without falling into abject poverty and into the workhouse. While hardly by design, workhouse foul wards catered primarily to young women.

Unfortunately little more is known about the identities of most of these patients. None of the parishes examined here recorded marital status on a regular basis. However, when we consider that such a large portion of female foul patients were considerably younger than the average age at which English women generally married in the period,[120] we can safely assume that many must have been single. Moreover, this assumption finds further support when we consider that the decision to marry was in large part contingent on a stable financial position. Considering the poverty of these women, one must conclude that, realistically, not only were many of them too young to have been married, most were also too poor.

The only patients who can be positively identified as unmarried were single mothers. As we have seen, mothers of illegitimate children and foul patients each represented significant financial burdens for the parish; and for that reason they drew more intense scrutiny from churchwardens. Some single women were unfortunate enough to find themselves both pregnant and infected, sometimes simultaneously. Needless to say, such women were not popular with eighteenth-century workhouse committees. For example, Mary Holman became infected and applied for medical care at St. Margaret's workhouse in September 1727. One week later the churchwardens arranged to have her treated in the Kingsland outhouse.[121] She returned to stay in the workhouse three years later for a short period in the winter of 1730/31.[122] By November she was pregnant and admitted to the workhouse, the minutes claiming that she was "Bigg with a B[astard] C[hild]."[123] On January 6 the churchwardens ordered an enquiry to discover the father. One week later the committee issued a rather draconian response, ordering "that Mary Holman when well of her lying inn of a bast[ard] child be sent to ye house of correction."[124]

The aforementioned Sarah Howard's case illustrates that St. Sepulchre also punished women who utilized parish resources both when infected with the foul disease and when pregnant. It will be remembered that St. Sepulchre's committee began denying Howard assistance when she arrived at the workhouse pregnant, having already been treated for venereal disease two years earlier.[125] The committee meted out similar treatment to Mary Green. A bastardy enquiry was ordered to discover who impregnated Green in September 1732, and a hospital petition was arranged to get her into a foul ward seven months later.[126] But she was denied further relief when she reapplied to the workhouse in July 1733.[127] Her illness continued until late-1734 when the committee finally agreed to admit her and her child. They continued to live in the workhouse until November 1735.[128] She was finally discharged in November, but only after the churchwardens issued a stern warning that essentially cut her off from any further relief by threatening to send her to Bridewell if she returned.[129]

As suggested, it seems that in each of these cases the woman's pregnancy demanded much more of the churchwarden's attention and ire than did their venereal infection. It is clear that in the eyes of parish officials, these women were bastard-bearers first and foul patients only second. However, the fact that each had also been treated for the foul disease could have only hurt them in the eyes of the parish officials charged with their care. First, their infection rendered them an even greater financial burden than ordinary single mothers, as did the possible infection of their children. More importantly, infection could also be grounds to criticize further the nature of their behavior and speculate on their character. In both Sarah Howard's and Mary Holman's cases their venereal infection came several years prior to their pregnancy. This gave workhouse committees a longer history of premarital sexual activity, which might well have influenced the subsequent relief decisions.

In most cases, one cannot know the actual sexual history of foul patients. Mary Holman and Sarah Howard were both sexually active out of wedlock over the course of several years. Either may have been prostitutes, but one cannot hazard a guess. In most cases we are left quite in the dark as to how these men and women came to catch the pox. However, there is one female foul patient/bastard-bearer for whom an enquiry survives. Importantly, her story differs from those of the women just outlined. Joan Rumbold was twenty-one when she arrived at St. Luke's workhouse infected and pregnant in March, 1757. She gave birth to a son, John, and in the following February mother and son were sent to the Lock Hospital for treatment. They returned from the Lock in mid-March and stayed in the workhouse until late-April. The workhouse sent them to live with Joan's father, but it did not work out and the two returned to the workhouse after just one week. The workhouse took them in, and after two months, the committee agreed to keep John and arranged a domestic service position for Joan in Brompton. They sent Joan to her new job with a new pair of shoes.[130]

Unlike the cases described above, there is no mention of punishment, nor threats to cut Joan off from future aid. The evidence given in Joan's bastardy inquiry may illuminate why it was that Joan seems to have received more lenient treatment than did either Mary Holman or Mary Green. According to her bastardy enquiry Joan became sexually active at the age of nineteen with a man named John Phillips. The two carried on a sexual relationship for three years. John passed on his infection to Joan at some point and deserted her once she became pregnant. Joan believed that John would marry her, and continued to wait for him to come for her while she was in the workhouse. While not all of her contemporaries would have approved of Joan's behavior, sex between betrothed couples prior to marriage was common and to a large degree tolerated in eighteenth-century England.[131] St. Luke's officials were somewhat sympathetic to Joan's case because Joan caught her infection from, and became pregnant by, a single man whom she reasonably expected to marry.

Contrast the nature of Joan's situation with that of women like Mary Holman and Sarah Howard. Because of their prior venereal infection, these women could not fall back on the standard story that many pregnant women (like Joan Rumbold) likely related at their bastardy examinations; namely, that they had been virgins, seduced by men who promised marriage and later deserted them. Thus, although venereal infection was not as important as bastardy to eighteenth-century churchwardens, it could colour one's sexual history and thereby affect how one was perceived and treated by parish officials.

It is worth considering the admissions process and life in the foul wards from the perspective of the inmates. This study has stressed the importance of privacy to early modern venereal patients. The same dynamics governing hospital admissions seem to be at work at workhouses, too. Just like the governors of St. Bartholomew's and St. Thomas's, churchwardens complained that "foul" applicants tried to conceal the nature of their illness. With the same caveats in mind concerning how we read such accusations, it seems as if workhouse applicants similarly used deception as one of their strategies when negotiating the admissions committee. Take the case of St. Sepulchre inmate John Rawlins. On several occasions the records explicitly state that Rawlins was diagnosed with the foul disease.[132] But it is interesting that Rawlins cites a variety of other ailments and symptoms when applying for aid over the course of several years. In January 1729/30 he claimed to suffer from a "Rheumatism" and bad eyesight. He applied again in April and the following January, complaining about his eyesight.[133] He then applied for a hospital petition on June 13, 1731. But four days later the workhouse master alerted the committee that Rawlins's illness was in fact the pox and that he was attempting to obtain a hospital petition under false pretenses:

> Mr. Clement reporting that John Rawlins had confessed to him he was affected with the foul disease it is ordered therefore that he may be got into St. Thomases Hospitall as a foul person and not otherwise. . . .[134]

The extent to which this was a true "confession" we cannot know. But Rawlins applies for relief so regularly, one gets the sense that there is some deft negotiation taking place. The following year Rawlins reapplied, but now he could not pass his condition off as something else, and the churchwardens granted him a petition for St. Thomas's foul ward.[135]

Such a wide array of symptoms accompanied the single diagnostic category "the venereal disease" that it was not difficult for early modern patients to seek care by citing various symptoms that did not necessarily indicate the presence of the pox. And again, this same range of symptoms also means that we cannot always assume that patients believed that they had the foul disease. Discrepancy between an applicant's diagnosis and that of the workhouse surgeon may well have been common. But for those who wished to deceive, there were many generic symptoms unrelated to urinary or genital health to hide behind. One of the most common conditions complained of by venereal patients was some sort of pain in the lower extremities, often diagnosed as a "sore leg." Elizabeth Sprigg applied for relief throughout the winter of 1729/30 and avoided calling her condition "foul" on four separate occasions. Each time she applied for relief she highlighted that she suffered "a lameness of her legs."[136] She had to wait a month for a vacancy in St. Thomas's Hospital and when she finally returned to the workhouse for her hospital petition the infirmary staff deemed her foul. Since she was willing to enter a foul ward she may have finally accepted the diagnosis.[137]

The fear of undergoing a settlement enquiry may have also influenced applicants who believed themselves to be foul to conceal the nature of their condition. Settlement enquiries added to the public nature and general intrusiveness of workhouse medicine. Not only did foul patients have to admit the nature of their condition to the men who comprised workhouse committees, but this could also lead to further investigations involving all kinds of acquaintances. Parochial officials might interview family members, neighbors, and present or former employers. Such enquiries opened the possibility that these people might learn the reason why the applicants sought entrance to the workhouse, and thus discover the details of their medical condition.

Moreover, venereal patients' reputation for guile seems to have led parochial officials to be particularly thorough when researching foul applicants' settlement details. Advice manuals stressed that when inquiring into the details of foul applicants' settlements the workhouse master "is not to depend on the informations of the paupers themselves" but rather research the details as thoroughly as they could by interviewing all those who might provide relevant information.[138] The suspicion that foul patients could not be trusted at their word, that they were more likely to deceive—which may have been well founded—meant that they faced more strict settlement enquiries than most other applicants. There is no way to know how discreet various churchwardens and overseers of the poor may have been when conducting such enquiries. But it is hard to imagine that these lay investigators protected such details or adhered strictly to an ethic of medical confidentiality when it is not even clear that such an ethic yet existed within the medical profession itself.

Those applying for parish relief may have had reasons other than modesty to conceal the nature of their illness. Workhouse committees also gave out a fair amount of incidental relief. Patients hoping to pry a few shillings out of the committee, while remaining out of the workhouse, did well to avoid mentioning the pox. As we have seen, churchwardens sometimes limited the amount of aid they were willing to grant to venereal patients who returned on more than one occasion. Henry Truslove and his wife were no strangers to the churchwardens; they appeared before the St. Sepulchre workhouse committee on at least twenty-five occasions between 1734 and 1752. Henry was diagnosed as foul as early as April 10, 1735; his wife, Jane, applied on his behalf and obtained for him a hospital petition.[139] The Truslove family clearly tested the patience and generosity of St. Sepulchre's committee, requesting financial assistance for one reason or another throughout the next year.[140] In 1736 the Trusloves requested financial assistance due to illness in September and again in October. On both occasions they avoided using the term "foul" and cited other maladies. In September Jane claimed that the couple suffered from "an ague and fever" and in October that they labored under a "fever." They were granted two shillings, six pence in September, but denied in October.[141] In December Henry decided to change his strategy and twice applied for relief on account of his "very bad leg," one time claiming to have dropped timber on it.[142] On at least five later occasions the Trusloves applied for relief claiming that they were "very ill," "sick," suffering from another "fever," "very ill of a consumption," and finally lame with another "sore legg."[143] The Trusloves were clearly experienced at negotiating the old Poor Law system. In addition to protecting one's privacy and/or reputation, denying that one was infected with the pox may have also afforded paupers a better chance to wring much needed money or medical care from their local churchwardens.[144]

Once in the infirmary, it seems that foul patients shared the general opinion of workhouse life: they often hated it. It is not entirely clear whether infirmary patients were made to work while under their course of mercury. Those who stayed in the house after their treatment certainly did, and many foul inmates resisted the labor regimen that workhouse masters tried to impose. Thus one finds entries like the following that records the aforementioned John Rawlins's refusal to work:

> John Rawlins having for his misbehaviour been discharged the hospital sooner then he need have been where he had been salivated for the foul disease, and having behaved himself in a very bad manner, and refusing to work in the house the committee ordered him to be discharged tomorrow.[145]

While Rawlins resisted house rules until he was thrown out, many foul inmates simply chose not to stay. Just like the general inmates that Hitchcock described, foul patients often responded to workhouse conditions by voting with their feet.

Some made that decision rather quickly. For example Ann Bell ran away from St. Luke's workhouse after only fifteen days in the foul ward. However, it seems more common for foul patients to stay slightly longer before going over the wall. Many of St. Luke's foul patients, like Mary Lewin and James Lamityer, remained for about a month before making their escape.[146] It is likely that such patients used the workhouse to obtain the course of mercury that they could not afford (the standard length of which was roughly four-to-six weeks) and then quickly made for the exit before the master could set them to work.

Unfortunately, mercury tended not to cure them, and many escaped inmates were forced to return to the workhouse and humbly reapply for admission once their symptoms returned. Mary Piper ran away from St. Margaret's workhouse after spending a little over two months in the foul ward during the winter of 1731/32. Yet she was back in the house by December, and again readmitted as "sick" in March, 1734.[147] Sometimes workhouse committees were not so forgiving. John Ashton entered St. Sepulchre's foul ward on August 12, 1750, and waited for the committee to arrange a petition for St. Thomas's. He left the house after only two weeks and does not seem to have actually entered St. Thomas's.[148] But when his symptoms returned the churchwardens refused his second application, citing his previous escape.[149] Paupers who ran away or refused to comply with house rules risked damaging their relationship with their local churchwardens. However, when faced with the conditions and discipline of the workhouse many inmates, foul or otherwise, chose to take that risk.

As these last cases illustrate, foul patients commonly found themselves returning to the workhouse for help on numerous occasions. In fact, the frequent return of foul patients to the workhouse marks a notable aspect of foul patients' experiences. Some paupers who suffered from the pox can be found returning to the workhouse seeking relief dozens of times. The aforementioned Henry Truslove applied for relief on twenty-seven separate occasions, while John Ralwins appears in the records no less than forty-seven times between 1734 and 1744. Often such paupers stayed just long enough to receive care and recover their strength before setting out on their own again, only to drag themselves back to the workhouse several weeks or months later when their symptoms reappeared. Two final cases demonstrate the pattern. These two cases have been chosen for their typicality. Not only do they demonstrate that patients returned to the parish with great frequency, but they further illuminate the overall experiences and strategies of poor foul women in eighteenth-century London. The uncertainties of retrospective diagnosis prevent me from declaring emphatically that this pattern illustrates the long-term effects of syphilis. Regardless, many of the patients like the following two women were clearly not well. Their poor health, whatever its nature, was an ongoing struggle.

Margaret Cock entered St. Luke's workhouse at the age of fifteen in September 1746. The register states that her parents were too poor to care for her. Her mother died shortly after Margaret entered the workhouse, and Margaret's father was forced to put her three younger sisters, Hannah, Ann, and Grace, into the workhouse as well.[150] The churchwardens arranged an apprenticeship for young Margaret in November, but she seems not to have stayed. She returned to her live with her father, but he was still in no position to support her, and in April 1748 she returned to the workhouse.[151] Now sixteen years old, she was considered "able to work" and discharged. Again she returned to her father; again this proved unacceptable; and again she returned to the workhouse in December 1749. The house kept her a mere five days before putting her back on the street to find work.[152] Within ten months she was back at the workhouse door, infected with the pox.[153] She entered the infirmary and was under the care of the house surgeon for eighty-two days. In the meantime the officers arranged a domestic service position for her following her "cure." She left for this position on December 24, 1750. Sixteen days later Margaret's symptoms returned and she was back in St. Luke's foul ward.[154]

The surgeon now decided that her case required hospital treatment, so the parish sent Margaret to St. Thomas's on January 17. She underwent the standard four-week regimen and was back in the workhouse by February 21.[155] She may have needed to recover, or they may have had trouble finding work for a recent foul patient, because Margaret remained in the house until mid-July before she was sent to another domestic service position.[156] Like clockwork Margaret returned ill to the infirmary in early October. She stayed in the house until spring, but in April she finally decided that she was sick of St. Luke's workhouse. On April 22, 1752, the clerk recorded that she "got away over the wall."[157] Despite her disdain for the institutions in which she had now spent a significant portion of her adult life, and despite her desire for independence, poverty and illness drove Margaret back to the workhouse by the end of October. On the thirty-first of the month Margaret humbled herself before the churchwardens and requested another petition for admission to St. Thomas's foul ward, which they granted. Within two days she was on her way for a second stint in the Southwark hospital.[158] She was still just twenty-one-years old. She underwent another four-week mercury regimen and was back in the workhouse on December 7. Over the course of the winter the churchwardens again arranged a service position for her. On February 10 Margaret left for her new job and finally disappears from the records.[159]

Elizabeth Wyatt was also a St. Luke's parishioner and similarly well known to the churchwardens. She first entered the workhouse at the age of sixteen because of a "swelling in her neck" in June 1750. She entered the house several times due to illness during the next two years. After each stint in the infirmary Elizabeth was sent to a domestic service position.[160] Elizabeth seems to have remained well until she returned to the house at the age of twenty-three. Merely described as "sick," she entered the infirmary in October 1757. Over the course of the next twenty-four months Elizabeth would enter the infirmary on four seperate occasions, each time claiming to be "sick" or claiming to have the "itch."[161] Each time she was discharged to a service position, with the exception of an occasion on February 24, 1759, when she ran away. Finally, less than a month after her last discharge, Elizabeth returned to the workhouse on December 4, 1760, and admitted that she had the foul disease.[162] By now she was twenty-four years old. The surgeon tried to treat her over the course of five months, but finally decided that she needed hospitalization. She was sent to Guy's Hospital on April 9, 1760.[163] She returned from Guy's to the workhouse in mid-June, and decided that seven months of institutional care was quite enough for her. She ran away on June 28, 1760. She could not have made it very far, because just two days later she dragged herself back to the workhouse, the register stating she was "not well from being foul."[164]

Clearly, Elizabeth had a hard time deciding between the two rather bleak options of either living in the workhouse or roughing it on her own. After just five days back in the workhouse she reaffirmed her earlier decision and escaped into the night for the third time.[165] By December illness brought Elizabeth back to the workhouse, but this time she told the committee she suffered from an "ague." The officers arranged a petition, and she spent just under one month in the hospital. She returned to the workhouse in February 1761 and stayed until early April.[166] The master may have kept a close eye on Elizabeth, because she did not run away. But she clearly did not wish to stay in the house indefinitely. She requested to leave, and the officers granted her wish on April 8.[167] Predictably, Elizabeth lasted until October before she had to return to the infirmary. Ill yet again, she entered the infirmary on October 10 and likely endured another in a long series of salivations. She stayed precisely one month and again requested to leave on November 10, after which point she does not reappear in the records.[168]

Clearly, both Margaret and Elizabeth badly needed the medical services provided by the workhouse. Neither seemed content to be there and both ran away. Yet each young woman found themselves returning to seek readmission. This dynamic marks a crucial component of the experience of poverty and the pox in eighteenth-century London. Elizabeth and Margaret were typical workhouse foul patients: single women who were infected before the age of twenty-five. By the time Margaret was just twenty-one she had been in the parish infirmary four separate times and in St. Thomas's foul wards twice. By the time Elizabeth was twenty four she had been in the infirmary on seven separate occasions and salivated in two different hospitals. Regardless of their disdain for the workhouse, they needed it.

Moreover, each felt that they could legitimately claim assistance there, and the actions of the churchwardens illustrate that they, too, felt that these women had a right to medical care. Each was well known to the parish, so neither faced settlement enquiries, at least as far as the records indicate. In Margaret's case, the churchwardens were not willing to support her while they considered her healthy and able to work. But once illness struck the workhouse provided medical care—in spite of the fact that neither of the women seems to have been an exemplary inmate. Both Elizabeth and Margaret may well have attempted to conceal the nature of their illnesses when reapplying for help, even though it may have been futile to do so considering how frequently they had been diagnosed as "foul." Both women exhibited similar strategies. They both tried to negotiate the poor law system to obtain the care they needed; but they both tried to do so on their own terms, choosing when they entered and when they left the infirmary. For the most part the parishes treated them not unlike other young women for whom they had become responsible. Churchwardens tried to place young women in a situation in which they could at least earn their own subsistence and hence be no longer a burden to the parish. For single women in eighteenth-century London this almost always entailed domestic service.

However, not everyone sustained such resolve when faced with the realities of what the pox meant for poor urban women. Try as they might to put their experiences of the pox behind them and start a new life, this could be exceedingly difficult to do. A final, tragic case of a woman very much like Margaret and Elizabeth drives home the dire circumstances faced by the plebeian women who came to be diagnosed with this disease. Frances Gardner entered the foul ward of the infirmary in the workhouse of St. George Hanover Square in March 1796. She underwent salivation without event. Proclaimed "perfectly cured," she was discharged in April. However, she showed no signs of optimism for her prospects. Her hopes of starting a new life at a new job seemed dim. Sarah Alexander, a nurse in the infirmary, reported that for several days "she appeared to be very low spirited and she went from the workhouse in that low spirited way about 6 minutes before 2 o'clock in the afternoon." That night, April 21, Frances hanged herself.

The nurse's deposition to the coroner reveals that Frances had cause to be pessimistic about her prospects. The churchwardens may well have tried to arrange a job for her, but as Frances found out, her medical history—and therefore her sexual history—hindered her chances. Two weeks earlier, a woman who was supposed to recommend Frances for a job came to the workhouse to tell her that, in light of her illness, she would provide no such recommendation. The deposition reads "that about a fortnight ago a woman called upon her at the said workhouse and told the deceased that she could not recommend her any more as she had lost her character by having that disorder. . . ." For all that we know about aristocratic and bourgeois credit in the period, we should not make the mistake of thinking that the wealthy cornered the market on reputation. Women like Frances Gardner may have had little else. The pox took even that from her.

It need not have been so. Had she the money for private treatment, her medical and sexual history might have remained a matter for her and her doctor alone. She might have had the chance to start anew. But Frances could not procure treatment out of the public eye, and now her chances of making an honest living were seriously damaged.[169] We have already seen in the Introduction that the pain accompanying the pox and salivation may have contributed to the despair that led to some of the suicides recorded by the Westminster Coroner. As Macdonald and Murphy have demonstrated, so too did poverty and shame.[170] The depositions given at Frances's inquest reveal that Frances's despair went beyond mere physical pain or the doubt in her chances for recovery. The pox—the disfiguring, stubborn, and stigmatized pox—heaped additional sociocultural and economic burdens on its victims, hardships felt particularly acutely by people like Frances Gardiner. Perhaps more than any other of the stories related here, the desperate actions of people like Frances speak volumes about the profound, wide-ranging, and long-term effects of the foul disease on plebeian folk in the long eighteenth century.

London workhouses took on a significant medical role during the eighteenth century. While they were not originally designed to do so, workhouses came to house infirmaries as a response to the overwhelming numbers of parishioners who were driven to seek aid due to illness or injury. These local poor law infirmaries developed into an institutional health-care system for the poor that existed beneath and in conjunction with those larger London hospitals that have received substantially more scholarly attention. Workhouse infirmaries comprised a key component to the city's overall medical charitable network that also included the royal hospitals and the smaller specialist hospitals supported by voluntary subscription. These various institutions did not exist in isolation. They interacted on a daily basis, passing patients back and forth. Workhouses often acted as conduits into larger hospitals for people who otherwise lacked the funds and support network needed to gain admission. Moreover, workhouses received patients after their courses of treatment and often housed them during their period of recovery. Understanding how workhouse infirmaries fit in to this multilayered system is crucial to the overall understanding of contemporary institutional medicine. However, a great deal of work remains to be done on these institutions and the more general topic of medical provision at the parish level in this period.

By the first third of the century both paupers and churchwardens had come to accept that paupers with legitimate settlements were entitled to parish assistance when they fell ill. The contemporary stigma of venereal disease was not so powerful as to overcome that belief. No workhouse infirmary refused venereal patients outright. That said, churchwardens clearly tried to limit their responsibility for foul patients. Usually they used settlement enquiries to pass as many foul patients as possible to other parishes. But this practice seems to have been driven more by economic concerns than moral ones. Members of eighteenth-century workhouse committees exhibited seemingly schizophrenic attitudes toward foul patients: at one moment discriminatory, at the next sympathetic. This was so because foul patients challenged the very concept of what it meant to be a "worthy object of charity." Some contemporaries felt that foul patients deserved their misery. But the face-to-face nature of eighteenth-century parochial charity ensured that workhouse committee members frequently saw first hand the debilitating effects of this disease. When faced with that reality, many early modern churchwardens came to view foul patients as legitimately needy.

As for the patients, evidence shows that the majority were young, single women. They seem to have had a complex relationship with workhouses. When pressed by the combined effects of poverty and disease, such women needed them. Indeed they felt entitled to demand medical care. Yet they simultaneously show evidence of loathing the circumstances in which that care took place. As in the hospitals already described, the nature of care in workhouse infirmaries was entirely public. Foul applicants faced groups of their socially superior, male neighbors who comprised workhouse committees, and who often ordered enquiries into their lives that involved family, friends, and employers. Just like contemporary hospitals, workhouses failed to provide the anonymity that venereal patients craved and which could be purchased in the London medical market by those fortunate enough to do so. The women and men who sacrificed their freedom of movement, donned numbered parish uniforms, and subjected themselves to house discipline and labor in order to obtain a bed in a workhouse foul ward must have done so only after exhausting all other available medical options. By and large, in eighteenth-century London women had fewer options than men. For that reason, these institutions at the bottom of the medical ladder became institutions that cared primarily for women. The gendered financial realities in eighteenth-century London determined that women outnumbered men in parish infirmary foul wards, even though men outnumbered women in the foul wards of all other hospitals in the city. The fee policies of the royal hospitals and the parishes' obligations under the English Poor Law combined to produce a gendered landscape of London hospital provision.

Workhouse records also begin to illuminate aspects of the patient-experience and patient's strategies that other sources cannot. First, it becomes painfully clear that for many people battling the foul disease was not a brief encounter, but often a long struggle and a vicious cycle. Patients like Margaret Cock and Elizabeth Wyatt resubmitted themselves to mercury regimens again and again, each time hoping that this salivation would be their last. When their symptoms subsided they set out hoping to make a fresh start at a new job, only to find themselves desperately ill a short time later. Sadly, this dynamic was extremely common. Moreover, it forced these people to return to the workhouse infirmary that they often loathed.

Illness in general was one of the main factors that forced people to go to the parish. For many who lived close to the margins, remaining healthy and capable of earning could make all the difference. Many poor Londoners, especially women, found themselves no longer in a position to support themselves thanks to the debilitating physical and financial effects of the persistent, enduring, and often incurable foul disease. But the evidence also shows that many poor people refused to give in when faced with infection. They set about using what meager resources were available to them. They tried to wring funds out of their churchwardens. When their illness worsened they used the parish to get themselves into a hospital, free of charge. If they instead found themselves in a workhouse infirmary, they often stayed just long enough to complete their treatment before making their way to the exit at the first opportunity. Even though many would be forced to return, their repeated attempts to conceal their illness and to escape the workhouse after treatment demonstrate a strong desire on the part of many foul patients to retain their autonomy and their dignity. For some, however, like Frances Gardner, the obstacles proved insurmountable.

5

THE FOUL DISEASE AND MORAL REFORM? THE LOCK HOSPITAL

In the summer of 1746 London hospital provision for the poor with venereal disease entered a new phase. Surgeon William Bromfeild placed advertisements in the *London Evening Post* and the *Daily Advertiser* inviting donations for a new charitable hospital for impoverished patients suffering under the disease. Within six months his new charity had solicited enough support to launch the venture officially. The new London Lock Hospital opened its doors in January 1747. Unlike the royal hospitals supported by rents, or the workhouses supported by the public poor rates, the Lock was a private endeavor. That hospital would stand at the center of London, indeed of British hospital venereology for three centuries.

Why did Bromfeild open the Lock if there were already so many institutions treating the pox? If hospitals were not shunning venereal patients, as standard historical accounts have held for so long, then why did Bromfeild and his supporters embark on this endeavor? It was clearly not launched on a whim. The administration struggled mightily, but successfully, just to keep the hospital afloat during the difficult early decades. The hospital survived the turbulent eighteenth century and eventually found the stability to last all the way to 1952. This is an impressive record. We must ask what drove the founders? Why was this hospital, which seemed to offer services already offered by many other London institutions, considered necessary by the late 1740s? London charities were many and the competition for benefactions was stiff. How did the Lock Hospital convince English benefactors to support its mission? It is worth considering what was unique about the Lock Hospital and what niche it filled.

If it was not to provide care that was otherwise lacking, perhaps its mission was linked to a wider reforming agenda. Along with the idea that the Lock represented new tolerance, the idea that a reform program drove the charity has been one of the central assumptions of the scholarly work done on the Lock.[1] It is often assumed that the Lock Hospital can be counted among those reforming, disciplinary institutions that grew out of the early eighteenth-century moral improvement campaigns characterized by such groups as the Society for the Promotion of Christian Knowledge (SPCK) and the Society for the Reformation of Manners. Social historians have done well to depict the anxiety caused by the problems of urbanization and the responses of those who, like the SPCK, sought to maintain social and religious order through various reform efforts. A venereal disease hospital seems to fit this model perfectly. After all, who stood more in need of reform than poxed paupers? The Lock Hospital, with its name that seems to imply confinement and harsh discipline, is often assumed to have been born primarily from the desire to institutionally isolate this identifiable group of sinners and set about changing their ways.

This view is too simple. The evidence shows that, while this picture holds some truth for the Victorian period described by Walkowitz and others, it is premature to read it back into the mid-eighteenth-century.[2] The moral reform of patients was not a primary impetus behind the original foundation of the Lock, nor did it not constitute a main function of the hospital during its first three decades. Instead, moral improvement became important to the Lock Charity only after a group of Evangelicals took control of the hospital's administration in the 1780s. Part of the confusion about the Lock's early mission may stem from its own promotional literature, much of which derives from this later period. By that point, the mission had shifted. However, the tendency to project this back to the earlier decades has obscured the early history of the hospital. Moreover, there was a disjunction between the public face that the Lock tried to portray in its fundraising literature, and the private reality of what actually transpired on the wards. An exploration of that gap yields new insights about the charity, which changed over time.

When it finally did take up the cause of reform, it is telling that the charity focused its campaign exclusively on women. When in 1787 the Evangelicals finally built the hospital's sister institution, the "Lock Asylum for the Reception of Penitent Women," they transformed what had been a primarily medical institution and launched it on an ambitious reforming campaign. This really marks the point at which institutional VD care took up a much clearer role in moral policing. "Lock" hospitals and "Lock" wards all over Britain would take up this asylum model and would soon hold a central position in the well-documented Victorian efforts to police poor women's sexuality.[3] At first, the Lock Hospital had largely resembled the other London hospitals that treated the pox. Yet by building a separate Asylum linked to the cause of saving fallen women in the late 1780s, the Lock evolved into something quite different from London's other hospitals that treated VD. The other crucial novelty here is that the Lock's mission at this point became strictly gendered. What attempts we do witness in earlier hospitals to police foul patients did not target one particular sex. Changes in gender ideology may have played an important part in this shift. The clear gendered campaign of the Lock Asylum grew out of the increasing importance placed on domesticity characteristic of the well-documented late eighteenth-century gender ideals trumpeted by London Evangelicals.

However, there is again a danger in overestimating the Lock's novelty, because even here the Lock cannot claim to be first. A short-lived and historically rather invisible institution called the Misericordia Hospital beat the Lock to the punch, linking the moral reform impulse of the penitentiary to the medical care mission of the venereal disease hospital more than a decade before the Lock Hospital opened the Lock Asylum. Regardless, the move to utilize the penitentiary approach typified by endeavors like the Lock Asylum was a watershed ushering in modernity for hospital venereology. It is a useful historical coincidence that as the Lock Asylum was just getting off the ground revolution broke out in Paris.

The story of the foundation of the Lock Hospital has been ably told several times. Donna Andrew included the Lock in her study of eighteenth-century charity and focused specifically on the Lock in a subsequent comparative article.[4] Linda Merians explored the Lock Asylum for Women in her contribution to *The Secret Malady*,[5] and the charity received its own institutional history when Sir David Innes Williams published his useful overview of two centuries at the Lock.[6] These studies agree that it was surgeon William Bromfeild who was the driving force behind the institution and largely responsible for getting the hospital up and running.[7]

Bromfeild (1712–1792) came from a London family with a long medical tradition.[8] He took up his father's profession and eventually attached himself to John Ranby, Sergeant Surgeon to George II. Bromfeild became a member of the Barber Surgeon's Company, lectured in anatomy, and published his *Syllabus Anatomicus* in 1736 and his *Syllabus Chirurgicus* shortly thereafter. He appears to have been successful from quite an early age. He cultivated his connections to the royal family and eventually became the surgeon to the Prince of Wales sometime in the 1740s. He married at twenty-three, and by thirty he had become a governor of St. George's Hospital, where he was elected Assistant Surgeon two years later, in 1744. St. George's was one of the few hospitals in London that claimed to exclude venereal patients according to stated policy (although evidence suggests that the policy was not rigorously enforced).[9]

Perhaps because he now worked in a hospital without foul wards, Bromfeild soon hit on the idea for the Lock. Bromfeild was able to put up the handsome sum of £350 to purchase the plot of land near Hyde Park Corner upon which the Lock would be built. Although this sum was large, such investments were not unheard of in the eighteenth century. As Lindsay Granshaw has shown, establishing specialist hospitals was a strategy used by many eighteenth- and nineteenth-century practitioners set on carving out a particular niche for themselves within the competitive London market.[10] However, Bromfeild did not allow his ambitious scheme for the Lock Hospital to interfere with his post at

St. George's. He remained on the surgical staff there until 1780, helping to sustain a relationship between the two institutions that would remain close throughout the century. Many donors gave to both charities, and, like Bromfeild, members of the medical staffs moved between both hospitals.[11] The Lock allowed St. George's surgical pupils to study there, and the two hospitals commonly passed patients back and forth. The close working relationship between the two institutions challenges the notion that St. George's stringently opposed VD care. To the contrary, St. George's passed venereal patients to the Lock Hospital, where members of their own medical staff often treated them. In a sense, Bromfeild's Lock Hospital became a sort of outhouse for St. George's Hospital, only a few hundred yards up the Pimlico Road.[12]

Bromfeild convened a meeting on July 4, 1746, to establish the charity.[13] He invited interested parties to join him at St. Martin's Library to discuss the venture. By the time the hospital opened on January 31, 1747, it had sixty-seven subscribers and benefactors. However, a core group of about fifteen men clearly ran the charity during the initial planning stages during the second half of 1746. The identities of these men suggest that at the outset the Lock's medical purpose may have outweighed its role as an institution of moral reformation. In addition to Bromfeild, the main decision-making team in the early days consisted of two physicians, an apothecary, another surgeon, and several men described by Merians as coming from "business, military, and political circles."[14] Bromfeild's court connections helped to raise funds from a good number of aristocrats, including his patron, the Prince of Wales, and the Duke of Ancaster.[15]

Bromfeild and his colleagues chose to employ an administrative structure used by many contemporary voluntary charities.[16] Unlike the royal hospitals that had the core financial support of guaranteed annual income from rents, smaller hospitals like the Lock relied on voluntary subscription. Like other voluntary charities, the Lock granted governorships to those who donated a specific sum. With one's donation came the right to participate in the hospital's administration. Those who gave five pounds per year became governors, and those who gave fifty pounds became governors for life.[17] Governors could attend and vote at all meetings. They could sit on committees such as the Thursday afternoon admissions committee, and most importantly they could nominate patients.

Unlike applicants elsewhere, patients at the Lock could not get in without first having obtained a governor's nomination. The General Court stipulated that "no patient be admitted but who brings a recommendation in writing signed by a Governor."[18] Governors were told to recommend only one patient at a time, and not to nominate another until the first had been discharged. The size of one's donation seems to have affected the success of one's nominees. The General Court ruled that in cases of conflict "a Preference be always given to those who subscribe the largest sums."[19] The charity established a medical staff consisting of two of St. George's surgeons, Bromfeild and Thomas Williams, apothecary James Bromfeild (William's brother), and two physicians who left the day-to-day patient care largely in the hands of the surgeons.[20] The early governors made the arrangements quickly, and the doors of the hospital opened in January 1747, six months after the initial meeting. A newspaper advertisement was ordered inviting nominated patients to come to the hospital at ten in the morning on January 31. The minutes of the General Court for the first year reveal little about treatment or the patients. The early rules hardly constituted a substantial disciplinary regimen. Only two specific rules were mentioned. Men were forbidden to enter female wards, and patients discharged for "irregularity" were forbidden from reentering the hospital.[21]

As Andrew and others have detailed, the charity got off to an inauspicious start, financially speaking. The administrative records indicate that funding remained a nagging problem for the charity throughout its first two decades. The hospital was designed to run on governors' subscriptions. However, the Lock's budgetary reports indicate that governors' subscription dues fell well short of the charity's annual expenses. In fact, the monies paid by men to become governors never met the Lock's financial obligations in any year during the eighteenth century. Its balance sheets reveal that in most years subscriptions amounted to less than half of the charity's annual expenses.[22] It quickly became clear that the charity needed to supplement its subscription income in order to survive. During the early years the charity counted on various fundraising events to close the gap between its subscription income and its expenses. It arranged with musicians for benefit concerts, with members of the theatrical community for benefit plays, and invited the wealthy to attend fundraising dinners and breakfasts. Such tactics were common among voluntary charities.[23]

Andrew has rightly described these extra fundraising activities as "particularly vital."[24] The figures support her characterization. In the first year for which such figures are available (1751/52) the charity nearly matched its subscription income of £401.2.0 by raising £334.9.0 through additional fundraising events. In certain years the charity actually brought in more money through concerts, plays, and dinners than it did through governors' subscriptions. For example, in the fiscal year 1756/57, the charity raised £601.2.6 by way of subscription. Yet, a benefit concert yielded £450, while a dinner raised £195.7.0, and a play, £28.8.0, for a total of nonsubscription income of £673.15.0. Two years later the charity's entertainments raised even more cash. In a year that saw subscriptions bring in merely £395.1.0, the various fundraising events provided the charity with more than £1,100.[25] It should not be surprising, then, that the governors discussed the logistics of such events extensively. Governors and members of the medical staff were constantly assigned to wait on the likes of Handel or David Garrick to persuade them, yet again, to donate their services.[26] In spite of these efforts, the Lock continued to struggle throughout its first few decades. Paupers with venereal disease simply did not tug at prospective benefactors' heart-strings quite as effectively as did foundlings or expectant mothers.[27] Subscriptions were often in arrears and by 1765 the charity owed more than £1,000.[28]

The charity got a much-needed boost in the 1760s, when its chaplain, Martin Madan, proposed to establish a chapel adjacent to the hospital.[29] The idea for the chapel came in the midst of financial turmoil. At a meeting of the Annual Court, May 21, 1761, the governors reflected on the hospital's financial woes. None of their usual fundraising schemes seemed to offer a solution. Bromfeild then drew attention to Madan's plan for a chapel:

> William Bromfeild Esq. reports, that he had reflected with great concern, on the ... [hospital's] debts, and that he had talked with the Reverend Mr. Madan on that subject, who told him if the Governors would give leave for the chapel to be enlarged, by a building to be erected on part of the present garden, he would get the same compleatly finish'd for the performance of Divine Service, without any expence to the charity; And after the deduction of the necessary sums to be laid out annually, he would give the net produce to the governors of this hospital, towards the support of the patients.[30]

The charity moved on the proposal, and the plan worked. The chapel quickly boosted the charity's finances. In addition to the pew rentals, which generated the most revenue, Madan also sold hymnals, published and sold sermons, and collected sacrament fees. It took until 1765 for the chapel to turn a profit. But in the fiscal year 1765/66 when governors' subscriptions only totaled £652, the chapel contributed over £947 to the hospital's operating budget. This trend continued. Over the next twenty years the chapel generated an average of over £935 for the hospital each year. During the same period governors' subscriptions averaged only £708 annually.[31]

The importance of the chapel to the charity cannot be overstated, and as Williams has shown, Madan's importance on the board of governors grew as a result of the chapel's success.[32] Bromfeild, who had put up significant start-up capital and whose noble connections were still invaluable to the charity, remained for the moment the hospital's dominant personality. But Madan gained a reputation as one the most dynamic preachers in London's growing Evangelical movement. With that reputation came a sizable and wealthy following who paid to hear him preach each week. As Madan became nearly single-handedly responsible for generating the bulk of the charity's funds, he quite naturally assumed a much more important leadership role in the administration.

But it is worth stressing that the chapel was founded to make money for the hospital by preaching to London's well-heeled, not to reform the hospital's patients. From the outset, the Lock Hospital had employed a chaplain, although it did not have great luck with its initial choices. A Reverend Williams, the first Lock chaplain, lasted less than a year before the governors discovered that he had embezzled funds.[33] His replacement, one Reverend Hughes, also quickly wore out his welcome. In 1749 he was fired because he failed to visit the patients.[34] Hughes's replacement was Martin Madan. However, Madan similarly seems to have devoted less than his full energies to reforming patients. He made more of an effort than his predecessors to visit the wards, but he seems to have done this infrequently. He read weekly services that the patients could attend, but it is not clear that attendance was mandatory. For example, a 1754 list of rules for patients conspicuously fails to discuss religious instruction of any kind.[35] The rules read quite like those for other contemporary hospitals, and thus do not embody a terribly strict disciplinary regimen. First of all, the Lock Hospital was not locked. Regulations stipulate that patients could leave the hospital, so long as they arranged permission. Otherwise, patients were merely instructed to heed the medical staff's instructions, "behave quietly and decently" while under care, and not to gamble or quarrel while in the house. These rules are disciplinary, to be sure, but they are no more so than any other contemporary London hospital, and they do not indicate that patients submitted to any particularly strict religious or moral instruction.[36]

Madan performed services for the patients and tried to visit them in their wards through the 1750s, but by his own admission he did so irregularly. Once the Lock Chapel project began in 1760 he focused even less attention on the actual objects of the charity. From that year he increasingly concentrated on the wealthy flock that came from all over London to hear him preach, to the detriment of the patients' salvation. In fact, it seems that by 1760 he had stopped visiting the wards altogether. To make up for his absence he revised an edition of John Reynold's *Compassionate Address to the Christian World*, which he placed in the wards for patients to read themselves. Madan's preface illuminates his attitudes towards the hospital. The reality of life on the wards of a venereal disease hospital proved too gruesome for him. Like several ministers who would follow, Madan admitted that he avoided the wards because they made him sick. In his Preface to the *Compassionate Address* he told patients:

> The cure of your disorder is of such a nature, as renders it often impossible for me to converse with you in private, and a stay of any long continuance in the wards, tho' the House where you are is so neatly and carefully kept, is what I have attempted, but cannot bear.

Madan simply could not stomach the reality of the wards. The *Compassionate Address*, promised Madan, "will be your private visitor in my stead."[37] The governors would later admit that visiting the patients in the wards "became so exceedingly offensive, that so long ago as the year 1760, Mr. Madan found it impracticable," and that from that point on "the office of regularly visiting the patients in their wards was discontinued."[38] Patients might have had access to a bit of religious instruction throughout the 1740s and 1750s, but the evidence hardly indicates that the governors made their spiritual salvation a primary goal in this period.

The belief that moral reformation constituted one of the main purposes of the Lock Hospital stems in part from the charity's own literature. By the end of the eighteenth century, as we will see, the Lock's annual fundraising pamphlets did stress that the charity sought to reform patients through Christian instruction. However, an analysis of the hospital's publications during the period 1746–80 shows that during its first decades the charity did not emphasize moral reformation when soliciting money. Instead, governors solicited support by stressing three main arguments. They portrayed their hospital as the patients' only hope; they appealed to the public's Christian duty of charity; and, perhaps most interestingly, they cast the patients as helpless and blameless victims. It is worth examining each in turn.

It is interesting, given the rather wide range of hospital provision for foul patients in eighteenth-century London, that the Lock portrayed itself as the only hope for the London foul. It was simply not in the charity's interest to highlight for potential benefactors that there were a number of other institutions that offered similar treatment. Instead, the Lock's governors tried to create the impression that their charity treated patients that other hospitals rejected. They made this case repeatedly. It seems to have had a significant impact on the historiography, which has tended to emphasize this line uncritically. Consider the following description taken from the Lock's 1749 annual report.

> Persons labouring under the Venereal Disease, are more destitute of relief than any other objects of public charity. . . . The county hospitals exclude them from their charity . . . [which] obliges numbers to resort thither in hopes of relief. But provision being made but for very few, only a small part of the many miserable objects which arrive here daily, can receive any assistance.[39]

The annual reports remained almost identical every year. One can find this particular passage reprinted verbatim a full forty years later.[40]

So, too, was another passage that similarly stressed the Lock's uniqueness. Near the end of each annual report readers were reminded that governors supported the Lock "because, the public infirmary, in

Westminster, as well as St. George's Hospital, entirely exclude such objects."[41] It is ingenious rhetoric. Rather than detail the list of hospitals and infirmaries that *did* treat the pox, it was more effective to mention two that did not. If one read only the Lock's fundraising literature one might well believe that the Lock was the only hospital in the city to which poor venereal patients might go. Of course there was a kernel of truth to the Lock's claim. That some voluntary hospitals did exclude foul patients certainly meant that the options for these patients were more limited than for other ill people. As we have already seen, hospital beds were at a premium. Moreover, we should not undervalue one element of the Lock's service that was unique; its beds were all free.

Of course the governors realized that venereal patients were not model recipients of charity. They faced opposition from those who argued that they brought their misfortune upon themselves through vice. The governors and chaplains confronted this issue head-on and begged prospective donors to remember their Christian duty to be charitable, in spite of the obvious shortcomings of their hospital's clientele. Madan's inaugural sermon in the Lock Chapel, entitled *Every Man our Neighbour*, exemplifies this rhetorical strategy. Madan based the sermon on the biblical lesson that one should love his neighbor as he loves himself. Madan sets up his plea by stressing that all people constitute our neighbors, thus we must love them regardless of their faults. He tackled the criticism that a hospital to relieve the pox might act as an encouragement to vice by arguing that Christians should find reasons for doing their neighbors good, rather than "inventing . . . excuses for withholding our charity."[42] Madan continued to employ this argument for years to come.[43]

Despite the logical strength of the argument that all Christians should be charitable, neither Madan nor the governors were naïve enough to base their entire fundraising campaign on this assertion.[44] The Lock competed for donations with charities like the Foundling Hospital and the Lying-in Hospital, which could base their requests on sympathetic images of orphaned infants and poor mothers. And the Lock was clearly losing. So they took a page out of their competitors' books and tried to cast their patients sympathetically too, not as wanton sinners responsible for their own misery, but as helpless victims infected through no fault of their own.

The gendered component of this construction is clear. The Lock administration stressed that its clientele were mainly women and children. Reading the annual accounts, or Madan's sermons for that matter, one might well imagine that the Lock specialized in treating such patients, for they dominate the discussion. Throughout the century the *Annual Account* stressed the "many poor creatures, labouring under this disease, [who] are in themselves no ways culpable, as it may have been occasioned by bad husbands, diseased parents, suckling children born with the disease, or children that have imbibed it from their nurses."[45] This trend is especially notable, because much of the prevailing discourse surrounding the disease in this period tended to present women as the *cause* of the disease, rather than its victims.[46] However, the Lock's literature shows that the manipulation of gendered imagery could work both ways. When applied to the cause of fundraising, it was thought appropriate to reverse the traditional gendered depiction of the disease. Roguish husbands become the culprits here, and the women they infect, the children they beget, and the wet nurses that their children subsequently threaten, now elicit sympathy.

This line of argument also formed a major component of Madan's sermon literature. Who, Madan demanded, "could be deaf to the cries of an unfortunate mother labouring under the most excruciating torments of the disease with this additional anxiety of mind, that her new born infant, is a joint sufferer with her[?]" Madan casts the husband in the role of cad; were not the mother and child sentenced to this fate "by the wickedness and folly of one, who by the laws of God and man was appointed their protector?" The rhetorical questions continued as Madan demanded of his flock: "Should the injured mother and helpless babe be denied relief, because the one is the wife of a profligate husband, the other the offspring of a cruel and unnatural father?"[47]

Madan highlighted the same themes in other sermons as well.[48] This rhetoric proved to play a major role in the Lock's construction of its own public image throughout the eighteenth century. At some point around 1770 the hospital must have endured some public attack, because in that year it published a pamphlet entitled *Some Reasons Tending to remove the Prejudices against the Lock-Hospital*, which is a clear response to an earlier salvo. That attack must have criticized its patients as unworthy of benevolence, because in response the Lock cast itself as "the friendly Asylum of the injured, deserted wife, Infected . . . by her VERY HUSBAND, designed by Heaven her Protector!" Moreover, it challenged readers to consider the "helpless harmless babe sucking poison from its mother's breast," and demanded "who will say that these are not objects highly deserving our kindest assistance, our warmest patronage?"[49]

When hitting the theme of the helpless victim, Lock governors seized upon one of the oldest popular beliefs about the disease. Since the Renaissance some believed that sex with a virgin could cure the pox. It is difficult to know how widely this belief was held. Winfried Schleiner could only find a handful of medical references to the practice throughout the sixteenth and seventeenth centuries.[50] By the eighteenth century medical writers condemned the notion outright, though they rarely spent more than a few lines dismissing it, suggesting perhaps that the idea was not widely believed. There are relatively few children that one can identify entering the Lock Hospital, although this may be a function of the records, which do not give ages. Moreover, given the major problems of children infected either congenitally or through nursing one must be cautious before presuming that all infected children had been raped.[51]

Nonetheless, Randolph Trumbach takes the practice very seriously, suggesting that child rape for the purposes of affecting a cure was emblematic of the violence that attended the new patterns of sexuality that he sees emerging in the eighteenth century.[52] Evidence from court records suggests that he is right, and that the notion of attaining a cure through intercourse with a virgin may have sustained wide plebian currency in eighteenth-century London. A cursory look at rape trials reported in the *Old Bailey Sessions Papers* between 1714 and 1759 indicates an overwhelming connection between rape cases involving young girls and venereal transmission.[53] Of the forty-six rape trials reported in this period that include a clear accusation of venereal infection, 84.7 percent involved female victims younger than sixteen years old. Only seven such cases involved adult women (two of whom were sixteen and seventeen). The remaining thirty-nine cases involved girls younger than sixteen, whose average age was 8.9 years. Children in rape trials—regardless of disease transmission—actually make up a slight majority (57.6 percent) of victims of alleged rapes testifying at the Old Bailey in this period, itself a phenomenon that deserves further research.⁵⁴ But the far greater incidence of accusations of disease transmission in cases of child rape is telling and demands explanation. In rape cases involving adult women, accusations of venereal transmission occur just 13.2 percent of the time. By contrast, accusations of venereal infection occur in more than half (54.1 percent) of all rape trials involving children.⁵⁵ More research is clearly needed here. But that venereal infection could seem so concentrated in cases of rape committed on young girls suggests that we may have to reconsider the continuing strength of the myth of the "virgin cure" in the eighteenth century, just as Trumbach suggests.⁵⁶

On the one hand it may be possible that some of these cases were the results of the belief in practice, perpetuated by infected men desperate for a cure. Given the struggles of foul patients that we have already witnessed, and the evidence that the disease could drive some to lie, others to steal, and still others to suicide, we have to consider the strong possibility that some may have resorted to this desperate and tragic act. On the other hand the myth itself may have influenced how contemporaries perceived cases in which adult men had sex with female children. Year after year publications like the Lock Hospital's *Annual Account* included language like this:

> [S]everal children from two to ten years old have become patients in this hospital from ways little suspected by the generality of mankind. . . . It is a received opinion with many of the lower class of mankind, both males and females, that when infected themselves if they can procure a sound person to communicate the disease to, they can certainly get rid of it.
>
> And from this principle the most horrid acts of barbarity have been frequently committed on poor little Infants, tho' these vile wretches have by experience been convinced of the absurdity of such vulgar notions, yet this requires the utmost publication to prevent such unheard of cruelty and Inhumanity for the future.⁵⁷

What could be more shocking to eighteenth-century bourgeois sensibilities?

Medical practitioners of all stripes testified in rape cases, discussing the physical effects of the act in order to prove (or disprove) penetration or the use of force. Frequently medical testimony also reported aspects of genital damage suffered by women and girls, which sometimes bore similarities to known symptoms of venereal infection (ulcers, presence of discharge, etc.). It may be that the fear of infected rogues preying on virgins predisposed those inspecting minors to interpret such symptoms as signs of transmission of the foul disease. In other words, the myth of the infecting scoundrel may have been part of the cultural understanding of pedophilia in the period and may help explain the frequency of accusations of venereal transmission in these cases.[58]

Regardless of what was really happening, such stories were fundraising gold. At least, the Lock governors must have thought so, because the stories are everywhere in the official pamphlet literature[59] and in Madan's sermons.[60] Of course, it would be wrong to portray the Lock as simply mercenary. On at least one occasion the governors contributed towards the prosecution of a man named Edmund Thirkill, accused of raping and infecting a five-year old girl named Mary Amelia Halfpenny treated in the hospital in 1764.[61]

Clearly the governors of the Lock Hospital felt compassion for the married women, wet nurses, and infants who arrived at the Lock Hospital. But we must consider their discussions of these innocent victims within the context of fundraising. The fact of the matter is that the Lock Hospital was not dominated by women and children. Throughout the eighteenth century the majority of its patients were men, frequently above 60 percent. Moreover, when the governors did keep records of the marital status of female patients, fully 87 percent of the women treated at the Lock were single, and few identifiable children entered the hospital in any given year.[62] In light of such figures, the images of innocent women and children that dominate the fundraising literature stand out. Given that such images do not accurately depict the vast majority of patients that the hospital actually treated, it seems clear that these images were crafted to elicit sympathy, and hopefully, money. This is hardly surprising. Raising money for the Lock Hospital was difficult.

Returning to the issue of reforming patients, it is still more informative to consider how the governors did *not* set about raising funds. During the period 1746–80 they based none of their fundraising campaigns on promises of moral reformation. The absence of those topics is even more conspicuous in Madan's sermons, where one might sooner expect to find discussion of the subject. It is thus hard to conclude that the Lock Hospital saw reform as a primary goal when the administrative records do not give the indication that such reform actually took place, and when the fundraising literature and sermons make no promises to that effect. In fact, the *Annual Account of the Proceedings of the Lock Hospital* indicates that during these first decades the governors felt that such reformation was to be left to each individual patient. Early editions of the *Annual Account* underscore that anyone who returned for treatment a second time would be refused. As elsewhere, patients got one chance to mend their ways, but the governors made it clear that it was truly up to them to do so. In 1747 the *Account* stated, "It is our Part to relieve the Distressed, theirs to amend their Lives."[63]

Other evidence demonstrates this gap between the public face of the hospital that the governors tried to advertise and the reality of the wards that they tried to conceal. This evidence further limits the notion that the hospital focused on patient-reform in the early decades. For example, the hospital consistently kept the public, especially prospective wealthy donors, from viewing the patients or the wards. They specified that the porter should escort any wealthy visitors away from the medical wards immediately. "If any Persons of Fashion Distinction or Quality shall desire to see the house," the porter was instructed to "immediately show him into the board room & acquaint the Matron therewith that she may attend them."[64] The wealthy were welcome to come to the Lock, but they were not welcome to see the harsh reality of what really went on there. The hospital also kept the doors and windows shut in order to conceal the wards from public view. In 1764 Bromfeild complained that patients loitered by open doorways. He did not base his complaint on a medical argument, that this practice might hinder their recovery in some way. Rather he warned that the sight of patients standing out front was so disagreeable that it might "prejudice ladys and gentlemen (who frequently pass by there) in their good opinion of the hospital, and [that] the charity might loose considerably thereby." The governors ordered the doors locked and the patients kept out of sight.[65] This was not the only such instance. In 1772 the board received a complaint that the male patients leaned out the windows and spat down on chapel-goers.[66] (They were being "salivated," after all.) The board ordered the windows locked before, during, and after service. The problem continued. Madan would later complain that patients hung out the windows and "behave indecently, in sight of people, passing & repassing the road." Once more they ordered the doors or windows shut.[67] These incidents betray a conscious effort to conceal the unpleasant reality of the wards rather than any effort to reform it.

The Moral Reform Movement

Yet reformation of the patients eventually did become a central concern at the Lock. In fact, the place of patients' salvation within the overall mission of the hospital became part of a fierce debate that transformed the charity in the 1780s. The debate began as a personal conflict between the Lock's main medical personality, Bromfeild, and its chaplain, Madan's replacement, Edward Charles De Coetlogon.[68] The relationship between Madan and Bromfeild seems to have been amiable for much of the time. However, there was an occasion when Bromfeild's son Charles (who was also a surgeon and who eventually replaced his father) was reprimanded for insulting Madan.[69] In addition, in 1770 Madan and Bromfeild clashed over the policy governing the cost of chapel admission tickets.[70] So David Innes Williams may well have been right to suggest that there was some tension between the surgical and religious figures on the Lock administration.[71]

The first salvo of the conflict came in 1780. In May a special court of governors convened to hear charges that Bromfeild had appropriated medical supplies from the hospital for his private practice. Four people testified that Bromfeild frequently came to the hospital and filled his carriage with medicine, linen, and instruments. They even claimed that he dropped off his personal laundry to be cleaned. Despite the testimony of several witnesses, the governors circled their wagons and exonerated their well-known founder.[72] The records do not reveal who brought the charges. Bromfeild likely believed that it was De Coetlogon, because he leveled a sort of counter charge just two weeks later. At another special meeting of the General Court, Bromfeild complained that the chaplain was not fulfilling his duties because he failed to visit the patients on the wards. The court concurred that the chaplain should visit the patients, and moved to consider what reward he should receive for this service.[73] They asked both Bromfeild and Madan to consider the motion, and each responded with a lengthy letter.

Madan's came first, and it supported De Coetlogon. Madan's letter again reveals that the purpose of the chapel, as viewed by those who ran it, was not the spiritual salvation of patients, but rather the financial support of the hospital. "The success and prosperity of the Chapel," Madan reminded them, "must be of the utmost consequence to the [hospital]." A talented minister was crucial to a successful chapel, and therefore one ought to be "genteely and handsomely rewarded."[74] Madan considered the attendance of patients to be an "additional requirement" for the chaplain, on top of his primary duties to the congregation. He therefore suggested that De Coetlogon receive a raise of fifty pounds annually in order to hire an assistant to visit the patients. Madan used the financial clout of the chapel to punctuate his letter:

> When therefore I consider this & the number of subscribers which have been brought to assist the charity with their annual donations through their respect & affection for Mr. D.[e] C.[oetlogon] as well as the income from the tickets which is procured by his labours I think if £50 p[e]r ann. were added to his salary, he then might afford to pay his assistant, and matters thus settled, would I trust & hope, turn out for the benefit of the charity. . . .[75]

Bromfeild responded with a strongly worded five-page report in which he lamented that patients lacked proper religious instruction because the chaplain neglected to visit the wards. Moreover, Bromfeild complained that Madan had "misapprehended" the matter. "Instead of considering the visiting of the sick patients in their wards as an *additional* duty of the chaplain," Bromfeild argued, "I am inclined to think . . . that it was the *first* and *most indispensable* part of the Minister's obligation." Since visiting the patients was not a new duty, Bromfeild argued it deserved no new remuneration. He did not mince words, citing the "evident neglect on the part of the present officiating Minister in this line of his duty." Bromfeild researched the salaries of other hospital chaplains, and presented a report that detailed how they already received lower salaries for, in some cases, heavier workloads. Bromfeild suggested that a select committee be appointed to investigate the issue.[76]

Bromfeild may have overestimated the strength of his position on the board of governors when he suggested the Select Committee. The following week, 157 governors convened to elect the twenty-one member committee. The election was fiercely contested. Three nobles, the Dukes of Ancaster and Manchester, and the Earl of Dartmouth received all 157 votes. Madan received 156. After these four, seventeen men were elected to the committee with either 79 or 80 votes, barely edging out another slate of seventeen who received just 77. Bromfeild was among those who were two votes shy.[77] There were clearly factions, and Bromfeild's had clearly lost.

The Select Committee's report leaves little doubt that the election signified a sound and embarrassing defeat for the hospital's founding father. The committee issued a thirteen-page report that disavowed Bromfeild's assertions.[78] It asserted that Bromfeild "started from false premises and was led into serious error," exaggerating the facts and misleading the board. Conversely, the report exonerated De Coetlogon. The Select Committee claimed that visiting the patients in the wards had never expressly been a component of the position, and that when Madan had done it, he had done so as a volunteer. Therefore, his precedent of visiting the patients was not binding to make it an official duty for his successors. The committee concluded by advising the governors to support De Coetlogon and issue him a raise if necessary. Again, they underscored their point by reminding the governors of the crucial revenue raised by the chaplain. Their report indicates how crucial the opinions of the congregation had become. The flock at the chapel liked De Coetlogon, so he was unassailable. The committee argued that they must not allow a "trifling addition of salary" to "displease that congregation or hazard so great an annual Income as they subscribe."[79]

The rebuff stung Bromfeild. He resigned his governorship and cut all ties with the hospital. He tried to extract some revenge by convincing noble benefactors not to pay their subscriptions.[80] However, this tactic seems not to have had much effect. The governors responded with advertisements in the popular press urging subscribers to attend the hospital and see for themselves that all was in order. Moreover, they tried to broker a peace. But the surgeon's pride could not be rescued, and he refused to have anything more to do with the charity. The last record of Bromfeild's words in the administrative records is the following embittered reply to the governors' attempt at bridge-building. Any attempt at reaching rapport was "fruitless," said Bromfeild, it not being in my power to negotiate for a single proposal that might be made for reconciliation, as that must depend wholly on the noblemen and gentlemen who thought themselves insulted and ill-treated at the last general court held at Hickfords Room. As for myself I never had but one object in view, which was the relief of those unhappy poor persons, who are the general inhabitants of the hospital—And as I hear the charity flourishes since I totally withdrew myself from the meetings of the Governors, I am extremely happy and shall be glad to see Mr. Hall at any time provided the Lock Hospital is not the subject of conversation.[81]

And so the founder of the Lock Hospital disappears from the record.

The clash between the surgeon and the chaplain is important for two reasons. First, such a public rift brought the core issue of the conflict into the open. As a result, the charity was now forced to confront the neglect of patients' religious instruction. Second, it marked the ascendance to power on the board of governors of greater numbers of clergymen. Soon after the rift between Bromfeild and De Coetlogon, the charity came to be controlled by a group of Evangelicals, sparking a period that Williams has accurately described as the "Age of the Evangelicals."[82] During the last decade and a half of the eighteenth century when these men controlled the charity the cause of moral reformation became much more central to the charity's overall mission.

Despite countering Bromfield's attack on De Coetlogon, the Select Committee had to respond to the core accusation that no minister had visited the patients, which everyone knew to be true. They instructed De Coetlogon to arrange for a minister to visit the patients in the wards three times per week.[83] But finding a minister proved difficult. De Coetlogon convinced a minister named Illingworth to take the job, but after two attempts to visit the wards he quit. Like Madan before him, the experience made him nauseous. De Coetlogon then turned to a minister named Goodhall. He, too, seemed overwhelmed by the task. He declined, citing his "fearful apprehensions" of completing "so disagreeable a business." De Coetlogon begged the board of governors for their assistance in what was becoming a difficult task.[84] The two ministers also submitted their own letters each declining the post. Goodhall declined politely, citing his health.[85] By contrast, Illingworth submitted a long and revealing letter that detailed the difficulties of offering instruction in the wards and gave his advice on how this might be improved. Illingworth considered his attempts to meet with patients "in vain." He suggested that the hospital arrange for a minister to preach to the patients collectively in the chapel. He believed that patients might be positively affected by the solemnity of the setting: "This might through the Blessing of God, more engage their attention by being brought into a place of divine worship; their minds being more abstracted from the disagreeable objects which necessarily strike them in the wards, which are unavoidable." His discussion of this proposal reveals that patients were not at this time attending service with the congregation. Moreover, his words indicate how the stigma attached to the foul disease forced the segregation of patients from the general congregation. He suggested that patients might enter the chapel as long as they did so long in advance of the congregation. If patients used the chapel several days before the congregation arrived, "nothing so offensive could remain" behind. So powerful was the aversion to these patients that any space they might inhabit needed to be aired out in order to cleanse it from the lingering contamination. Illingworth's letter also reveals how the patients responded to the sudden intrusion of a minister into the wards. He described how patients purposely resisted face-to-face religious instruction. Group instruction in the chapel was superior, Illingworth asserted, because "they [the patients] would not have their beds to hide themselves upon, nor such opportunities to evade attention."[86]

Despite these difficulties, De Coetlogon finally found someone willing to visit the patients for the thirty-pound salary, one Reverend Twycross. He visited patients only twice per week, not three times as the governors had hoped. But it does seem that patients began attending services again in the chapel.[87] The Select Committee quickly moved to inform the public of this change. In August the committee resolved to amend the text of the *Annual Account of the Proceedings*, the charity's fundraising pamphlet that had remained virtually unchanged for decades. The bulk of the text still remained the same. Prospective donors continued to read about the helpless women and molested children whom the governors put at the center of their appeal. However, the Select Committee now drafted a new paragraph that read:

> A clergyman of the Church of England reads prayers and preaches in the Chapel, three times every week, to a large congregation; when the officers,

servants, and patients attend, he (or his assistant) likewise attends the patients, twice every week in their wards; and performs other religious offices as the necessity of their cases require.[88]

This marks the first time that the charity openly appealed for funds based, at least partially, on the promise that the hospital would work towards the moral reformation of the patients. It was 1781.

Undoubtedly, the charity moved to put a minister on the wards and publicized that move in order to head off a public relations disaster in the wake of the highly public Bromfeild/De Coetlogon conflict. However, evidence shows that it began to show real commitment to the cause of spiritual reformation from that point forward. In doing so the governors proved to be very much of their time, for historians of charity have pointed to the late eighteenth century as a point at which personal visitation became much more prominent in voluntary relief schemes.[89] The following year the weekly board moved, and the General Court resolved, to have the weekly visitors (the governors assigned to inspect the hospital) look into the effects of clerical visitation. Specifically, the inspectors were told "to enquire particularly into the good effects of the constant visitation of the Patients in their Wards by the Chaplain or his Assistants; that such signs of Penitence & Reformation in the Patients occasioned thereby, may be noted in the Minutes, and reported from time to time, to the general court." Moreover, the governors requested that Twycross himself report on his progress saving patients.[90] Twycross reported that he had met with some, albeit modest, success. Twycross could not report any specific success stories, but he could say that his labor did not seem "entirely in vain." "There is great hope of penitence and reformation in several women patients," he claimed,[91] a passage that may indicate that the governors had already begun to concentrate their reformation efforts on the female wards. This was encouraging enough to convince the governors to employ two clergymen on a permanent basis. One would tend to the congregation in the chapel while the other focused on the patients in the wards.[92]

Similarly, the following January saw the governors issue another public statement that acknowledged that the charity now held a dual mission, part medical, part spiritual. In a letter to select nobles whom the charity hoped to reach, the charity's secretary, Jalez Fisher, gave a short history of the charity. He did not stray far from the script of the *Annual Account of the Proceedings*, again citing infected wives and molested infants. But to that standard appeal he added:

> But the Governors have another argument to urge in favour of this charity, which is, that they have the greatest reason to believe many of the patients admitted into this hospital have not only been cured of their disorder, but by the preaching in the chapel, and other spiritual offices performed by the Chaplain & his assistant, been brought to a proper sense of their former evil courses, and from that time become useful members of society.[93]

The administrative minutes do not indicate who these "many" reformed patients were. Considering Twycross's modest report about his early success preaching in the wards, it is likely that Fisher was exaggerating.

But it makes little difference. One does not find this kind of statement in the Lock Hospital's literature before the 1781. At this point the campaign to reform patients was only getting off the ground. But the charity was now clearly committed to the cause of turning patients into "useful members of society" for the first time in its history. Moreover, within a few years the charity was able to locate a chaplain who was firmly committed to this new moral mission. In 1785 it hired Thomas Scott, who announced "that he would feel a peculiar satisfaction in visiting the unhappy Patients in the House"—at which the governors proudly, and probably with some measure of relief, proclaimed that "the patients will be again visited with zeal & attention"[94] By 1788 patients could be found on Sundays attending divine service in the chapel, though they were segregated from the congregation in balcony rooms, concealed from all view by blinds.[95]

This shift in focus at the Lock was undoubtedly the result of the influence of members of the Evangelical movement who came to hold significant positions on the charity's executive. Andrew, Williams, Trumbach, and Merians have all correctly identified their influence on the history of the Lock.[96] Further, they have also described the reciprocal importance of the Lock Charity and its chapel to London's Evangelical movement. Many prominent Evangelical ministers preached at the Lock, and William Wilberforce is even said to have been converted by one of De Coetlogon's sermons.[97] This shift in membership of the board of governors coincided with a shift in the overall mission of the charity. Andrew stresses the importance of Evangelicalism to London charity in the last two decades of eighteenth-century, especially its influence on establishing the primacy of moral improvement in the purpose of many London charities, setting the period after 1780 off as a distinct era in English philanthropy. For Andrew, the Lock (and other charities during this period) began to represent the ideals of Evangelicalism, which held firmly that "positive change could only come about as a result of prior moral improvement."[98] Now charity would be given only "in the manner most conducive to the improvement of manners and morals."[99]

It is indeed informative that when this shift took place the Lock focused its campaign exclusively on women. The clearest example of the Lock's new commitment to reform was the establishment of its sister institution, the Lock Asylum for the Reception of Penitent Women, which received women after they had been treated in the hospital and submitted them to a further regimen of moral cleansing and discipline. The Asylum did not arise *ex nihilo*. It grew out of cultural assumptions about "Fallen Women" that were quite prevalent by the 1780s. Susan Staves sees the fallen woman in need of salvation as so common in eighteenth-century literature as to make it something of a literary trope. Writers never seemed to tire of describing the notorious details of young women seduced into prostitution by wicked bawds who manipulated them.[100] Tony Henderson concurs that the view of prostitutes as victims of seduction became so entrenched that a generic model of the descent into prostitution can be constructed that accurately fits most of the literature on the topic from the second half of the century. Above all, such portrayals depict prostitutes as victims and follow a plot line much like Hogarth's *Harlots Progress*.[101]

Of course the Lock's own fundraising literature drew on and contributed to this growing mid-eighteenth-century tendency to present women as helpless victims in need of salvation. An example of the influence of this outlook at the Lock comes from the pen of Madan himself. Madan's solution to the problem of prostitution did not involve a proposal for the Lock Asylum. Instead, he offered *Thelypthora, or a Treatise* on *Female Ruin* (1780), a defense of polygamy. He argued that potential prostitutes could be saved from their cruel fate by marriage. Madan's hypothesis could only work so long as men—that is, wealthy men—be allowed to marry several different women, thus extending the benefits of domesticity to more and more women, thereby protecting them. This engendered a massive rhetorical rebuttal by the many people offended by this idea, which hurt Madan's reputation.[102]

Less controversial were suggestions that a charity be formed to put into practice the agenda of saving those whom Staves has called Britain's "Seduced Maidens." Of course, this had first happened in 1758 when Jonas Hanway founded the Magdalen Hospital for penitent prostitutes. As Stanley Nash and Sarah Lloyd have described, the Magdalen was England's first fallen-women's asylum, and it would prove the model for the scores of female penitentiaries that would follow suit throughout the nineteenth century.[103] Indeed, it provided the model for the Lock Asylum as well.

However, while there can be no doubt that the Magdalen Charity provided a model for the subsequent Lock Asylum, it is worth pointing out how the Asylum represented a pronounced shift, for before 1780 the Lock Hospital saw itself as quite distinct from the Magdalen, with its own mission and its own goals. At its outset the founders of the Magdalen approached the Lock's administration to propose forming official links between the two charities. In their attempts to reclaim prostitutes the Magdalen governors understood that they would confront the pox regularly. They hoped that the Lock would establish a special ward for the reception of women sent from the Magdalen. It is telling that Bromfeild responded to this proposal coolly. He formed a committee to discuss the matter, which quickly and firmly decided against the idea. However, Jonas Hanway, as a leading member of London's charitable community, was not to be dismissed out of hand. Moreover, Hanway had already presented to the Lock the donations of twelve Magdalen governors.[104] So Bromfeild had to reply delicately. He responded to Hanway that the Lock Hospital would gladly treat some patients received from their fellow charity, but he politely declined to forge any official connection between the two charities. The Lock would neither set up a separate ward for penitent prostitutes, nor appoint a special chaplain to attend them.[105] The Lock did treat patients sent from the

Magdalen house for a few years, but they received standard treatment. And the Magdalen's governors had to continue paying annual subscription fees in order to use the Lock's services, just like everyone else. However, in 1764 the Magdalen account fell into arrears, and the secretary of the Lock had to write to his counterpart at the Magdalen the following letter:

> Sir,
> Upon Sarah Baunton's Application to the Board this day for admission, the Governors present order'd the books to be search'd for the number of patients recommended from the Magdalene, & they find that in the year 1762 they receiv'd nine into the House, & cured one as an outpatient, and in the year 1763, Seven Patients receiv'd, and in this year in proportion, and upon likewise searching the Subscription Book, they can only find three Guineas Subscribed by Jonas Hanway esq. as a particular consideration for the expence this House is at upon the Magdalene Account. The Board has therefore order'd me to let you know that they intend to report this to the next Genl. Court, and in the mean time beg to be excus'd from receiving any more Magdalene Patients. . . .[106]

The early history of the relationship between the Lock and the Magdalen again underscores the significant differences noticeable at the Lock before and after 1780. During its first decades the Lock had ample opportunity to commit to the cause of reforming of prostitutes. They clearly declined that opportunity. Before 1780 the Lock had a mission all its own, and it left to Jonas Hanway the task of saving London's fallen women.

The Misericordia Hospital

Hanway did that and more. Rebuffed by the Lock, and forced to pay the admission fees demanded at the royal hospitals,[107] he opened an entirely separate VD hospital, which has received little attention beyond passing mention by his biographers.[108] The Misericordia[109] Hospital appears to have been short lived, and its history must remain obscure because little documentation has survived. But a brief sketch is possible. It opened its doors on November 19, 1774, and is listed in the *Medical Register* until 1783,[110] but it likely faded sometime thereafter. All of its scant surviving literature comes from the period 1775 to 1781.[111] Situated in Goodman's Fields in the East End, the same drive to open institutions like the London Hospital, specific to that part of town, may have played a role in its foundation.[112] The architect of the original idea may have been William Blizzard, surgeon to both the London Hospital and the Misericordia.[113] It is even possible that the Misericordia may have taken up the vacated building that originally housed the Magdalen in Goodman's Fields before it moved to its new home in Southwark in 1769.[114]

Regardless, geography was hardly Hanway's main concern. If the Lock would not link its medical mission to his reforming vision, then he would do so himself. Hardly "lacking in novelty," as one commentator has suggested,[115] the Misericordia fused the medical and the moral in a much more comprehensive way than the Lock was willing to attempt during its first four decades. As Herbert Jones's inaugural sermon announced, the Misericordia's goal was not merely patients' physical recovery "but also, to afford them instruction in righteousness: to bring them from darkness into light, from the power of Satan to the living God."[116] Moreover, Hanway's eighty-page *Account of the Misericordia Hospital* (1780) demonstrates that he saw the new hospital in the same light as the Magdalen charity, and that he constructed it along similar lines.

Hanway appropriated to the new hospital the very attributes of the Magdalen that Miles Ogborn sees as central to its modernizing role—a focus on the rational, self-governing individual, with stress laid on solitude, prayer, and self reflection.[117] The Misericordia, not the Lock Hospital, was England's first institution to fuse venereology and the penitentiary. The Misericordia fits neatly into the wider penitentiary movement of the 1770s described by Michael Ignatieff, which sought to establish "total institutions," replacing arbitrary gross physical punishment with a rational regimen of solitude, work, and prayer. Of course, Hanway's central role in that wider movement makes this hardly surprising.[118] When Blizzard, also noted by Ignatieff for his involvement in the penitentiary movement,[119] approached him to attach himself to the institution as its treasurer, Hanway stipulated that he would only lend his support provided that the hospital expressly attend the soul as well as the body, calling for the institution to cement "religion and policy . . . [in] their natural union."[120] Hanway's significant clout as a fundraiser meant that his influence must have been considerable.

At a time when the Lock still debated whether the chaplain needed to visit patients, the Misericordia's minister played a central role. Patients were ordered to be in prayer always: think of Christ "when you rise, when you labour and when you rest."[121] However, as Ogborn has suggested for the Magdalen, reform at the Misericordia hinged on the internal.[122] The chaplain had a role, but reform had to come from within. Hanway stressed the rational capacity of patients, which brought with it the hope of, and the responsibility for, self-government and self-restraint. "You are a reasonable creature," Hanway scolded inmates.[123] "It is presumed that you will be so watchful of your thoughts, and attentive to your actions, that every hour of your life may give proof of your sincerity."[124] Hanway hoped that their illness would have brought inmates to a point of destitution and misery at which they would become finally amenable to such self-governance.

> Can the Ethiopian change his skin, or the Leopard his spots? And most true it is that some evil habits are seldom corrected without extremity of misery: But That extremity having reached you, it may be good for you to be thus tried. It is ground for hope that you will not return like the dog to his vomit: and the proper use you may make of [society's] suspicionis to be the more watchful of yourself.[125]

Other than a constant attention to prayer and the standard demand that strong patients help the hospital staff with daily chores, the regimen in the hospital is unclear. Without administrative records we only have a sense of Hanway's *ideals*, not the day-to-day reality of life on the wards. As Ogborn suggests for the Magdalen this is useful in itself, but not the whole story.[126] However, the house rules do give an indication that the Misericordia acted more decidedly than the Lock on its reforming mandate. The chaplain was instructed not just to read divine service, as at the Lock, but to "admonish the patients in their several circumstances," to distribute reading materials and prayer books, to read to patients, and to preach directly in the wards.

As at the Magdalen, the Misericordia strictly monitored patients' communications with those outside. They forbade the porter from carrying any message whether written or verbal in or out of the hospital unless properly screened. Nor were visitors allowed in unless similarly vetted.[127] Just as at the Magdalen, patients were warned against profane or immoral conversation. And in an attempt to cut more completely patients' ties to their own pasts, patients were further forbidden from even discussing their personal histories. On threat of discharge patients were warned "[not to] mention a word relating to the vicious ways of life you have been in."[128] Patients had to reject their past completely. This began from the moment they entered the hospital, at which point they were forced to make declarations of their intention to repent. Not just to the hospital staff, Hanway made it clear that he expected patients to make such declarations to their family and friends. Hanway asked if they did not "owe it to God and your own soul to make application to such persons, as may patiently attend to your humble suit, when on your knees you implore forgiveness with tears of contrition, before them and before God."[129] It was only to those who made such declarations that Hanway would turn his benevolence. Hanway also initiated a scheme to provide patients who had properly repented with letters of recommendation to present to their families or future employers when they left. With a notable paternalism Hanway would then speak on patients' behalf, vouching that

> Her promises of amendment of life are . . . very strong and I apprehend no less sincere. The committee have questioned her particularly and find there are [circumstances] which take off from the malignity of the guilt, otherwise I should not have made this application in her behalf.[130]

Hanway presents this letter as the brass ring for which all Misericordia patients should strive. Such a letter may have proven practical for helping discharged patients to find work, and we cannot doubt the importance of this, given how public knowledge of infection could seriously damage one's employment prospects, as Frances Gardiner too sadly discovered. However, such a letter may have also helped to weld the character of "foul patient" more permanently to one's identity. Rather than leaving the hospital behind and moving on with a clean slate, patients bearing such a letter would now wear that identity in a much more fixed and lasting way, carrying it with them wherever they went. With true Enlightenment hubris, Hanway believed in the power of institutions like the Misericordia to transform individuals. Patients would reject their past, abandon it, reinvent themselves through cold prayer, work, and sober self-reflection, and carry forth with a new identity. Hanway's optimism notwithstanding, it is not an identity that many patients necessarily wanted.

However, Hanway leaves evidence that Misericordia Hospital may not have become quite as effective a penitentiary as he may have wished. Despite its goals, its function as a medical hospital may have kept it from performing the fuller reformative role of the Magdalen. For example, Hanway's account of the patients treated from 1774 to 1780 points out that that some patients who had given particularly strong evidence of repentance were conveyed to the Magdalen Hospital after their treatment. In other words, they were transferred to another penitentiary to undergo fuller reformation than the Misericordia could actually offer. However, of the 1,465 patients treated to that point, Hanway estimated the number shipped to the Magdalen to be "about 40," or a scant eight patients per year.

The hospital's sizeable outpatient program also meant that the majority of patients never actually faced the Misericordia's intensive moral regimen. Just 355 patients actually entered the hospital; fully 1,110 were outpatients.[131] The chaplain was instructed to meet with outpatients when they came for their appointments with the medical staff, but given the significant weight that Ogborn places on the importance of institutional space to the penitentiary's impact, we must consider the experiences of those patients who never actually inhabited such spaces as entirely different.[132] It is also worth pointing out that while the Misericordia seems to have represented the one locale in which men experienced an approach to reform generally reserved for wayward women, Hanway's rhetoric repeatedly stresses the Misericordia's female inmates, indicating that, as at the Magdalen, women were the primary focus of its campaign. The attention to solitude, prayer, and self-reflection established in the Magdalen Hospital and later found in similar female penitentiaries, may have been, for a very short time, administered to men as well. But Hanway's *Account* does not indicate his strong interest in the reform of promiscuous men.

The story of Hanway's Misericordia Hospital sheds some new light on the changes witnessed at the Lock Hospital after 1780 and the eventual foundation of the Lock Asylum. When faced with a chance to forge a link with Hanway and the Magdalen, the Lock rejected the opportunity to place the reformation of fallen women at the center of its mandate. Yet twenty-five years later the Lock set the very same goal for itself, and essentially replicated the Magdalen in its own Lock Asylum for fallen women. Clearly the charity and its aims had changed over time. It is also important to keep the new Misericordia in mind when considering the Bromfeild/DeCoetlogan clash of 1780. That clash threatened to damage the Lock's public image by broadcasting its utterly lackluster commitment to patient-reform. The Lock governors must certainly have known that prospective donors considering various charities would have seen the Misericordia as an option that offered the added benefit of moral reformation that the Lock simply did not provide. In other words, the presence of the Misericordia may have increased the level of competition that the charity felt, and this may well have contributed to the greater emphasis placed on reformation, and the efforts to publicize that new emphasis, that we witness after 1780.

But undoubtedly the shift in personnel tells the story. The Evangelicals who came increasingly to control the Lock's administration were cut in exactly the cloth of those who, as Ignatieff shows, advocated the penitentiary model most vigorously and, as Andrew shows, led the movement to establish moral improvement as a cornerstone of late eighteenth-century charity.[133] There can hardly be doubt that they proposed the idea for the new asylum in the spring of 1787.[134] Consider their significant presence in the Lock Asylum's administration: at the Asylum's first meeting were Wilberforce, several central members of the Clapham Sect such as Henry and Robert Thornton, Charles Middleton, William Morton Pitt, and several Evangelical ministers including De Coetlogon, his new assistant, Scott, Reverend Basil Wood, Reverend Hadley Swain, and Reverend Henry Peckwell.[135]

The Evangelicals who seized control of the Lock Charity also had a powerful vision of gender relations. The men who launched the Lock Asylum were at the vanguard of the movement in the last decades of the eighteenth century to create what feminist scholars have termed separate spheres for men and women. Leonore Davidoff and Catherine Hall's now standard formulation argues that women came to be confined increasingly to the domestic realm and excluded from the political and economic world of public life.[136] Seeking to lay the groundwork for Victorian gender relations, Davidoff and Hall point out that writers of conduct books and other proscriptive literature increasingly stressed women's passive, gentle qualities while highlighting their domestic and reproductive tasks, aimed at persuading women to accept their roles in the home as wives and mothers, leaving the world of work to men, and creating a world of starker division between men and women's realms. Central to this endeavor was an intensification of the importance of female domesticity.

This model has come under criticism, the general thrust of which has been that women were not actually excluded from eighteenth-century public life to the extent that the separate spheres model suggests. Rather, they remained quite active in the public sphere and central to cultural production. Moreover, the home, by contrast, was hardly the exclusive domain of women, as men performed functions in the so-called private sphere, too. Thus, as an expression of the lived experiences of eighteenth-century men and women, the model is somewhat too stark. Furthermore, the increasing emphasis on domesticity, which critics readily acknowledge was indeed present in the late eighteenth century, was not entirely new, for similar expressions can be found earlier. Nor was it the sole model for gender relations in the period.[137]

These points notwithstanding, there is no denying that the attitudes on which the separate spheres model is based were present and strongly championed in the last third of the century. It may well have been that women's expanding participation in the public sphere of the eighteenth-century city generated anxieties in middle-class men who reacted by stressing the need to limit women to home and hearth as one way to control a world they saw changing rapidly around them.[138] So the *ideology* of separate spheres was alive and well, and, what is important here, it was fostered by precisely the same men who took control of the Lock in the 1780s. The Evangelicals and members of the Clapham Sect were some of the strongest proponents of new arguments for more intense gender division, female subordination, and the increased importance of women's domesticity and moral purity.[139] While not everyone in eighteenth-century London adhered to the theories that Hall and Davidoff make their focus, the men who ran the Lock Hospital clearly did. Imbued with Enlightenment optimism, confidence in the power of institutions, and faith in the individual's ability to change,[140] the Evangelicals who took the helm at the Lock in 1780 set about crafting a new institution to impose an internalized reformation on the hospital's female patients, molding modern, reformed, self-reflective Christian wives and mothers out of the clay of prostitutes and wayward girls. Class is also crucial here, for the Asylum represented a clear response to anxieties that Anna Clark claims the middle classes felt toward the plebeian "sexual crisis" that they perceived in the late eighteenth-century city.[141] The Lock Asylum was thus one more instrument in the campaign that Ruth Perry characterized as the "colonization of the female body for domestic life."[142]

At least, that was the plan. The purpose of the Lock Asylum was to offer women who had showed signs of "sincere repentance" in the hospital an opportunity for further religious instruction and vocational training that would help them to secure employment. In keeping with contemporary tendencies to portray women as helpless creatures in need of defense, the founders resolved "to maintain and protect them till they can be restored either to their friends or to the community at large, in a way of Industry according to their ability."[143] In the main, they hoped that they could keep the Lock's women from resorting to (or returning to) prostitution. They realized that the economic realities for single women in London left many women without significant alternatives. So the Evangelicals planned to use the Asylum to train women for careers in domestic service. Of course, the ultimate hope was that its inmates might one day become marriageable. The absolute pinnacle of success for their new institution, as defined in their own terms, would be cases in which the Asylum transformed streetwalkers into wives and mothers. Indeed the Asylum's fundraising literature advertized just this promise.

One of the oft-repeated "success" stories lauded in the Asylum's early fundraising literature told the story of a woman returned to the safety of a domestic setting, in this case returned to her father:

> [H]er father . . . willingly received her, and some months after expressed his entire satisfaction in her conduct, and his gratitude to the charity in the strongest terms; his daughter having (as he said) taken care of his household affairs, ever since her return, in the most prudent and commendable

manner; and had at that time a prospect of being married and settled to advantage, and the completion of his wishes on her behalf.[144]

The text was amended two years later: "*She has since been married and was expecting.*"[145] Even if she were fictional, she represented the epitome of the Lock's new mission: she performed proscribed domestic tasks in a controlled household, and having proven herself worthy and reformed, she readily assumed her proper role as a (re)productive member of society.

The Asylum opened its doors on July 5, 1787, to four women, Ann Cox, Sarah Parsons, Susannah Parker, and Ann Ray.[146] It is not an overstatement to assert that it institutionalized the double standard. For while both men and women still received physical care in the Lock Hospital, men were now released after their course of mercury while women were transferred to a second institution for further discipline. Men's treatment involved a single stage, cure the body; women's treatment was now a two-stage process, cure the body then the soul. To support this new mission the Asylum generated new rhetoric. Merians is certainly right to criticize the biased language of the Asylum's fund-raising pamphlets, which now presented prostitutes as a frightening threat.[147] "A common prostitute," benefactors were told, "is an evil in a community not dissimilar to a person infected with the plague; who, miserable himself, is daily communicating the contagion to others, that will propagate still wider the fatal malady."[148]

The Asylum's founders justified their institutionalized double standard in the name of economics. Women presented a greater threat to society than men, because their limited employment options made their return to prostitution likely. Men, they claimed, could return to their jobs and homes. Thus once they received their mercury "all that can be expected from us hath been done in order to their reformation."

> But most of the women are of that class whose misery and baleful influence have been noticed, many of them have no method of subsistence, but by prostitution; and can procure no lodging but in an house of infamy. These have scarce any alternative, but starving or a prison on the one hand, or returning to their former practices on the other.[149]

The founders believed that only if a serious enough impression could be made upon the women while they were institutionalized could they withstand the lure to return to a life of sin. For the Lock this represented a new spin on the now well-worn saving-women trope. The Lock had always broadcast that they saved women. However, the women the hospital had originally claimed to save were not "fallen." They were allegedly the innocent victims of dastardly husbands, not of bawds who had seduced them into prostitution. The difference was a subtle one, but it has important ramifications for gauging the perceived role of the charity over time. By 1787, the Lock Charity had fallen into much closer step with the Magdalen Hospital, and now put forth its own version of the fallen woman that it promised to rescue.

Moreover, the Lock Asylum promised to save society *from* them. Throughout the century there remained remnants of a line of thought that linked female physiology to the generation of the disease. The institutionalized double standard represented by the Lock Asylum emerged from a culture in which one could still find those who believed that the bodies of promiscuous women could literally generate the pox. Since the seventeenth century some doctors had warned that female promiscuity resulted in an unnatural mixture of different men's seed within the overheated and unclean wombs of sexually charged women. Doctors frequently described how such a preternatural mixture putrefied within the womb and transformed into the poison of *Lues Venerea*, which unchaste women went on to spread.[150] Such theories became especially influential in England between roughly 1670 and 1750. Their influence waned after 1750 when arguments about the New World origin of the disease led many to posit that only the bodies of non-European women in exotic climates possessed the requisite corporeal heat to transform seed to pox. But remnants of the theory remained in the second half of the century as doctors continued to speculate on the damage done to promiscuous women's genitals by the constant "frictions" of their repeated sex, which could cause genital ulcers that might prompt venereal infection.[151]

Mary Spongberg's study of nineteenth-century prostitution has demonstrated that these etiological theories of spontaneous female generation of the disease provided important ideological support for the Victorian asylum movement and campaigns to police prostitutes.[152] The belief that women could actually generate new strains of the pox *ex nihilo* had always offered an important rationale for scrutinizing female sexuality. In some ways the Lock Asylum merely represents an institutionalized manifestation of a much older policing discourse. Certainly, for those in the late eighteenth century who still believed in the danger of putrefied mixed seed, or in the pathology of the womb, or in the degenerative effects of repeated sexual frictions, a female asylum aimed at policing plebeian women's sexuality must have seemed like a vital public health development. Moreover, there was a long tradition that warned that female physiology obscured symptoms, leaving lingering suspicion about whether or not women were ever truly cured. Writing just as the Lock Asylum was getting off the ground surgeon Jesse Foot warned that women, despite treatment, might remain contagious for more than a year after visible symptoms disappeared, making a more pronounced quarantine of female patients seem like solid medical common sense.[153] The Asylum's own rhetoric casting prostitutes as infectious wandering plague victims makes clear just how profound the sense of contagion was.

Nonetheless, the relationship between theories of venereal pathology and the mission of the Lock Asylum was probably quite indirect. First, the late eighteenth century seems to be a valley, not a peak, for theories on spontaneous female generation of the pox. By this point many were convinced that only tropical women's bodies could generate the pox, thus freeing European women from responsibility. Second, there is no smoking gun linking such theories to the foundation of the Asylum. Nowhere do the governors refer to the danger of spontaneous female generation of the pox, nor does a review of the publications of the Lock Hospital's medical staff reveal that they adhered to any of these ideas.[154] Still, the shadowy danger of the womb continued to lurk beneath the surface of medical theory in the late eighteenth century and lent important support to the agendas of institutions like the Magdalen, the Misericordia, and the Lock Asylum.

The Asylum's records for the period 1787–1800 suggest that the Asylum drew heavily on the penitentiary model set by the Magdalen and Misericordia. We have seen that the rules for patients in the Lock Hospital were hardly unique, and did not constitute what one might consider a strict disciplinary regimen beyond what one witnessed at most other contemporary hospitals. By contrast, the "penitents" in the

Asylum faced a regimen quite similar to that in the Misericordia described above. Inmates' labor was aimed at preparing them for domesticity, as women learned to spin and sew. Moreover, the Asylum sought the same internal, self reflective reformation pioneered at the Magdalen. Inmates received religious instruction every morning and every evening, in addition to attending divine service in the chapel. They could not wear their own clothing, but wore uniforms. No games were allowed, and the chaplain forbade any books not explicitly approved.

Note also the solitude, such a conspicuous element of the Magdalen and Misericordia penitentiaries: women were not allowed to speak, save in the presence of the matron.[155] Here, too, the women were cut off not just from each other, but from the world outside. The matron inspected all their personal belongings and censored all letters that came into or left the house.[156] The rules that restricted women to the premises and which kept them from communicating with outsiders were aimed at creating a sort of moral quarantine in the Asylum. The founders expressed fear that women might be enticed back into their former lifestyle if they came into contact with their former associates. The annual *Account* stipulated that women must be "preserved from ever seeing or conversing with their former abandoned companions."[157] And importantly, women had to demonstrate that they had internalized their reformation and that they "appear sensible of the Moral Evil of [their] former practices, and give hopeful evidence of sincere repentance." The matron was ordered to "keep a very watchful eye over their moral and religious conduct" and to "make a weekly report of their behavior to the Board." Thus the board actively monitored inmates' behavior on an ongoing basis.[158]

For the Lock, this was new. It marked a radical departure from the comparatively lax approach to discipline during its first four decades. Taking the long view, this approach was new for London, too. We must remember that London had already had an all-female VD hospital for two centuries, the Kingsland Hospital. Yet the Kingsland provides no evidence of any systematic attempt to save its women. The Lock Asylum was not the first asylum for fallen women. Nor was it the first attempt to apply the penitentiary model to a VD hospital. The Misericordia deserves that special honor. But the Misericordia did not last; the Lock Asylum did. It would stand as the general model for institutional VD care in the new century. The combined efforts of Jonas Hanway at the Misericordia and the Evangelicals at the Lock changed the face and the function of hospital venereology. As the nineteenth century dawned, VD hospitals began to play an increasingly central role in broader social policing campaigns, campaigns that gathered steam in the Victorian period.

Nonetheless this reform agenda did not emerge until four decades after the Lock Hospital's foundation, and this still leaves us with the question posed at the start of this chapter. For if William Bromfeild did not establish the Lock to affect a specific moral reformation or social policing task, why did he? What purpose was the Lock supposed to serve that made it seem necessary in the late 1740s? It is to that question that we will turn in the next chapter.

6

RETHINKING THE LOCK HOSPITAL

Extremely high incidence of the pox in the mid eighteenth century necessitated more ward space for London's venereal poor. This may seem strange since the royal hospitals had large operations, and parishes treated an increasing number of patients within workhouses. By the 1740s there were more hospital beds for the city's poxed poor than ever before, and thus the Lock's emergence in that decade still confuses. Understanding the Lock's original purpose ultimately rests on understanding how it related to the pre-existing network of medical institutions. When we consider the Lock within the wider context of the city's medical welfare system, it becomes clear that its purpose was to fill a particular niche within that system. Despite the growing number of beds in London's foul wards a particular portion of the city's population remained neglected, and the Lock Hospital tried to fill the gap.[1] It is only by considering the Lock in relation to the broader social welfare network that its unique character may be understood. Once understood, we can better compare the experiences of its patients to those we have already encountered.

The need for more hospital beds for venereal patients was acute in the early to mid eighteenth century. Demand continued to outstrip supply when it came to hospital beds for the pox. By the time William Bromfeild convened a meeting to found the Lock Charity, London had witnessed the establishment of yet two other major hospitals that cared for venereal patients, Guy's Hospital in Southwark, and the London Hospital in the East End.

Guy's Hospital, established in 1727 by famed benefactor Thomas Guy, originally intended to care for "incurables"—that is, those that other hospitals refused because their conditions seemed too desperate to cure successfully. Considering this original mission, it is surprising that foul wards came to house a significant portion of Guy's patients. Unfortunately, the vast majority of Guy's Hospital's administrative records were not available for consultation at the time of this study.[2] Therefore, it cannot be known exactly why the foul disease, which was not considered incurable in the eighteenth century, came to occupy such a prominent place in Guy's Hospital. The official histories of Guy's Hospital, like the general histories of St. Thomas's and St. Bartholomew's, do not say much about the foul disease.[3] It is possible that Guy's specialized in severe cases of the pox that other institutions considered beyond care, but that will not be known until the damaged records are repaired and made available again.

Regardless, some available evidence demonstrates that those in charge of Guy's perceived the need for more hospital beds devoted to the pox. Thankfully, admissions records survive similar to those from St. Thomas's. Of the 1,814 patients admitted into Guy's Hospital during the twelve-month period January 1–December 31, 1760, at least 409, or 22.5 percent, entered the foul wards.[4] That figure may even underestimate the true number of foul patients in Guy's.[5] Whether or not Guy's founders originally intended to take in venereal patients, they soon devoted significant ward space for that purpose,[6] and by the second half of the century Guy's Hospital could be counted among the many hospitals that provided significant venereal care as one of its primary purposes. However, it is important to note that unlike some of the other voluntary hospitals, Guy's charged admission fees. Predictably, foul patients paid more than twice their clean counterparts: clean patients paid three shillings, six pence, foul patients seven shillings.[7]

The London Hospital also offered venereal care prior to the Lock. Founded in 1740 along lines typical of contemporary voluntary hospitals, the London Hospital emerged in Moors Fields to offer medical care to the suburban poor who crowded into the city's East End. A. E. Clark-Kennedy has ably told the story of the London Hospital, which was very much the brainchild of surgeon John Harrison.[8] Harrison and the London's governors geared their benevolence towards the working poor, regularly specifying that the hospital focused on "Poor Manufacturers, Sailors in Merchant's Service, and their Wives and Children."[9] The London's governors had initially planned to exclude venereal patients, but this did not last long. Within a few months of opening, the board already instructed the medical staff to bend the rules to expand treatment to the poxed, "but not so as to admit them into the House."[10]

The hospital's considerable outpatient service seemed to provide a solution. The governors could extend their charity to these patients, while still exercising a strict separation between them and clean patients. However, salivation was an intense operation, and doctors considered certain cases too serious to handle through outpatient care alone. So John Harrison argued to allow inpatient venereal care in August 1742 and several governors concurred that "it was very fit and proper to make some provision under this Charity for such poor persons whose cases require Salivation."[11] It must have been a hot topic, for the minutes record that "many debates" followed. When they finally called the question the governors resolved that venereal patients "were proper objects of this charity, and ought to be received and provided for accordingly." Harrison drew on the existing model set by St. Bartholomew's and proposed that the hospital arrange a separate house for venereal patients, noting that they could not be accommodated them in the hospital without imposing on other patients.[12] What else could such a house be called besides a "Lock"? The governors fixed up their Lock in time to begin receiving patients in September 1742.[13]

They advertised that they now accepted foul patients in March, 1742/43.[14] The text of their advertisement indicates that it was indeed the large numbers of foul applicants that inspired this change of policy.[15] The London Hospital also followed the royal hospitals' lead by creating a separate fee structure for foul patients. They offered free care to general patients, but demanded ten shillings, six pence per week from foul inpatients, and the same amount as security for good behavior of venereal outpatients, the latter to be refunded upon returning thanks for their cure.[16] Eventually these fees rose to three guineas.[17] The governors expressly endeavored to make the new Lock "bear its own expence without burthening the other branch of this charity." They even proposed, rather naively, that the Lock might provide revenue. If they kept nine patients in the Lock at a time, they imagined they could cover operating costs, and if they kept fifteen patients they believed it could even provide a surplus income of a hundred pounds per year.[18] Hope springs eternal.

The "Lock" at the London Hospital was an experiment, and it did not survive a decade. Specific patients can be found in the records throughout the 1740s but not in such a way as to ascertain the scope of the operation with any confidence.[19] The London also used its Lock to house nonvenereal patients, including infectious patients and those with small pox.[20] This may have happened with increasing frequency as the decade wore on, because in 1747 the hospital ordered that "the Title of the Lock upon the Facia of the Infirmary be taken out, Other persons being admitted therein besides those whose Cases require Salivating."[21] This certainly suggests that a stigma accompanied the term "Lock" and that governors moved to ensure that nonvenereal patients did not suffer from a misconception of their diagnosis. On occasion, governors complained when the nonvenereal patients whom they had nominated ended up in the Lock. The Lord Bishop of Oxford issued such a complaint on April 26, 1748, stating that one Francis Flight, his nominee, lay in a "noisome" room. Responding as politely as they could under the circumstances, the governors wrote back that Francis was in fact in the Lock. Although Francis did not need salivating—he had a broken hip—the governors regretted informing the Bishop that Francis nevertheless was "so Dirty and offensive that we could not properly lay him anywhere else."[22]

In either 1749 or 1750 the governors reversed themselves, and they again treated venereal cases as outpatients only. One can find records of patients paying three guineas to be salivated as late as November 15, 1749. However, by May 1750 one finds foul applicants like Hannah Devereled refused. The House Committee wrote to her nominating governor to inform him that "agreeable to the Rules of the House she cannot be continued, but will be relieved as an Out-Patient."[23] Curiously, neither an official decision nor a debate on the policy change emerges from the administrative minutes. But by October 1750 one can find the policy expressed in plain terms: "Persons afflicted with the Venereal Disease be received by the surgeons, only as Out Patients."[24]

Thus ends the brief history of the London Hospital's "Lock" house. After 1750, we can assume that venereal patients such as Jane Gallicote and Elizabeth Wheeler continued to receive outpatient care,[25] but the London Hospital's dispensary leaves no records to gauge the extent of this practice. We can be sure that the hospital itself never reversed its decision. There survive quite good patient registers, covering 1760 as well as 1791, and 1792, which actually indicate each patient's diagnosis.[26] Not a single venereal patient appears in any of these registers. Given the clear diagnostic terms then in use—"foul" "venereal," "poxed," etc.—as well as the phrases used to describe treatment such as "under a salivation," one can say with confidence that venereal patients did not enter the hospital in these years. Someone raised the issue of taking venereal patients in the hospital again in 1774.[27] The proposal cannot have garnered much support. The House Committee resolved to postpone the discussion a fortnight, at which time it resolved to postpone it again. The committee considered the proposal two weeks later and for a third time resolved to postpone its decision.[28] The proposal then disappears, seeming to have died a committee death. The records of the other administrative committees show that it never made it to their desks.[29] Considering that Hanway's Misericordia Hospital had just opened just its doors in East London in November of this year, the board may well have felt the initiative unnecessary.[30]

The evidence of ever-expanding hospital provision for venereal patients at institutions like Guy's, the London, the Lock, and the Misericordia raises the question of whether we can see this increasing number of foul ward beds as an indication of rising rates of infection.[31] Methodological obstacles and the inability to equate the foul disease with syphilis prevent a definitive answer, but theories about possibly changing patterns of sexuality in the period make the question worth considering. Two theories have been advanced, both of which see penetrative sex on the rise in the eighteenth century. Hitchcock has argued for a sexual revolution of sorts, with an increase in penetrative sex at the expense of other forms of nonpenetrative sexual activity as the century wears on.[32] This view is grounded empirically in the rise in illegitimacy that the century witnessed.[33] Unwanted pregnancies increased markedly, and Hitchcock argues that this occurred because sex came more and more to mean penetration.

Randolph Trumbach is sympathetic to such a reading. However, Trumbach seeks to recast the sexual playing field in an even more dramatic way. Trumbach argues for a gender revolution first, which in turn led to a sexual one. In this model, a third gender, the sodomite, was recognized for the first time in the eighteenth century. In response to the appearance of a new, homosexual third gender came a new kind of male heterosexuality, which Trumbach portrays as reactionary and excessively predatory. One of the results of this new sexual economy was a new pattern of prostitution that served the increased needs of men for sex with women in order to prove their status as heterosexuals through penetrative (heterosexual) sex. Just like the rise in illegitimate births, a change in the pattern of venereal infection could serve to shed light on patterns of sexuality. Even given the nonsexual means of transmission, mainly congenital transmission and breast-feeding, rates of venereal infection would seem to offer a good measure for sexual activity.

Thus it is tempting to see the rise of the Lock Hospital, coming as it did on the heals of Guy's, the London, the increase in workhouse foul wards, and the birth of the Misericordia, as evidence of ever rising rates of infection. Certainly, if Hitchcock is right, more penetrative sex would mean more sexual infection, and this can help to explain why London seemed to need more and more hospital beds for the pox as the eighteenth century wore on. Venereal disease plays a central role in Trumbach's argument, and he indeed argues, in part from Lock Hospital data, that the rates of venereal infection were on the rise in London after 1690, especially among common folk.[34]

The evidence from hospitals may serve to support the view that sexuality changed in the eighteenth century, but this has to be taken with some caution. First, while I have argued for extremely high incidence in the eighteenth century, the pox seems to have been omnipresent throughout the seventeenth century as well. The records of Bart's Hospital, especially, demonstrate that not only was the disease widespread well before the 1690s, but that it was rampant among the poor. Trumbach's vision of a new pattern of venereal infection begins from a position that presumes the disease to have been largely limited to the rich in the seventeenth century. Second, while the eighteenth-century increase in hospital beds might suggest more infection, it may only represent an increased tendency towards institutionalized care, as a result of the more general institutionalization represented by the voluntary hospital and workhouse movements. There is no way to argue with absolute certainty that new hospital beds reflect an increase in raw demand.

Also, while the eighteenth century may well have witnessed more penetrative sex and more resulting VD, there is one aspect of Trumbach's picture that must remain open to question, even though it certainly does find some support. Generally, Trumbach presents a pattern of infection in which a small minority of women (who are prostitutes) infected a wide portion of the heterosexual male population.[35] This new pattern of prostitution is itself constructed partially out of Lock Hospital data. When gauging the number of prostitutes in London he bases his estimates in part on the size of the female patient population at the Lock, presuming that virtually all the single women in the Lock Hospital were prostitutes, a questionable assumption.[36] However, when taken together the various institutional records seem to indicate a more even distribution between the numbers of men and women diagnosed with the pox in the eighteenth century. The royal hospitals and the Lock Hospital tended to treat more men than women. But (as we saw in chapter 4) parish records show a large population of infected women receiving treatment there. Therefore, it would appear (very roughly speaking) that nearly as many women as men received treatment for the pox in eighteenth-century institutions.

This could mean one of two things. More balanced rates of infection could speak to the common practice of intercourse as part of general courtship.[37] There is significant evidence from bastardy enquiries that plebeian women often engaged in premarital sex, especially with their co-workers in domestic service.[38] Some of these encounters may have involved some form of coercion, such as a false promise of marriage, or even outright rape. Still, given what we know of eighteenth-century courtship, we should not assume that all single women in hospital foul wards were streetwalkers; it is likely that many contracted the disease by engaging on some form of consensual, or at least non-mercenary, sex. By contrast, if Trumbach is right and prostitutes were the dominant vector for transmission, than it may well be that prostitution formed a much more common aspect of plebeian women's lives than we have perhaps imagined. If the majority of men caught their infection from prostitutes, the hospital records examined here would suggest that the sheer number of prostitutes in London may have been larger than even Trumbach suggests.

Finally, while the data on VD in hospitals may contribute some new evidence for historians of early modern sexuality to consider, it has to be borne in mind, once again, that the early modern diagnosis "foul" is a shadowy piece of historical evidence. We simply cannot know how often the diagnosis "venereal" actually stood for a sexually transmitted condition. If incidence of the foul disease was rising, it may only represent an increased tendency to render that diagnosis. One can speak with confidence that the rising number of illegitimate babies in the eighteenth century all resulted from penetrative sex; one cannot make such confident assertions based on the VD data. However, medical historians might do well to work in the reverse. To the extent that a portion—albeit unknown—of "venereal" cases *were* cases of syphilis or gonorrhoea, then we should hardly be surprised that a changing sexual culture that produced more illegitimate babies also witnessed more disease. Even if the timing of Trumbach's chronology is open to debate, it is impossible to doubt his overall characterization of the impact of the foul disease in eighteenth-century London: it was extremely pervasive and tremendously damaging. He is right to characterize it as the "harsh legacy" of the sexual culture produced by a rapidly expanding urban environment.[39]

That harsh legacy produced the significant need for yet another institution devoted to treating the pox in the middle of the eighteenth century. Even if we must refrain from suggesting a rigorously documented increase in "rates of infection," it is clear that there remained a significant, perhaps increasing, portion of the plebeian population that still stood in need of care, despite the dramatic expansion of hospital foul wards. When considering the need that still remained, it is helpful to consider which particular kinds of people remained who needed the services that the Lock could provide. In short, a particular demographic of foul patients still lacked care, and they essentially necessitated the new hospital. The group that made the Lock Hospital necessary was the population of itinerant poor who had left the countryside to seek employment in London.

It is well known that London's population grew dramatically over the course of the period of this study. Estimated at 200,000 in 1600, London nearly tripled over the seventeenth century, reaching 575,000 by 1700. The boom continued. By the time the Lock opened its doors in the middle of the century London was home to 675,000 people. By 1800 the metropolis bulged to just under a million. Put simply, the city grew by roughly 500 percent over the course of the seventeenth and eighteenth centuries.[40] That growth was overwhelmingly the result of migration from rural England. The rates of mortality in the capital were devastating, as the number of deaths outpaced the number of births year after year. Yet the growth of London raced ahead. In order to sustain that growth in the face of those mortality rates, demographers estimate that an average of seven thousand new people had to move into the city from the countryside each year.[41] That was an unprecedented number of people for an early modern city to absorb year after year. The process of absorption was imperfect, and it put an enormous strain on London's social welfare network, which, we have seen, relied heavily on parochial relief.

Virtually all of those who migrated to London left behind their parishes of settlement. Many obtained new settlements in London parishes by working, marrying, or serving an apprenticeship therein. But others were not so lucky, as parishes were not quick to recognize newly arrived paupers. The presence of ever more unsettled paupers on the streets of London posed a problem that elicited a range of responses, including harsher policing of vagrants.[42] Those who could not arrange new settlements had no recourse to the medical resources offered by workhouse infirmaries in London, so vital to the London poor. The mobility that came to characterize early modern England ensured that London became home to a sizeable group of migrants who no longer resided in the parish in which they could claim relief. By and large, these are the people the Lock Hospital specialized in treating. It is worth noting that by the time the Lock opened its doors migration patterns had subtly changed. In the eighteenth century new Londoners had usually migrated a shorter distance than their sixteenth- and early seventeenth-century predecessors.[43] Thus they lived closer to support networks of kin and parish upon which many likely relied in bad times, and we must assume that returning home was one survival strategy that London paupers utilized when the going got tough. However, some could not do this, and others chose not to do this. Perhaps their illness made such travel impossible. Perhaps they feared facing the scrutiny of their home community when they returned with the pox. Perhaps they feared that the churchwardens back home would not be so welcoming and that they might well face protracted resistance if they returned seeking care. Or perhaps they felt that their community could not provide the same quality of care that London could provide. For regardless of their somewhat closer proximity to their home parish, many eighteenth-century migrants who contracted the pox remained to seek assistance in the city.

It is unclear whether Bromfeild and the original governors consciously intended unsettled paupers to be the core objects of their new charitable initiative. It is unlikely that they did. The literature produced at the time of the Lock's foundation specifies neither the Poor Law nor unsettled paupers. The minutes of the original governors' meetings and the earliest surviving annual report of the charity merely stipulate that the intention of the charity was to relieve persons afflicted with venereal disease. In fact, initially the Lock Hospital likely admitted patients from London parishes. For example, the original recommendation form left a space to indicate the patient's parish. The form reads:

> Gentlemen
> To the Governors of the Lock Hospital I desire you will admit into your Hospital the Bearer ----[blank]---------------of the Parish of ----[blank]---- (if h__ Case intitles h__ to the Charity) being well assur'd -----[blank]----- is a proper Object, and am
>
> Your Humble Servant.[44]

This text leaves some room for debate. The mention of the patient's parish could have merely referred to the parish in which she resided, not the parish in which she was settled. Moreover, the phrase assuring that the patient was "a proper object" may indicate that governors looked into their nominees' settlements. It almost certainly did not refer to a medical diagnosis. In other words, governors were not assuring the hospital that their nominee was a proper object because she suffered from the proper ailment. It is much more likely that the governor's assurance indicated that the nominee was sufficiently impoverished and devoid of resources to warrant the hospital's charity (in addition, of course, to being properly deferential). So the space in the form for the patient's parish could have been to let the admissions committee know that a patient's parish of settlement was *not* in London, but somewhere far away. It is possible that governors may have taken patients' settlement status into account at this early point, but this is far from clear.

Communications with the parish in which the Lock Hospital was located, St. George's Hanover Square, indicate that the charity did treat some patients sent by that parish's overseers during the charity's first decade. For example, in May of 1757 the surgeon at the parish's workhouse accused the Lock Hospital of discharging several parishioners uncured whom he had subsequently to treat. Bromfeild attended the next meeting of the parish's weekly committee to refute the accusation.[45] The Lock's report on the matter reveals that it had commonly accepted patients from the parish, claiming that they promised "to oblige the Gentlemen of St. George's Parish by taking in all their venereal paupers that offer."[46] Indeed, St. George's paupers can be identified in the records during this period. On May 26, 1759, the governors ordered that an infant named Walter Cameron be admitted "upon producing a request from the proper officer of St. George's Hanover Square."[47] So the hospital did treat some paupers settled in its own parish.

However, after its first decade the charity changed its policy and began to charge parishes for treating their paupers. The Lock Hospital stands apart from most other contemporary hospitals treating the disease for explicitly not charging admissions fees. Yet after 1760 they began to charge fees from one group of patients—those sent by parish officers. In 1761 one finds entries recording the collection of a fee of two guineas each from the churchwardens of Hurst (in Berkshire) and of Hammersmith.[48] These two entries are brief and unrevealing. But later entries demonstrate that the governors had by this time ceased to view people with parish settlements as proper objects for their hospital. Beginning in the 1760s the struggling charity began to refuse to take on patients that they considered a parish's financial burden. If parishes wished to get their parishioners into the Lock they now had to pay.

The Lock's responses to settled applicants clearly display this (likely new) attitude. For example, on April 19, 1764, Thomas Mayo, the master of the workhouse in the parish of Abbington, in Berkshire, petitioned on behalf of an infected couple, John and Ann Rauce. The petition was not successful, even though they had arranged a nomination from a governor. The board of governors ruled that they were "not proper objects of this charity." "The parish of Abbington are certainly lyable to maintain & take care of them as their poor, and therefore [we] cou'd not admitt them on the recommendation of any Gov[erno]r, it being contrary to the institution of the charity."[49] The board agreed to treat the husband and wife, but only after Mayo signed a promissory note for four guineas, two guineas per patient. This demand outraged the parochial officials in Abbington. One of its officers sent a bitter letter criticizing the charity for demanding payment. Moreover, he requested that he be allowed to pay one additional guinea to bring the total to five guineas, the current subscription rate for a governor, this so that someone representing the parish could become a governor and nominate parishioners for free treatment in the future. He was probably surprised to receive the following reply that denied this seemingly fair request.

> The Gov[ernor]s present are sorry you shou'd so far misunderstand the intent of this charity, as to imagine that the nobility & gentry who support it by Voluntary Subsc[riptio]ns ever meant to receive people from parish workhouses who by the laws of the land have or ought to have very sufficient Provision the Design of the Gov[ernor]s was to relieve those who are destitute of any known Relief whats[oever] who by distance from or ignorance of their settlements must perish with[ou]t the assistance of this hospital, therefore whenever parish officers have brought objects here they have always willingly comply'd with the standing order of this house w[hi]ch is that a benefaction of Two Guineas be paid on the account of each patient by the parish he belongs to . . .[50]

Surely angered by this, the parish officers refused payment. In November the Lock threatened a lawsuit against Mayo to make him honor his bond. When their threat failed they filed suit the following January.[51]

The incident is telling because it marks the first time that the governors made this argument and the first time they used this language. It was far from the last. The governors now honed their definition of what constituted a "proper object" of the charity. They now sought only to relieve "those who are destitute of any known Relief whats[oever] who by distance from or Ignorance of their settlements" would otherwise be without recourse to treatment: in other words, those without a London settlement. It is difficult to know for certain whether this represented a new policy. There is some reason to believe that it may have predated the 1760s. The reply to Abbington asserted that this had always been their policy, calling it a "standing rule" with which other churchwardens have "always complied." However, even if it was on the books prior to 1761, the Lock's records do not indicate that it was enforced during the 1750s.

This episode was merely the first of many such incidents recorded. Soon after the Abbington episode the governors developed a rather formulaic response to these requests, which confronted them regularly. Five months later the governors refused to admit a woman named Ann Judd, a parishioner of Richmond. The secretary reported that she "appear[ed] on Examination to be a pawper of the parish & to be sent up here at the Expence of the parish, [and] was by the Board on account thereof, (according to the Institution of this Charity) Deem'd not to be a proper Object thereof, & consequently cou'd not be admitted on the Recommendation of any Gov. whats[oeve]r." In language quite similar to the previous response to Abbington—and which would be repeated virtually verbatim throughout the rest of the century—the secretary responded to the churchwardens in Richmond thus:

> [T]he original institution of this charity & the design of the Gov[ernor]s (who support it by voluntary subscriptions) was & ever has been to relieve those poor objects who are destitute of any known relief whatsoever who by distance from [or] ignorance of their settlements must perish without the assistance of this hospital, & not meant or intended to receive people from parish workhouses who by the laws of the land have or ought to have very sufficient provision & care taken of them....

However, the secretary went on to explain that a special arrangement could be made in those cases when a pauper had been brought to London from a distant parish. They, too, could pay two guineas per head to get their paupers into the Lock.[52]

This particular case is revealing because the board also wrote to Pryse Campbell, the governor who recommended Judd, to ensure that he not recommend any settled paupers in the future. Reiterating many of the same phrases, Secretary Stamford informed Campbell that his nominee could not be admitted "without deviating from or breaking thro' the Establish'd Rules & Orders" of the charity because Judd held a legal, known settlement. Stanford went on to repeat his statements about the "original intention" of the charity vis-à-vis settled paupers.[53] The secretary sent virtually the same letter to Sydney Stafford, when that governor recommended one John Sale, legally settled in Wimbledon.[54] This policy remained throughout the rest of the century. Moreover, they enforced it. The records indicate that the charity demanded payment— always two guineas per patient—from at least twenty-five separate parishes. Certainly there were more cases, but the records for the 1790s tended to leave out the name of the parish, and recorded only payments received.

Distance seems to have made a difference. The charity tended to make exceptions and admit parish patients for a fee when they had traveled to the hospital from a parish outside London. The language used in such instances indicates that they were truly exceptions. When Secretary Stamford responded to such parishes he made the point that the hospital considered these special cases. In words that would be repeated to parochial officials countless times, Stamford stated, "when Parish officers have brought or sent objects here out of the Country, the Gov[ernor]s (rather that they shou'd be sent back or depriv'd of their cure) have made it a stand[in]g rule to extend their charity so far as to receive them on the parishes giving two guineas . . . but not otherwise."[55] A list of the parishes from which the Lock accepted settled patients (for a fee) lends support to the assertion that the governors made this exception only for parishes outside central London.

The twenty-five identifiable parishes that paid to send patients to the Lock during the eighteenth century were: Abbington, Hammersmith, Richmond (Surrey), Henden (Middlesex), Arsley (Bedfordshire), Long Ditton (Surrey), Upton (Berkshire), Dartford (Kent), Clapham (Surrey), Deal (Kent), Great Missenden (Berkshire), Newland (Gloucestershire), Ham, Sutton (Kent), Horton Kirby, Airencester, Walton St. Lawrence, Windsor, Clewer (Berkshire), Hurst, Putney, Darenth, Leatherhead, Hayes (Middlesex), and Kingston. Nowhere on this list are any of the major parishes that one would expect to find sending patients to a central London hospital. Where on this list is St. Giles-in-the-Fields, or

St. Sepulchre, or St. Margaret's Westminster? By 1760, and perhaps earlier, the Lock governors considered paupers settled in those parishes someone else's responsibility. The absence on this list of major London parishes is not entirely surprising. Viewed from the perspective of London churchwardens, the Lock was an expensive option. The two-guinea fee at the Lock was more than double the ten shillings, six pence charged at St. Thomas's or Guy's, and even when the per diem costs at those hospitals are factored in, the fees charged at the Lock were excessive.[56] The governors at the Lock must have been aware of the fee schedules at the other hospitals in town. Their decision to demand more expensive admissions fees than the royal hospitals was likely aimed at dissuading London churchwardens from sending foul patients to their hospital. The strategy worked. St. Thomas's foul wards were filled with patients from central London parishes; the Lock's records indicate that London parishioners rarely ever entered.[57]

The governors clearly came to view their hospital as fulfilling a specific task within the larger network of London's healthcare institutions. The law clearly obligated parishes to relieve their own sick parishioners. In the eyes of those who administered the Lock Hospital, those who could claim parochial relief in London were not destitute. Therefore, they were not proper objects of their charity, which had quite limited resources. This may have been the policy from the outset of the charity; the governors who enforced the policy after the 1760s certainly claimed that this was the "original intention" behind the hospital. However, the founders never mentioned the policy, and it does not seem to have been enforced in the 1750s. So it is just as likely that the governors came to view the project of relieving the unsettled over time, only after witnessing so many of the people who came to the Lock and who had nowhere else to go.

This does not mean that no one with London settlements ever entered the Lock Hospital. As usual, rules were made to be avoided. The records of a single parish, St. Luke's Chelsea, reveal fifteen different parishioners who left that parish's workhouse and entered the Lock Hospital between 1756 and 1800.[58] It was impossible for the weekly admissions committee to perform settlement examinations on all the patients who entered the hospital. Many individual governors likely thought it their prerogative to nominate who they pleased, regardless of settlement. So doubtless many settled patients slipped through rather large cracks in the system.

But unsettled Londoners who fell ill must have clearly presented a major problem, for the City of London also began paying to hospitalize such people in the late 1780s. Despite the efforts of the Lock Hospital to care for such folk, St. Thomas's recorded a significant number of patients admitted for whom the City of London agreed to pay all fees. Patients entering both clean wards and foul wards supported by the City begin to be a common feature of the admissions registers after 1788.[59] St. Bartholomew's financial records similarly begin to record monies received from the City to support patients in 1789.[60] The City supported no small number of patients. At St. Thomas's in 1789 and early 1790 it paid to support just over 7 percent of all patients in the hospital. Moreover, issues of gender, poverty, and sexual infection seem again to have influenced who had to rely on City authorities to get a hospital bed. Just as foul women were heavily dependent on parish authorities, so too were they more reliant on the City when they fell ill. While foul women only represented 7.3 percent of the entire patient population, they accounted for 17.5 percent of patients sent by the City of London. And while the City supported roughly two men for every woman generally, 55 percent of its foul patients were female.[61] These patients were likely people deemed vagrants by the city, who may have been too ill to imprison or return to their own parish, but who may have faced those fates once they recovered their health. Whatever the case, the records of the royal hospitals support the claim that the problem of the unsettled sick remained a serious dilemma for eighteenth-century authorities.

Whether or not governors at the Lock recognized this in 1747, it remains clear that after 1760 they did what they could to steer the charity's resources more exclusively towards treating those migrant paupers who could not otherwise obtain treatment within a healthcare system based largely on the Old Poor Law. The medical social welfare system that had evolved was simply not equipped to handle massive migration. The migration that brought thousands of new people to London each year created a new underclass, namely, those who lacked a settlement they could realistically rely on in time of need. The Lock Hospital came to care primarily for these people. In that sense, it is safe to say that the

Lock Hospital emerged neither from new enlightened tolerance, nor from reformist zeal. It was a product of demographic shift.

Given the Lock's unique character, it is worth considering the characteristics and experiences of its patients. Many were likely impoverished migrants for whom the dream of better economic times in the city never materialized. Without a settlement in town by which to claim relief, migrants without sufficient funds to hire a doctor or even to pay the admissions fees at city hospitals saw the Lock as their main source of care once they came to believe that they needed hospitalization. Many such migrants who were women may well have turned to prostitution, but we cannot ascertain what portion of the Lock's patients were prostitutes. Tony Henderson's study of London prostitution has suggested that a sizable portion—around 60 percent—of London prostitutes were not born in London, but migrated.[62] Considering this, it may well be that the female wards at the Lock Hospital witnessed more prostitutes per capita than the female wards of other infirmaries in the city, which Trumbach suggests.

The gender breakdown at the Lock resembles that of the two royal hospitals. The admissions and discharge records show that men outnumbered women and accounted for up to 60 percent of all patients. The figures fluctuate within a range. Taken at ten-year intervals, men represented 53.5 percent of all patients admitted to the Lock in 1760, 60.2 percent in 1770, 53.2 percent in 1780, and 61 percent in 1790.[63] These data confirm that the fundraising literature that stressed married women and children as the Lock's main clientele was misleading. For example, during the twelve months of 1765 only three children and two married women entered the hospital.[64]

The only surviving data on patients' ages suggests that the majority of women to enter the hospital were young. Surgeon John Pearson's surviving 1798–99 casebook (discussed in the Introduction) details the cases of 86 patients, all of them women.[65] Pearson recorded their ages. In line with the average ages of foul women in St. Luke's workhouse (23.8), the average age of the Lock's women was 23.3 years. Importantly, this also fits with Henderson's data on the age of prostitutes in the city. So the Lock's women would have shared not one but two characteristics of the demographic sketch of prostitutes. Members of both groups would have often hailed from outside the capital and would have been roughly the same age, lending slightly more evidence in support of Trumbach's assumption that the Lock treated many prostitutes, although I would still caution against too firm a generalization. No data has survived that indicates the ages of men in the Lock up to 1800. However, a casebook by surgeon John Ritchie survives for the period 1813–14. While it covers a much smaller sample, it does detail twenty-one men's cases.[66] It suggests that the ages of the Lock's male patients resembled the ages of men in parish foul wards. Just as in workhouses, the Lock's men appear to have been older than their female counterparts, with an average age of over twenty-eight years.

These casebooks indicate that patients did not immediately run to the Lock Hospital upon contracting the pox. Most applied to the Lock only after many months, or in some cases years, battling the disease. We know that entering an early modern hospital was a last resort. So we should not be surprised to find that even these patients, with little in the way of resources, initially struggled to cope with the disease on their own. Both Pearson and Ritchie believed that the length of a patient's ailment was an important factor in prescribing a treatment; the date when the patient believed that they initially contracted the pox was almost always the first detail recorded. On average, Pearson's female patients endured their pox just over six months before seeking admission into the Lock Hospital.[67] However, the influence of gender again appears great. Men coped with their infection for more than twice as long before applying for admission to the hospital. On average Ritchie's men fought their pox for well over a year—66 weeks—before they applied to the Lock.[68] Yet again this difference probably comes down to economics. These (older) men likely had greater resources with which to seek private medical care, in the hopes that they might avoid resorting to hospitalization. The pox itself was impoverishing. If we take application to a hospital like the Lock as evidence of destitution, then this evidence suggests that the pox drove women into abject poverty twice as quickly as men.

Ritchie's case histories include more detail in descriptive prose than does Pearson's casebook. His descriptions illustrate that these men did, in fact, frequently avail themselves of private medical care before applying to the Lock. All but one of his male patients for whom full case histories are given indicated that they procured some sort of private medical care before applying to the Lock Hospital.[69] Presumably the women who could afford to do so also attempted cheap self-medication during the six months that they tried to cope with the disease. Hospital records can say very little about the role of self-medication for the venereal poor. But the patients who entered the institutions studied here most assuredly would have tried to treat themselves before they applied for hospital admission. To the extent that this book has tried to explore the illness-experiences of the venereal poor, it must be stated in the clearest terms that an important component of that illness-experience, self-help, would have to be added to complete the picture under construction here. The evidence in Ritchie's casebook gives us one small glimpse at the experiences of plebeian patients before they applied to a hospital.

Once it became clear to folks that they needed the services of a hospital, those who could not count on the support of a nearby parish had to seek a nomination from a governor. This stipulation, perhaps more than any other, distinguished the experiences of patients at the Lock Hospital from patients at the other hospitals covered by this study. To this point I have argued for the importance of privacy and the public nature of early modern institutional care. In all cases, venereal patients who entered a hospital or infirmary had to make a public admission of their infection before they could be admitted. In some cases this meant facing a hospital admissions committee, in others a workhouse master or local churchwarden. The nature of the charitable exchange at the Lock was even more pronounced. Before prospective patients faced the admissions committee they first had to solicit a nomination from a governor. These men who held the proverbial keys to the institution were often quite wealthy; some were even titled nobles. This meant that prospective patients at the Lock had another highly public step involved in their admissions process, one that involved clear dynamics of both class and power. In a position of rather dire need, prospective patients had to prostrate themselves before socially superior strangers, admit their infection, likely explain how they contracted the disease, and beg a governor's charitable mercy.

Unfortunately, institutional records do not yield significant information on these exchanges, which happened before patients ever arrived at the doors of the hospital. An investigation of the papers of individual governors might one day harvest a description of these meetings. Until then, we must assume that a significant amount of role-playing took place on both sides.[70] Paupers probably tried to cast themselves as repentant for their moral slips. Many may have used the strategy of the poxed villager, described by G. R. Quaiffe, who sought relief from the local curate by promising that "if he might have help it should be a warning for him as long as he lived."[71] Of course, many may have lied about how they contracted the disease, as they tried to cut the figure of the worthy object of charity. There was no denying their infection, for they were applying to a venereal disease hospital. But they could surely craft a story of how they contracted the disease, which might mitigate their guilt. Who knows how many prostitutes claimed to be wet nurses?

In addition to creative spin put on their medical and sexual history or the promises to reform that were likely extremely common, applicants may have also had to craft other aspects of their story. Governors probably questioned them on their finances, trying to ascertain whether they were truly destitute. Paupers probably had to prove they were ignorant of their parish of settlement, or else that they lived at a great distance and had good reasons for not returning to apply for relief. Governors on the other hand likely assumed the paternalist role that came with their station, casting themselves as charitable Christians and warning supplicants that they would not be readmitted if they contracted the disease again.

It is possible that some governors chose not to meet with infected paupers face to face. In such cases paupers may well have met with stewards or secretaries who acted as intermediaries. And it is possible that the network could have involved several degrees of separation between governor and pauper. Paupers, too, might have had someone acting on their behalf, contacting a governor or his representative in their stead. Of course, the nature of these exchanges would have differed from one governor to the next, especially after the Evangelicals became more prominent among their ranks in the 1780s and 1790s. Still, it is vital to understand that the patients who came to occupy beds in the Lock Hospital all had to negotiate this exchange as part of their admissions process. This process differed markedly from the admissions process at the other institutions analyzed here, and it comprised an important element in the illness experiences of patients at the Lock.

This book has also argued for the centrality of gender to the illness-experiences of foul patients. Here again women likely faced unique challenges confronting this admissions system. Voluntary hospitals stressed that governors should only nominate applicants who were morally worthy of benevolence. Given the different moral yardsticks for men and women in the period, we can only assume that governors understood male and female "worthy objects" differently. Successful applicants required the all-important skill of *appearing* worthy. Once infected with the pox, making the case that one was morally worthy of benevolence was a difficult argument to muster. This may have been an instance where the sexual double standard hit hard. While on some levels all paupers may have resented having to assume their deferential role in the morality play that hospital admission required of them, this was likely much more difficult for women to endure. That so many women in St. Thomas's chose to confide in other women when arranging their admission bonds indicates that the aversion to disclosing one's infection to a man was not unique to elite women. We have every reason to believe that poor women who had to solicit a nomination and face the accompanying interview process to determine their worthiness also resented the process.

A telling piece of evidence lends support to the claim. Even more obscured than the history of the Misericordia Hospital described in the previous chapter was that of a short-lived institution called the "Misericordia General Dispensary for the Venereal Disease." Despite its name it seems unaffiliated with Hanway's hospital. Located in the Strand, the Misericordia Dispensary, which offered exclusively outpatient care to venereal patients, seems to have grown out of the wider dispensary movement.[72] Founded by physician Loftus Wood in 1779, and listed in the *Medical Register* of 1783,[73] it, too, had a brief history of just a few years. Wood moved to Colchester in 1783, and the dispensary seems to have closed. There was one notable aspect of its operation. Aware of the particular sensitivity involved when women applied for assistance, the dispensary exempted women from its rule that all patients solicit a governor's nomination letter. In one of the few instances in which we see a reverse double standard at work, women's application process here was actually made easier:

> In order to rescue Female Patients from that Load of Shame and confusion, which, even in the most depraved State, they would necessarily feel, on applying at Gentlemen's Houses for Letters of Recommendation; all such shall be admitted to the benefit of this Charity without any Form of Recommendation whatever.[74]

Though we know little else about this institution, this rule alone helps to confirm that contemporaries were well aware of the sting of the disease's stigma for poor women, and the difficulties that the voluntary charity model posed for them. In the end, one reason the Lock Hospital treated more men than women may have been that women were reluctant to submit to the scrutiny that its admission process demanded, and thus fewer women may have applied. Certainly the governors of the Misericordia Dispensary recognized this possibility.

Even those who successfully negotiated the process and secured a governor's nomination were not always guaranteed a bed in the Lock. Like all institutions the Lock had a limited capacity. In some periods there were simply no available beds. At such times applicants were not refused outright; names went on a waiting list. However, in some periods, especially in the 1760s, the waiting list grew exceedingly long, and paupers faced quite lengthy waits. The 1760s were a particularly bad time for the English poor, for if grain prices are any indication, it was a period of crisis. It may be that the increased demands on the services of an institution like the Lock Hospital represent the trickle down effects of larger economic downturns, like the well-known economic crisis of the 1760s and 1770s that drove even more people into poverty.[75] For example, during the twelve months of 1765, 490 people applied to be admitted to the Lock Hospital. Of those, 186, or almost 40 percent, were placed on the waiting list. The average postponed patient waited more than two weeks before entering the hospital.

However, this figure tells only part of the story. Some patients waited as long as two months before being admitted. For example, Anthony Davis applied on November 21, 1765, sponsored by Lord Farnham. Despite his aristocratic nomination Davis was not admitted until January 16, 1766.[76] For some, such a long wait proved too difficult to endure. A significant portion of those postponed never actually entered the hospital. The aforementioned two-week average waiting period applied only to the 133 postponed patients who actually entered the hospital and for whom there exists a date of entry with which we can calculate the total length of their wait. The other 53 of the 186 postponed applicants, or over 28 percent, gave up and never entered the hospital. When placed on the waiting list applicants were instructed to attend the hospital each week to see if a place had become available.

Evidence shows that many abandoned hope after a while and stopped appearing. Perhaps some made the trip back to their parish of settlement; perhaps some sought admission to another hospital; perhaps some died. It is impossible to know what happened to these people, but the records show that many remained on the waiting list for a considerable time before they simply stopped showing up. For 40 of these 53 applicants dates are available that indicate either the last time they appeared in person or the point at which the governors officially struck them from the waiting list for failing to appear. On average, these patients waited for more than a month (34.8 days) before moving on. A month may not seem long upon first consideration, but time is relative; to these diseased and suffering people, a month was clearly a very long time. We will also remember that typical patients would have likely already battled the disease for as long as a year or more before applying, their symptoms worsening over time.

Moreover, many patients waited considerably longer than the one-month average. A woman named Rebecca Flower could still be found on the postponed list on July 27, five months after she had first applied on February 21.[77] Another aspiring patient, Richard Bland, was put on the waiting list on November 21, 1765. After waiting for almost three months (84 days) with no luck, he failed to appear and was struck from the waiting list on January 30, 1766. Probably failing to obtain any significant treatment on his own, Bland reapplied on May 1, 1766. He went back on the waiting list for four more weeks, but he again lost patience and disappeared from the records after May 29, having never entered the hospital.[78] Postponed patients occasionally received outpatient care while they waited, but these cases were rare. Only three of the 186 applicants postponed during 1765 received any care on an outpatient basis.[79] We are left to assume that the other 183 applicants were on their own.

Occasionally, postponed applicants solicited the assistance of their nominating governor in an attempt to speed up the admissions process. Angry governors sometimes complained that their nominees had not been admitted. While this may have greased the wheels of the bureaucracy, the board's responses always reiterated that patients would only be admitted in due turn.[80] The records furthermore indicate that patients were often refused outright if their nominating governor already had a patient on the waiting list.[81] Considering that most applicants probably endured their symptoms long before they applied to the Lock Hospital, and considering that they probably applied only after their conditions notably worsened, these lengthy waiting lists constitute an important factor in the experiences of those who sought access to the Lock Hospital. As elsewhere, a bed in the foul wards was not easy to come by.

Yet another revealing case emerges from the records of the Westminster Coroner's Inquests that demonstrates the desperation prospective patients sometimes felt as they lobbied in vain for a bed during this extremely crowded period at the Lock.[82] In November of 1762 Robert Munro had already been a patient in St. George's Hospital for at least a month. A variety of deponents later characterized Munro's general mood as "melancholy and dejected," "low-spirited," and "pensive." He had entered St. George's for a pain in his head, which apothecary William Dampier described as "intense," stating that it "baffled all . . . remedies." According to the medical staff Munro had suffered from the pox for several years. He had already been salivated in the Lock about two years previously, to no avail. He now sought to return to the Lock Hospital to give salivation one more try, but he had trouble getting a bed. He entered St. George's Hospital and became the patient of none other than William Bromfeild, who, we will remember, served at both hospitals. According to testimony, Munro placed his final hopes on securing a bed at the Lock.

One of St. George's porters, William McGregor, knew Munro for five months; his deposition tells part of Munro's story. McGregor reported that Munro confided in him that he had been suffering from the foul disease for about two years and that he complained of his inability to get a bed in the Lock Hospital. McGregor described him as exceedingly despondent, claiming that he "despaired of a cure." When asked why his hopes were so low, McGregor reported that Munro "replyed that he was disappointed of a place [at the Lock] which some gentleman had promised him." McGregor asked Munro whether he had gone to the Lock personally to enquire about a place. Munro responded that he had, without luck. Paraphrasing Munro, McGregor claimed that this turn of events pushed him to consider suicide: "[H]e had [been to the hospital] and did not expect a cure; that he was more ready to make a way with himself than anything else, and [he] appeared . . . to be very melancholy & dejected."

The despair that he would not get a bed in the Lock, and would thus never be cured of his pox, seems to have led Munro to take his own life. In the early morning hours of November 20, Robert slit his throat with a straight razor as he lay in bed. Bromfeild and the other surgeons attended him, but he could not be saved. More than one deponent cited Munro's severe disappointment at failing to get into the Lock Hospital, concurring with McGregor's testimony that "a gentleman" had promised to secure him a place, but failed to do so. Joseph Hall, St. George's clerk, put it this way:

> Sir, I am directed to inform you that one Robert Munro a patient in this hospital, did cut his throat here this morning; and is just now dead of the wound. The best account that can be given at present for his doing this act is that he despaired of being cured of his disease (for which he had been in the Lock Hospital before he came here) and have been in expectation of a place, which it is said had been promised to be given to him, he was acquainted, that he must lay aside all thoughts of it, for it could not be procured for him; and this it is thought was what determined him to put a speedy end to his life.[83]

Munro's is an extreme case to be sure, but as earlier discussions have demonstrated, it was hardly the first instance in which the pox contributed to a pauper suicide. While extreme, it serves to put a sad human face on the scores of names of those who applied unsuccessfully for a Lock Hospital bed, as well as the thousands of those whose names we do not know, who, throughout the century, tried to negotiate the health care system and failed, at this hospital and others. Given what we now know about the Lock Hospital, it is likely that Munro had neither the money to procure private care nor the resources of a London parish upon which to fall back. For him, the Lock really does seem to have been his last hope. Perhaps the Lock's fundraising literature, which presented the hospital in such terms—as the only hope for the city's poxed poor—was not so far off the mark after all. Somehow Munro networked to find a "gentleman" to act on his behalf, probably not a governor himself, but someone trying to solicit a nomination for him.

Who knows how hard he tried to get Robert a bed? Nonetheless, timing was not on their side, for in the early 1760s, space at the Lock was at a premium. Robert Munro was hardly the only one to get his hopes up to no avail.

Munro attempted to get into the Lock at the worst possible time. The waiting lists at the Lock were longer during this period, the early- to mid-1760s, than at any point in the century. For example, in 1760 the figures resemble the 1765 figures cited earlier. One-third of all applicants went on the waiting list in 1760 (128 of 383), and over 20 percent of those postponed never entered the hospital.[84] However, the numbers begin to drop by the end of the decade, and they remain lower throughout the rest of the century. In 1770 only 12 percent of all applicants (88 of 698) went on the waiting list. The list dwindled further to just eleven postponed applicants in 1780, none of whom waited longer than two weeks before obtaining a bed. And during the twelve months of 1790 no applicants had to wait. This trend resulted from the expansion of the Lock's capacity. Through 1765 the hospital only cared for an average of just over 450 patients annually. However, until the end of the century the operation extended, and during the period 1765 to 1800 the hospital admitted over 653 patients annually.[85] Of course, the period after 1765 marked the point at which significant funds from Madan's new chapel began to pour in, making expansion possible.

The administrative records verify that the governors worried about the lengthy waiting lists of the 1760s and moved to expand operations as a result. For instance, the board moved to open another ward for men. The minutes read: "Upon considering the Large Arrear of Postponed Patients on file, it was thought proper to open the inner Ward for Men Patients."[86] Of course, this means that lengthy waiting periods, during which time patients suffered without any treatment, constituted a more significant component of the patient experience for those applying to the hospital during the early part of the Lock's history. More importantly, this data also offers some explanation for why the Lock seems to have tightened its policy regarding patients with parish settlements in this very period, the early 1760s. It was only then, when the hospital's resources clearly could not provide for the throngs of paupers who applied, that the governors began strictly enforcing its rules governing settled patients.

While fewer patients faced long waiting lists as time went on, the admissions records indicate that prospective patients faced a different obstacle in the later decades of the century. A significant number of applicants were refused admission because the surgeon diagnosed their condition as "not venereal." No significant development in theory or new method of diagnosis can account for this trend. Rather, it should probably be attributed to the vagaries of individual surgeons. Yet again we see how the phenomenon of diagnosis was a fluid and contested process. In these cases we see a large number of patients who believed themselves to be poxed and acted on that assumption. They found a governor, secured a nomination, and applied to a hospital that specialized in venereal disease, only to have the house surgeon disagree with their diagnosis and block the door.

It is telling that the frequency of these incidents shifts over time. Neither in 1760 nor in 1765 were any applicants refused because they were diagnosed not to have the pox.[87] However, in 1770 more than one applicant in ten was diagnosed as "not venereal" and thereby refused. Fully 77 of the 698 paupers who applied that year were unsuccessful because of a disagreement over diagnosis. The same fate awaited many, albeit fewer, paupers in 1780 and 1790. In those years respectively 42 and 36 applicants failed to obtain beds because they were deemed "not venereal." The period after 1765 witnessed several changes to the medical staff at the Lock. Bromfeild's son Charles took over his father's surgical post in 1770, and John Pearson began his forty-year career as the hospital's surgeon in 1780, after Charles resigned in the midst of his father's clash with De Coetlogon and Madan. These latter surgeons seemed more scrupulous than their predecessors, more likely to negate patients' self-diagnosis. Whoever was responsible, the admission records indicate that a sizable number of applicants who thought themselves to be infected were refused after 1765 for diagnostic reasons.

Those who successfully negotiated the admissions process seem to have received rather standard treatment once in the hospital. Mercurial salivation, as elsewhere, remained the usual regimen throughout the century. Pearson's casebook demonstrates the point. Of the seventy-five patients in Pearson's casebook (for whom complete records survive[88]) seventy-one underwent salivation. Pearson noted the date on which salivation began, and he monitored the amount of sweat and spit (as well as urine and stool) that each patient produced. Ritchie's casebook further indicates that mercurial salivation remained standard at the Lock during the first decades of the nineteenth century.[89] Their casebooks also indicate that the surgeons visited the patients quite frequently. According to the dates of their entries, both Pearson and Ritchie visited patients at least every second day. Sometimes they saw patients daily. The patients in Pearson's casebook remained in the house for an average of 51.1 days. On average, those patients were visited by a surgeon on 29.9 of those days.[90] Therefore, Pearson visited each patient more frequently then every second day.

The casebooks suggest that patients tended to remain in the house just long enough to undergo salivation. Although it is incomplete, Pearson's casebook allows us to calculate the duration of twenty-four patients' salivation. On average, these patients underwent salivation for all but eight days of their stay in the hospital.[91] Generally, patients seem to have entered the house and prepared for salivation for a few days before the procedure began, some more than a week. However, this may have meant that doctors administered mercury for several days before actual salivation began. Patients endured salivation for about forty days,[92] during which time they continued to take mercury. They then stayed in the house for a very brief period before being discharged. Some patients left the very day that their salivation ended. All but two of Pearson's fully documented salivated patients left within four days of completing the procedure.[93] Pearson also published a textbook that reveals that he experimented with other substances besides mercury, such as nitric acid. However, he concluded that mercury remained the most reliable drug for the pox.[94] Other forms of treatment at the Lock included the use of bougies, thin instruments used to keep the urinary tract from closing due to ulceration. Their use is not surprising since they became a quite popular way to treat urinary disorders by the middle of the century.[95]

Finally, Pearson also illustrates how individual surgeons could have an impact on the patient experience at the Lock. The average duration of treatment increased after Pearson became the Lock's surgeon in 1780. When William Bromfeild ran the medical staff in 1760, patients remained in the house for an average of 40.6 days. Similarly, in 1770 when his son Charles took over, patients stayed an average of 38.2 days.

However, by 1780, the year that Pearson arrived, the average length of a patient's stay rose to 51.4 days. It remained virtually unchanged in 1790 at 51.2 days, and in 1798/99 the patients in Pearson's casebook stayed for an average of 52.4 days. So the nature of treatment remained similar throughout the century, but its duration changed over time. After 1780 patients who entered the house stayed for roughly a week and one-half longer than patients did earlier in the century. Probably this was a result of a new surgeon's opinion. However, it is also possible that in the 1760s and 1770s the medical staff responded to the pressure of heavy demand and long waiting lists, and tried to move patients along as quickly as possible.

As chapter 5 began to argue, discipline seemed less than stiff at the Lock during its early decades. Indeed patients at the Lock faced a regimen that may have even been somewhat more lax than patients at other contemporary hospitals. For example, the patients in the Lock Hospital did not have to perform any labor. While other hospitals and workhouses often made inmates work assisting the staff with general housekeeping tasks, Lock patients got off lightly. The governors argued explicitly against the practice of making patients work. For example, in 1768 the governors discovered that the staff made the patients wash the wards. They responded that such labor "retarded their cure" and "endangered their healths, to the discredit of this charity." They proceeded to ban the practice of labor by patients.[96] By 1783 the staff again tried to make use of the patients in the kitchen, and again the governors put a stop to it. On October 31, 1783, the General Court ruled to "confirm the regulations of the weekly board of the 17th Inst. respecting the patients not being suffered to do any work in the house, shop, or kitchen &c." The governors ordered that an extra maid be hired so that patients would not have to work.[97] Considering that mandatory labor and the instilling of a work ethic were major components of many eighteenth-century disciplinary/reforming initiatives, the absence of mandatory labor at the Lock is notable. Combine this information with what we already know about the scant religious instruction during the early decades and the overall regimen seems even more strikingly lax.

Conflicting evidence emerges on the issue of discipline after 1780. On the one hand, the administrative minutes suggest that hospital regulations may have became somewhat stricter in the years following the Evangelicals' ascendance. After 1780 the governors moved to tighten up hospital security. In October 1782 the governors passed several new disciplinary resolutions, bringing the Lock Hospital more nearly in line with the penitentiary model. They seemed especially concerned with keeping patients within the confines of hospital grounds, complaining that patients had frequently left the house without permission, and they prohibited the staff from sending patients out of the house on errands.[98] In 1787 the governors ordered that patients' visitors should be monitored more strictly, that patients' meetings with visitors should be supervised, and they now reserved the right to inspect parcels sent to patients in the house.

But the Evangelicals seemed to have aimed their scrutiny particularly at women patients. Just when the charity was focusing increased efforts on policing women in its new asylum, the charity also moved to govern the female hospital wards more strictly. Even though the rule concerning supervision of visitors was supposed to apply to all patients, the governors subsequently directed the matron to be particularly vigilant of the hospital's women. In 1788 the governors ordered:

> That during the time when persons are admitted to have access to the female patients, the Matron or some other person . . . be always present, to the intent that no improper conversation may pass between the female patients & visitors.[99]

However, the evidence of patients' own actions indicate that they resisted house rules and generally expressed dissatisfaction with the wards more frequently in the period *before* the Evangelicals took over the charity. Acts of resistance, either disobeying house rules or running away, were more common in the period before 1780. In 1770, sixteen different patients found the Lock Hospital unbearable and ran away, and four others were discharged for grossly disobeying house rules. Patients such as Jane Grimfield were thrown out for such offenses as "Fighting, and Swearing, and otherwise behaving ill in the wards and not conforming to the rules of the house."[100] However, such incidents were ten times more common that year than they would be a decade later. In all of 1780 just two patients ran away and none were discharged for disobeying regulations. Likewise, in 1790 only five patients escaped, and none were discharged for misbehavior. The behavior of the patients themselves indicates that they may preferred the conditions in the house in the era of the Evangelicals, despite the seemingly more stringent regulations. However, it is difficult to draw conclusions. One the one hand, the general conditions of the wards (e.g., cleanliness, quality of food) may have simply improved, and thus patients' greater compliance may not reflect their attitudes towards disciplinary rules in particular. One the other hand, patients may have resented the new rules, but it may have been more difficult to run away once more stringent monitoring commenced.

But regardless of what period one analyzes, the patients at the Lock simply did not exhibit particularly insurgent behavior. The twenty incidents of resistance in 1770 represent less than 4 percent of the entire patient population.[101] The handful of incidents in 1780 and 1790 represent far less. The fact of the matter is that these patients felt that they badly needed the services of the Lock Hospital. They had made the decision to apply for admission, and they probably realized that they had few options available to them in their diseased and impoverished state. Once in the house they found themselves occupying a bed that they had secured only by mustering considerable networking resources and enduring a difficult process of scrutiny. As at the royal hospitals, most seemed unwilling to risk these hard-won resources through insurgency. Moreover, the overall disciplinary regimen at the Lock seems hardly stricter than any other contemporary hospital. For a good portion of the century it seems that patients could leave the house when they wanted to, and they did not have to work. And we cannot forget, their care was free; the importance of this economic reality cannot be overestimated. Patients must certainly have known that patients in Bart's, St. Thomas's, or Guy's had forked over a goodly amount of cash to get the same care that they were receiving for free. And without parish support to fall back on, the Lock's patients surely knew enough not to rock the boat. In all, the Lock's patients leave us little evidence that they complained very loudly.

The women in the Lock Asylum are an entirely different matter. The asylum's penitentiary regimen of prayer, work, and solitude met with significant resistance.[102] Women in the asylum resisted house rules much more frequently than did patients in the hospital. The records of the Lock Asylum are quite incomplete—the entire decade of the 1790s is virtually undocumented.[103] However, its first two years are well documented. The records for 1787 and 1788 show that many women found the place difficult to endure, and left. According to the admissions and discharge records from July 1787 through to the end of 1788, almost 16 percent of all the women who entered the asylum ran away. Some left after a less than a week, while others remained for a few months before making for the exit.[104]

The charity's own published reports demonstrate that the success rate for the asylum was much lower than even this figure suggests. In the *Account of the Institution of the Lock Asylum* (1796) the charity gave figures on its rates of success. The asylum could not boast an impressive record. Against 94 cases of "successful" reformation (in which women were either returned to their family or placed in domestic service), the asylum had to admit 123 failures. Virtually half of all the women admitted to the asylum had run away or been expelled for bad behavior. Most of them, 112, ran away, either from the asylum or from the service in which they had been placed.[105]

Despite these disappointing figures, the governors tried to put a positive spin on the facts. As Linda Merians put it, "the necessity to put its mediocre results in a positive light was . . . crucial to the asylum's continued survival."[106] They did this with the rationalization that the few cases of true reformation were worth the efforts expended on all the women who rejected the asylum's mission.[107] Still, the governors could not deny that most women rejected their efforts to save them. The stricter regulations at the asylum likely drove many women to escape. One does well to remember that all the women who entered the asylum had already experienced institutional life in the hospital. One is forced to conclude that, despite slightly stricter rules in the hospital in the late 1780s and 1790s, life in the asylum was vastly different from life in the hospital. The women who rejected the asylum were not rejecting institutionalization in general; they rejected the asylum in particular. Fully half of those who had already stomached life in the Lock Hospital could not, in the end, bear the realities of life in the Lock Asylum. As the nineteenth century dawned the governors at the Lock had to admit begrudgingly that curing someone's body was a lot easier than saving their soul. Salivation, it seems was one thing, salvation quite another.

CONCLUSION

POVERTY AND THE POX IN EARLY MODERN LONDON

On the night of February 7, 1734, Elizabeth and John Byon lay in a rented room that they could barely afford. John was a fan painter and he paid two pence to Magdalen Jones for one night's lodging so that he and Elizabeth could get off the street. Elizabeth was extremely sick. John tried to explain away his wife's weakness by telling Magdalen that she was drunk. But when what a servant would later describe as "dismal groans" emanated from Elizabeth's bed, Magdalen came upstairs to check what was wrong. John was afraid that Magdalen would turn them out if she knew the nature of Elizabeth's illness, so he lied again. Changing his story he now said that Elizabeth had a cold. He had good reason to worry. Magdalen was not fooled. She ordered them out, and refused to allow them to lie in the house, despite the couple's pleas to let them stay until morning. "You told me your wife was only drunk, but she is rotten with the Pox, she shall not lye here, and so take your Groat again, and take your Time to dress her, and carry her out." Magdalen's husband became impatient for them to quit the house. Not allowing them to finish dressing, he forced them into the street. It was 10:30 at night, and it was February. John helped Elizabeth to a bench where he finished dressing her. They tried to move on but they had nowhere to go. Elizabeth made it just a few blocks before she collapsed in the street and died.

John and Elizabeth had lain for two nights on the pavement in front of a churchwarden's door, but their settlement details were contested. They sought relief in St. Ann's, but the officers would not acknowledge them as settled parishioners. St. Ann's officers refused to accept them without a pass from St. Giles. So they gave up seeking alms in St. Ann's and made their way across town and into St. Giles, where Magdalen Jones lived. They may have come into the parish too late in the evening to apply for relief. Perhaps they planned to apply at the workhouse in the morning. They had two pence that they may have begged while sleeping rough, so they sought meager shelter for the night. For the moment, if they could not get a bed in the workhouse or a chance to see a parish surgeon, at least they could get in from the cold. Or so they thought.

John brought a murder charge against Magdalen Jones, who stood at the Old Bailey accused of

> dragging [Elizabeth Byon] out of Bed, and forcing her into the street, between 10 and 11 at Night, when she was weak, and sick, and groaning, and leaving her there, almost naked, and not able to stand, so that with Cold, and Want of due Care, requisite to a Person in such a weak and sick Condition, she Instantly dy'd.

The court heard only the testimony of John Byon, a servant maid, and the city coroner. Byon and the maid told essentially the same story, save that the maid made the point to note that that Byon was "rotten with the Pox" just like his wife. The coroner merely gave a one-line testimony: "The Body of the Deceased was full of Blotches."

The jury acquitted Magdalen Jones.

We can only surmise that the jury essentially agreed that it was legitimate to shun the foul. Certainly the testimony of all witnesses bore out the accusation—nobody contested that Jones, as accused, had cast Elizabeth Byon into the cold winter street when she was obviously sick. There was no testimony challenging the charges against her. There was merely the chilling testimony of the coroner, verifying that Elizabeth was in fact poxed. One might think that this should have sealed Jones's conviction, proving that Elizabeth was indeed direly ill and thus Jones knowingly threw her into the street. Yet the jury members essentially sanctioned Magdalen Jones's actions.[1]

This book has tried to tell the story of people like Elizabeth Byon by focusing on the options available to them when they fell ill. It has done so mainly through the lenses of institutional records, a method with obvious limitations, but one that can still yield valuable insight. To draw the themes together, let us consider for the moment the patients' perspective: what did it mean to live in early modern London and contract what you thought to be the pox? As the preceding chapters have argued, the options available, and therefore one's ultimate illness-experience, would have varied immediately according to time and place and even further according to sex and station. The experiences of people like John and Elizabeth Byon differed significantly from those of their social betters, and also, because of the nature of their diagnosis, from the experiences of other plebeian people struggling with different medical conditions.

Consider first patients of means. Patients with money to spend encountered London's vibrant medical market, which provided a range of options for those afflicted with the pox. Physicians, surgeons, apothecaries, and empirics of all sorts all treated the disease. While most offered standard mercury treatment throughout the entire period, patients could also find a host of alternative remedies for sale. Frequently such treatments employed substances besides mercury, which by all accounts was rough to endure, while others offered more gentle forms of mercury treatment employing smaller doses of the drug. Many alternatives to salivation held the promise of treatment that was briefer and less conspicuous than lengthy salivation, which took weeks and brought perceptible side effects. The advantages offered by alternative treatments proved popular, as evidenced by their ubiquity in the market. The Montpellier method was just one of the more successful alternative treatments that offered patients greater privacy.

Patients shopping for treatment certainly held discretion as a primary concern as they consulted their many treatment options. The presence of a sexually transmitted disease so challenged the traditional doctor-patient relationship that it forced innovation in a number of ways, none more important than the increased emphasis placed on the provision of medical privacy. As a result, foul patients of means could shop around and select from a host of different practitioners who provided a range of different services in a variety of different circumstances. Shy patients could visit doctors under the cover of darkness or use backdoor entrances. They could receive treatment within the privacy of their own homes or obtain medicine and instructions on a seemingly innocent visit to a bookshop. They could even hide behind the anonymity of a letter and obtain drugs through the mail. However, gender further complicated the medical exchange, and under no circumstances was this more apparent than in the case of venereal disease. The market responded with the provision of female healers for shy women.

The ability to take advantage of these services was contingent upon wealth. Since entering a hospital in this period was very nearly a last resort, it is likely that people of even rather meager means initially tried to obtain medical care on the open market. Who knows how long Elizabeth and John Byon had been sick? It is likely that early in their illness they tried some of the cheap drugs provided by one of London's many empirics. Handbills for such drugs were everywhere, and they would have had little trouble locating a "doctor" happy to sell them a pill or powder sure to clear them right up. In fact, the real trouble when confronting the medical market would have been trying to distinguish one cure from another, considering the number of heralded treatments.

Scholars remain dubious about the efficacy of pox treatments that predate the age of antibiotics, mercury included. It is likely that most plebeian sufferers tried at least one, perhaps several, such remedies to no avail. Hospital records show that most patients had usually already tried their luck with a private practitioner or some form of self-medication before they applied for admission. However, not everyone could continue to avail themselves of these expensive options indefinitely. Certainly, the combined financial effects of mounting medical expenses and lost wages may well have driven people on the margins further into poverty. Those with little extra cash could not long continue paying surgeon's bills whenever their symptoms returned, especially if their illness rendered them incapable of working or if exposure of their infection cost them their job. The constant recurrence of symptoms marks an important component of the illness-experiences of this disease in the period, one that magnified the general economic effects that attended all forms of illness. For those whose "pox" was more than a simple urinary tract infection that might clear up by itself, illness could be cyclical, even perpetual. Considering this, we would do well to consider the pox itself as impoverishing, driving many further into a state of poverty, and forcing them to seek forms of public relief that they would not have otherwise needed. Evidence shows that gender cut to the bone on this count. Women felt the financially debilitating effects of the pox much more quickly than did men, who held out for twice as long before becoming destitute and having to apply to a hospital.

Throughout most of the seventeenth century care for such people was free at both St. Thomas's and St. Bartholomew's. However, this changed as the century drew to a close. Thereafter these institutions largely restricted their services to those with some resources to contribute towards their own care. St. Bartholomew's records indeed indicate that by the eighteenth century the royal hospitals' clientele came from a slightly higher rank than the absolutely destitute. It is not likely that John and Elizabeth Byon could have afforded the fees, which were not insignificant. The first step in the admissions process for patients here demanded that they arrange their own financing. Some patients could pay their own fees and post their own bonds. However, most could not. The majority had to come up with the necessary funding by relying on a support network of family and friends. Some who hoped to avoid disclosing the sensitive details of their condition to such people could arrange their financing through one of the many alehouse-bondsmen who would post their bond for a small fee. Yet again, medical discretion came with a price tag and was not available to everyone.

St. Thomas's admission registers demonstrate that gender again played a pivotal role. Many of the women who entered St. Thomas's foul wards relied on other women when soliciting funds for admission. This underscores the point made in chapter 1 that infected women were generally reluctant to discuss their situation with men. When they could, women chose to confide in other woman. Virtually one out of every four women in St. Thomas's foul wards got another woman to finance their hospitalization, making them ten times more likely to do so than men. Of course, patients facing economic difficulties could do what foul patients had been doing at hospital gates for two centuries—creatively construct their diagnoses. Many patients seem to have slipped into the clean wards where they would pay lower fees by citing alternative illnesses. However, there was much more at stake in this than mere admission fees, so patients also actively participated in negotiating their diagnoses to avoid the public labeling that accompanied the medical verdict "foul."

While some patients may have been able to arrange their financing more discreetly than others, once they entered the foul wards of Bart's or St. Thomas's all patients faced highly public care. Patients had to face hospital governors, members of the medical staff, and the weekly admissions committee, and in the eighteenth century they were on display for medical students. For most of the period patients at Bart's had to enter a former leper house. While these may have carried a greater stigma than the corresponding wards within St. Thomas's, their distance from the main hospital may have actually allowed for less stringent discipline. Patients there seem to have left the premises when they wanted, unlike the patients at St. Thomas' whose movements were more strictly monitored and where patients generally faced stricter discipline. At St. Thomas's they had to endure public whipping in the early seventeenth century, and in the early eighteenth century were forced to go before a minister just prior to and immediately following their salivation. This helped to merge spiritual purging and biological purging into a single process, a development not apparent at the other hospitals studied.

But there still remained many people, Elizabeth Byon for example, who could not afford the services of the royal hospitals. Some could not afford the fees; some lacked a support network. Quite simply, the royal hospitals became priced beyond some peoples' reach, especially as time went on and fees increased. Such people had to fall back on the medical resources available to them by their parish under the provisions of the Old Poor Law. These services varied from parish to parish. In the seventeenth century they often took the form of a dole, or a visit from a surgeon hired to care for the local poor. However, as the eighteenth century wore on, Poor Law medical provision in London increasingly took the form of the workhouse infirmary. It is not too much to say that patients resented, and indeed resisted, this option. Studies on workhouses and the evidence presented above have shown that paupers were reluctant to enter the workhouse, where foul patients incurred a double stigma. Not only did they carry an embarrassing sexual disease, but they had to enter a dreaded institution and frequently don the parish's pauper uniform.

Far more women than men had to rely on this medical option. In four separate parishes, women represented three-quarters of all foul patients. Moreover, St. Thomas's admission registers show that women in the royal hospitals were far more likely than men to have been sent there by their parish. More than one out of every five women in the foul wards came from a parish, while less than one in twenty men did. Gendered economic realities determined that workhouses, institutions for the very poor, were primarily female institutions. This point is particularly significant when we consider the poor conditions and disciplinary aspects of workhouses. These institutions, consciously designed to be distasteful, provided the context in which most poor women with the pox found their care.

Several factors distinguish the experiences of those patients forced by need to utilize workhouse infirmaries. Settlement enquiries provide a good example. Venereal patients were more likely than most other applicants to undergo background checks to determine their eligibility for relief. In some cases, parishes seeking to limit expenditure succeeded in passing the fiscal burden of poxed paupers onto other parishes. Some venereal patients might find themselves bounced from one parish to another before they could finally obtain an infirmary bed. Settlement enquiries could also further affect the issue of medical discretion. There is no evidence that churchwardens kept applicants' conditions a secret while soliciting details from their family, friends, and employers.

Workhouses were also unique institutions because patients might utilize them on a variety of occasions throughout their life, not merely when they were sick. Because parishes administered all kinds of welfare, the very poor went to the parish for many reasons. In some cases a pauper's history of relief could color how the parish responded to his application. For example, parish officials already knew the sexual histories of women who had mothered bastards, and this could hurt their chances when they applied for care while afflicted with a sexual infection. And the reverse was just as true: a stay in the workhouse foul ward might remain in the parochial memory and affect one's chances for future relief.

Despite this scrutiny, most foul applicants were relieved. This was so because throughout the eighteenth century, custom and to an extent the law obliged parishes to relieve those with a legal settlement. The poor were entitled to relief regardless of the nature of their condition. Thus it is not clear that workhouses could have rejected venereal patients out of hand, as did voluntary charities. It was by employing this entitlement that the London poor helped to medicalize workhouses. By simply showing up and demanding assistance when they fell ill, London's poor, especially its women, pressured parish vestries to recast their new institutions, transforming large portions of their new workhouses into infirmaries, something workhouse planners show no signs of having intended.

Despite this important system of local medical care, a significant segment of the population still could not access these resources, as Elizabeth and John Byon sadly discovered. The officers in St. Ann's refused to relieve them, claiming that they were another parish's burden, and allowing them to wallow on the workhouse doorstep for two winter nights before they gave up and moved on. John and Elizabeth tried to make their way back to St. Giles where they could employ their claim to settlement, hoping to trade in on one of the last resources at their disposal. But they did not make it. In theory everyone in England could rely on their parish of settlement. In practice the issue of settlement was muddy and contested, and from a pauper's perspective far from certain. While parishes had some obligation to relieve the sick poor, the vagaries of settlement often gave them the leeway they needed to exclude undesirable applicants, as the Byons learned. Moreover, migration put incredible strains on the system. As thousands of migrants arrived in London each year, some failed to establish a new settlement and thus found themselves at a great distance from the parish in which they could claim medical relief in times of crisis. In a sense, these people existed as a segment of the underclass even more destitute than those encountered in workhouse infirmaries. It may seem odd to consider the people lying in beds in eighteenth-century workhouse foul wards "winners," but such things are relative. Someone lying in St. Ann's workhouse in February 1734 had succeeded where Elizabeth and John Byon now failed. A viable parish settlement was a crucial resource for plebeian folk, and yet some Londoners lacked even this.

The voluntary Lock Hospital emerged because each year hundreds of people like Elizabeth and John still fell through the cracks in the health care system that had evolved under the Old Poor Law. It is not entirely clear that the founders initially recognized that unsettled migrants represented the biggest problem for the existing system. The early publications of the charity do not mention the issue. But soon after they opened their doors they began to stress that they reserved their beds for those without a local parish settlement. London parishes continued to try to get their patients into the Lock, where they might receive free care, but the Lock always denied such requests. The Lock emerged neither from new tolerance nor from the drive to reform. It was a product of demographics. London was booming, and its welfare network could not handle the growth. The Lock Hospital emerged to plug one of the rather large gaps in that system.

A sick pauper who wished to enter the Lock first had to secure a governor's nomination. This represented yet another exchange into which poor foul patients had to enter with their social betters. The class divide and the corresponding disparity in power in these exchanges could hardly be greater, the supplicants extremely poor, the governors quite well-off. It is probably safe to assume that paupers resented having to go through this process, especially women. We must assume that some amount of role-playing took place, and that many patients carefully crafted their stories about how they became infected, struggling to appear worthy of charity.

Since their settlement was unclear, perhaps John and Elizabeth could have qualified for a bed in the Lock. John was clearly practiced at crafting stories about the couple's health, giving Magdalen Jones two separate explanations for Elizabeth's sickness, as he tried to tell her what he thought she wanted to hear. Who can doubt that had he the chance to catch the ear of one of the governors of the Lock Hospital he would have similarly tried to play the role of the worthy object of charity with whatever creativity and guile he could muster. But catching the ear of one of the well-heeled men who ran the Lock required networking resources, what we would call connections. And, in point of fact, they did not stand a chance; Elizabeth died more than a decade before the Lock opened its doors. Their story illustrates the significant need for the Lock's services that still remained in the eighteenth century, despite the significant VD provision already mounted at other institutions.

Taking these various hospitals together has allowed us to get a glimpse of early modern London's multilevel healthcare system. In many ways, a study of almost any disease would have given a similar picture. The ill in London—whether infected with the pox or otherwise—faced a variety of options in the seventeenth and eighteenth centuries, starting with private care and trickling down through the royal hospitals, parish infirmaries, and voluntary hospitals. Of course this was hardly a "system" in the modern sense; rather it was a conglomerate of individual institutions that arose independently. There was no overriding strategy governing London's hospital development. Even the workhouses, which did emerge as a result of an identifiable workhouse movement, only cropped up in London one at a time, parish by parish. So the health care system, such as it was, was an amalgam of individual institutions, each with its own history, each developed to perform a particular service. But, I have argued, it is crucial to study these institutions in relation to one another, because they interacted almost every day. Moreover, as they developed, they defined themselves in relation to one another, and grew, almost like plants, to fill particular niches. Thus integrated, or perhaps entangled, they came to form a system, imperfect and ad hoc as it was. The Lock Hospital, which from the start cast itself against existing Poor Law provision, illustrates this point. We should not be surprised to find that other voluntary hospitals similarly defined themselves in relation to the existing Poor Law institutions.

The experiences of all these options differed dramatically according to both gender and class. The experiences of those who could afford private care could hardly differ more from the experiences of the poor who had to rely on institutional care. Poverty forced the latter to solicit care according to the ground rules set by those who established and administered early modern hospitals. Patients of means confronted a range of treatments, while the hospital poor increasingly faced debilitating salivation as their only therapeutic choice. Importantly, the circumstances in which treatment took place and the processes by which people procured treatment also differed dramatically. Patients of means usually received care within the privacy of their own home. By contrast the poor had to forego their freedom of movement and privacy to enter a hospital. We can be sure that those who entered hospitals were frequently driven there by the most dire symptoms and by cold fiscal need. People who clearly did not want to go public about their condition, only did so when they had no other choice. The poor were not without agency, and they tried in a variety of ways to negotiate the system with their dignity intact. But this was difficult to do.

Furthermore, evidence also shows that the experiences of women were unique at every turn. The fact that administrators, governors, churchwardens, doctors, and medical students were uniformly male magnified the power dynamics involved as infected women sought treatment. Such women were reluctant to expose themselves before the male medical gaze, but they had few other options. The fact that the marketplace provided female practitioners for wealthy women, but that poor women could not access this service in hospitals, serves to demonstrate how the dynamics of gender and class could converge. The presence of the Lock Asylum demonstrates the point in another way. The Lock's case for the reform of poor women hinged on the notion that poverty would force them to turn to prostitution. Its presence underscores how anxiety about female sexuality and anxiety about poverty conjoined in the eighteenth century.

Moreover, I have argued that the Lock Asylum is crucial to the larger history of hospital venereology. Attempts by hospitals to impose moral reformation on venereal patients were fairly limited and generally sporadic in hospital foul wards before 1780. St. Thomas's may have made more of an effort than other institutions to bind its moral mission to its medical one, resulting in public flogging in the seventeenth century, and a ritualized blending of purging and proselytizing in the early eighteenth century. But even here the evidence is spotty rather than consistent, and is somewhat anomalous when set against the longer history of the foul wards in the period up to 1780, when attempts at discipline were unimpressive. At Bart's outhouses and at the Lock in its first decades we are struck by mundane normality, where the discipline in the foul wards barely differed at all from what one found elsewhere in other hospitals in the period. Patients came in, received treatment, and left. Little attempt was made to change them during their brief stay. By contrast, the merger of the VD hospital with the penitentiary, first as an experiment in Hanway's Misericordia and then permanently by the

Evangelicals at the Lock, would be a portent of things to come, and marks the beginning of a new era in the history of venereal disease.

That development was also a product of shifting notions on gender. Feminist scholarship has established that the later eighteenth century witnessed the development of gender ideals characterized, at least in part, by the increasing emphasis on female domesticity and the call for the greater confinement of women to the household. Challenges to that doctrine have shown that that the ideal was hardly achieved, as women and men were not so easily confined to these separate spheres. But the force of that *ideal* was still clearly felt in the late eighteenth century. The Lock Asylum for Penitent Women, which sought to retool wayward women with the moral probity and domestic skills necessary to take their places as wives and mothers, was merely one attempt to set that ideal into action. The connection is beyond doubt, for the Evangelicals who founded the asylum were the very same men that Hall and Davidoff demonstrated were among the primary champions of the new domesticity.[2] The fact that it is only at the very end of the eighteenth century that we witness the first gender specific reform effort mounted in a VD hospital, after more than two centuries, suggests strongly that institutions were beginning to represent these shifting cultural sands.

There is no doubt that responses to the pox were already intense in the seventeenth century. The disease was already shameful. The eighteenth century may have witnessed some intensification of the reaction to the disease, thanks perhaps to larger changes in politeness and the effects of the civilizing process. But a more dramatic shift in hospital venereology had to await the 1780s. The Lock Asylum stands out not only for its much clearer campaign to employ a medical institution for producing moral citizens, but for the stark gendered focus of that campaign. There is nothing about the quite pronounced disciplinary efforts that we witness at St. Thomas's to indicate that they targeted either sex. St. Thomas's regimen was intense, and at points downright cruel. But it was gender neutral.

By the last decades of the eighteenth century, the women who entered the Lock Hospital had ventured too far out into the new public sphere in a dangerous way. Indeed, many were suspected of being "public" women, a noteworthy term for prostitutes. The Lock Asylum promised to reclaim them and prepare them for their place within

England's households. Thus a major sea change had transformed the Lock Charity from its humble beginnings. As the century drew to a close the charity bore a new commitment to social engineering, manifesting emerging gender ideals in brick and mortar. Situating itself at the front lines of social policing, the Lock Asylum now wedded medicine and social discipline in a bold new way. As we look forward to the nineteenth century and beyond we begin to see "Lock" hospitals and asylums all over Britain exercising social policing roles to a degree not witnessed in foul wards during the previous two centuries. But perhaps a bit more evidence that the Evangelicals' vision of the new domesticity was merely one vision among many can be found in plebeian women's reaction to their new endeavor. Many of the same women who had endured institutional life in the Lock Hospital, now rejected the mission of the Lock Asylum as they made their way out its windows and over its walls.

Institutions were changing. Gender was changing. Medicine was changing, too. The birth of the clinic, the rise of a new corporeal paradigm that saw bodies as uniform and rendered individual patient-narratives obsolete, forged new medical relationships that first emerged in hospitals. As patients lost their voices and increasingly submitted their bodies to the medical gaze in silence, they lost the ability to craft their own illnesses in their own terms. This was disempowering for all patients, but it may have been particularly costly for venereal patients. We have seen how the ability to negotiate the diagnosis was a crucial part of the survival repertoire employed by venereal patients facing hospitalization. The opportunities to contest their own diagnosis or to craft a story about their mode of infection now evaporated, as doctors gave them less chance to speak, preferring to mumble in Latin to their students rather than listen to patients' interpretations of their bodily signs. The loss of voice that Mary Fissell documented so well[3] may have stung more acutely in the foul wards than anywhere else. That said, there may well have been foul patients who embraced the new silent role imposed on them, preferring not to engage in the traditional medical exchange, which would have forced them to discuss their infection, something patients over two centuries show signs of resisting intensely.

This alternative reading of the effects of the birth of the clinic on foul patients also forces us to question whether hospital medicine was becoming less personal for VD patients in this period. The literature of the rise of clinical medicine suggests that medicine became less individualistic, less personal, and more anonymous. The evidence to support this view is strong. Yet there may have been important exceptions to this trend. If we look at the gaze of the clinician, medicine indeed looks increasingly impersonal as the nineteenth century dawns. But as we have seen, there were more than just doctors in early modern hospitals. Just as medicine was becoming more anonymous in its clinical gaze, hospital administrators were conversely looking more deeply at each individual woman. Consider the Lock Asylum or the Misericordia. Here we do not see care becoming less personal, but rather more intensely personal. The purveyors of these hospitals took a probing interest in the lives and the souls of each individual inmate. Just as patients may have faced fewer interviews from their doctors, the women in the asylum were facing increasing daily interviews from the chaplain, who questioned them about their fallen lifestyle and implored them to change their ways. But unlike the rise of clinical pathology, which affected all patients, this development targeted only women. If there is any evidence that venereal patients may well have preferred greater anonymity, it may well be found in inmates' common rejection of the new scrutiny practiced in the Lock Asylum.

Patients who entered the foul wards as the nineteenth century dawned entered vastly different institutions than their predecessors of two centuries earlier. They now entered hospitals produced by visions of starker gender separation in which new medical relationships rendered them voiceless. Patients must surely have continued to try to exert control over their own illness-experiences within these new institutional realities. But this may have been more difficult than ever.

The sheer ubiquity of venereal disease in early modern London underscores the importance of these findings. Venereal disease seems absolutely endemic in early modern London. By the late eighteenth century one would find almost 30 percent of all patients in St. Thomas's Hospital in its foul wards. No other single diagnosis accounted for more patients in early modern hospitals, save perhaps the ubiquitous and poorly defined blanket diagnosis "fever." The issues complicating early modern diagnostics and the dangers of retrospective diagnosis prohibit the scholar from identifying any particular case as certainly "syphilitic." But such methodological caveats hardly go so far as to deny the existence of syphilis in this period. It cannot be pinned down perfectly, but it is there, shadowy, lurking within all of this evidence. We can never know what portion of "foul" patients would today be diagnosed as "syphilitics." But it hardly matters. There is no denying that the foul disease was omnipresent in early modern London or that foul patients represented a major portion of the early modern ill. Regardless of the criteria for rendering the diagnosis, whether patients diagnosed themselves or were labeled by others, many termed "foul" were clearly very sick people suffering from dreadful physical symptoms. We must be careful not to let our methodological sensitivity prevent us from allowing that the bacteria *treponema pallidum* may have had a lot to do with it.

What can be said with certainty is that contemporaries evaluating the situation in seventeenth- and eighteenth-century London would have come to the conclusion that the foul disease was absolutely everywhere. In an era when infected people might resume an active sex life because they assumed (probably incorrectly) that they were cured, and in a period when condoms were rarely used, the pox was a fact of life that touched a much larger portion of the population than we previously imagined, rich and poor, men and women. Early modern society had to respond to such an endemic disease. It could not simply write off those infected with smug indignation, leaving them to fend for themselves because they deserve what they got. To the contrary, the dilemma of VD among the urban poor proved too large to ignore, and society devoted an extraordinary amount of resources to the problem. Analyzing how English society responded to that dilemma reveals a great deal about prevailing values. For instance, that paupers with the pox were considered worthy of benevolence reveals the continuing strength of Christian concepts of charity and strong notions of civic obligation throughout more than two centuries.

Unfortunately, the wide range of institutional responses to the challenge, while impressive, still failed to solve the problem. We cannot doubt that the combined efforts of St. Bartholomew's, St. Thomas's, Guy's, the Lock, the Misericordia, the London Hospital, and more than seventy parish workhouses throughout the city helped to ease the burdens of many hundreds of sick people every year. But stories of failure, like that of Elizabeth Byon, still remain. The stories of people defeated by the twin specters of poverty and the pox, who failed to access care and who, in some cases, chose to give up the fight rather than carry on, remind us that not even these impressive efforts were enough. And despite the significant evidence of charity, the continued segregation and in some cases rejection of the dreaded "foul" simultaneously indicates the revulsion and fear that forever accompanied this disease. In the end, the story of poverty and the pox in early modern London is a story rife with contradictions. It is marked by genuine mercy and stark prejudice, by real compassion and grim struggle.

NOTES

Archival materials in the notes are cited by the following abbreviations. The full references can be found in the Bibliography.

BL	British Library
CLA	Camden Local Studies and Archives Center
CLRO	Corporation of London Records Office
GL	Guildhall Library
LMA	London Metropolitan Archives
OBSP	Old Bailey Session Papers
RCP	Royal College of Physicians
RCS	Royal College of Surgeons
RLH	Royal London Hospital Archives
SBH	St. Bartholomew's Hospital Archives
WCA	Westminster City Archives Centre
WALM	Westminster Abbey Library and Muniments Room.

Introduction

1. Anon., *Some Reasons of a Member of the Committee, &c. of the Trustees of the Infirmary in James Street Westminster, near St. James Park, for his dividing against the Admission of Venereal Patients. In a Letter to a Lady* (London, 1738), 1–3.

2. I use the term "venereal disease" rather than the more recent "sexually transmitted disease" simply because the former was current throughout the period under study. I should add that the acronym "VD" is used in this book merely as a shorthand and is not meant to imply that the acronym was current in the seventeenth or eighteenth century.

3. See Johannes Fabricius, *Syphilis in Shakespeare's England* (London: Kingsley, 1994); the contributions to Linda E. Merians, ed. *The Secret Malady: Venereal Diseases in Eighteenth-Century Britain and France* (Lexington: University of Kentucky Press, 1996); Winfried Schleiner, "Infection and Cure Through Women: Renaissance Constructions of Syphilis," *Journal of Medieval and Renaissance Studies* 24 (1994): 499–517, and "Moral Attitudes towards Syphilis and its Prevention in the Renaissance," *Bulletin of the History of Medicine* 68 (1994): 389–410; Raymond Anselment, "Seventeenth-Century Pox: The Medical and Literary Realities of Venereal Disease," *Seventeenth-Century* 4 (2) (1989): 189–211; Marie E. McAllister, "Stories of the Origin of Syphilis in Eighteenth-Century England: Science, Myth, and Prejudice," *Eighteenth-Century Life* 24 (2000), 22–44; Bruce T. Boherer, "Early Modern Syphilis," *Journal of the History of Sexuality* 1 (1990): 197–214; Owsei Temkin, "On the History of 'Morality and Syphilis'" and "Therapeutic Trends and the Treatment of Syphilis before 1900" in his *The Double Face of Janus and Other Essays in the History of Medicine* (Baltimore: Johns Hopkins University Press, 1977), 472–4 and 518–24.

4. Robert Jütte, "Syphilis and Confinement: Hospitals in Early Modern Germany," in *Institutions of Confinement: Hospitals, Asylums, and Prisons in Western Europe and North America, 1500–1950*, eds. Norbert Finzsch and Robert Jütte (Cambridge: Cambridge University Press, 1996), 97–116 (see esp. 100–1). A major contribution on Italian hospitals is Jon Arrizabalga, John Henderson, and Roger French, *The Great Pox: The French Disease in Renaissance Europe* (New Haven: Yale University Press, 1997).

5. See David Innes Williams, *The London Lock: A Charitable Hospital for Venereal Disease, 1746–1952* (London: Royal Society of Medicine Press, 1996), 8–10. W. F. Bynum, "Treating the Wages of Sin: Venereal Disease and Specialism in Eighteenth-Century Britain," in *Medical Fringe and Medical Orthodoxy, 1750–1850*, eds. W. F. Bynum and Roy Porter (London: Croom Helm, 1987), 7; T. J. Wyke, "Hospital Facilities for, and Diagnosis and Treatment of Venereal Disease in England, 1800–1870," *British Journal of Venereal Disease* 49 (1973), 78–85; M. W. Adler, "History of the Development of a Service for the Venereal Diseases," *Journal of the Royal Society of Medicine* 75 (1982): 124–8.

6. While in no ways whiggish, Peter Lewis Allen has also described the period as extremely intolerant, claiming that "hospitals refused to admit syphilis patients." This is because "[the] disease was a moral problem so severe that it barred even charity itself. . . ." *The Wages of Sin: Sex and Disease, Past and Present* (Chicago: University of Chicago Press, 2000), 42–3 and 57–8.

7. Norman Moore, *The History of St. Bartholomew's Hospital* (London: C. A. Pearson, 1918); F. G. Parsons, *The History of St. Thomas' Hospital* (London: Metheun, 1932).

8. James Bettley, "Post Voluptatem Misericordia: The Rise and Fall of the London Lock Hospitals," *London Journal* 10 (1984): 167–75. A notable exception is Margaret Pelling, "Appearance and Reality: Barber Surgeons, the Body and

Disease," in *London, 1500–1700: The Making of the Metropolis*, eds. A. L. Beir and Roger Finlay (London: Longman, 1986), 95–105. Fabricius draws on Pelling in his brief discussion of hospitals, *Syphilis in Shakespeare's England*, 71–2.

9. Phillip K. Wilson, one of the few medical historians to comment on early hospital care at St. Thomas's asserted its novelty, stating, "unlike St. Bartholomew's, St. Thomas's treated venereal cases." See " 'Sacred Sanctuaries for the Sick': Surgery at St. Thomas's Hospital 1725–26," *London Journal* 17 (1992): 51, n.14.

10. Roy Porter, "The Patient's View: Doing Medical History from Below," *Theory and Society* 14 (1985): 175–98; Dorothy Porter and Roy Porter, *Patient's Progress: Doctors and Doctoring in Eighteenth-Century England* (Oxford: Polity Press, 1989), 33–114; Roy Porter, ed., *Patients and Practitioners: Lay Perceptions of Medicine in Pre-Industrial Society* (Cambridge: Cambridge University Press, 1985); Barbara Duden, *The Woman Beneath the Skin: A Doctor's Patients in Eighteenth-Century Germany*, trans. Thomas Dunlap (Cambridge, Mass.: Harvard University Press, 1991).

11. Mary Fissell's *Patients, Power and the Poor in Eighteenth-Century Bristol* (Cambridge: Cambridge University Press, 1991) is a notable example of how to capture the experiences of plebeian patients in the period (esp. 148–70).

12. Randolph Trumbach proposes that "Venereal disease in the sixteenth and seventeenth centuries had probably been contained in specific social groups like gentlemen and soldiers and sailors. In the course of the eighteenth century, however, the gentlemen's disease became the common disease of the London poor. . . ." See *Sex and the Gender Revolution*, vol. 1, *Heterosexuality and the Third Gender in Enlightenment London* (Chicago: University of Chicago Press, 1998), 196–9.

13. Edward Shorter similarly suggested that VD "remained at the margins of the population" before 1850, only slowly increasing among the general population at the end of the eighteenth century, and not spreading widely until the twentieth. Shorter did not peg aristocrats as the primary carriers, however, but rather the fringe groups of "soldiers, students and prostitutes." *Women's Bodies: A Social History of Women's Encounter with Health, Ill-Health, and Medicine* (New Brunswick, N.J.: Transactions Publications, 1991), 263–7. First published as *A History of Women's Bodies* (New York: Basic Books, 1982).

14. For example, see N. F. Lowe's "The Meaning of Venereal Disease in Hogarth's Graphic Art," in Merians, ed., *The Secret Malady*, 172–81.

15. A notable exception that focuses on London is Susan Lawrence, *Charitable Knowledge: Hospital Pupils and Practitioners in Eighteenth-Century London*, (Cambridge: Cambridge University Press, 1996).

16. To see the utility of this approach see the work of Sandra Cavallo, especially *Charity and Power in Early Modern Italy: Benefactors and Their Motives in Turin, 1541–1789* (Cambridge: Cambridge University Press, 1995), and "The Motivations of Benefactors: An Overview of Approaches to the Study of Charity," in *Medicine and Charity Before the Welfare State*, ed. Jonathan Barry and Colin Jones (London: Routledge, 1991), 46–62.

17. Adrian Wilson, "Conflict, Consensus and Charity: Politics and the Provincial Voluntary Hospitals in the Eighteenth Century," *English Historical Review* 111 (442) (1996): 599–619.

18. On the response to urbanization see Lee Davison, et al., eds., *Stilling the Grumbling Hive: The Response to Social and Economic Problems in England, 1689–1750* (New York: St. Martin's Press, 1992).

19. See the contributions in Tim Hitchcock, Peter King, and Pamela Sharpe, eds., *Chronicling Poverty: The Voices and Strategies of the English Poor 1640–1840* (London: Macmillan, 1997) and Jeremy Boulton, " 'It is extreme necessity that makes me do this': Some 'Survival Strategies' of Pauper Households in London's West End During the Early Eighteenth Century," *International Review of Social History* 45 (2000): 47–69.

20. James C. Scott, *Domination and the Arts of Resistance: Hidden Transcripts* (Yale University Press, 1990).

21. Ibid., xi.

22. W. F. Bynum described the Lock as "charged with a double mission of caring for disease and reforming morals," Foreword in Williams, *The London Lock*, vii. W. A. Waugh similarly claimed that "The Lock Hospital was founded on motives of reform." W. A. Waugh, "Attitudes of Hospitals in London to Venereal Disease in the Eighteenth and Nineteenth Centuries," *British Journal of Venereal Disease* 47 (1971): 147.

23. Mary Spongburg, *Feminizing Venereal Disease: The Body of the Prostitute in the Nineteenth Century* (London: MacMillan, 1997); Judith Walkowitz, *Prostitution and Victorian Society: Women, Class and the State* (Cambridge: Cambridge University Press, 1980); Linda Mahood, *The Magdalenes: Prostitution in the Nineteenth Century* (London: Routledge, 1990); Paula Bartley, *Prostitution: Prevention and Reform in England, 1860–1914* (London: Routledge, 2000). On the central role of VD policy to moral hygiene campaigns in a variety of modern contexts see the contributions to *Sex, Sin, and Suffering: Venereal Disease and European Society since 1870*, eds. Roger Davidson and Lesley Hall (London: Routledge, 2001) and Peter Baldwin, *Contagion and the State in Europe, 1830–1930* (Cambridge: Cambridge University Press, 1999), 355–522. For the Scottish case see Roger Davidson, *Dangerous Liaisons: A Social History of Venereal Disease in Twentieth-Century Scotland* (Amsterdam: Rodopi, 2000).

24. Kathryn Norberg, "From Courtesan to Prostitute: Mercenary Sex and Venereal Disease, 1730–1802," and Merians, "The London Lock Hospital and Lock Asylum," in Merians, ed., *The Secret Malady*, 34–50 and 136–8.

25. I have elsewhere explored how early modern medical discourse on the pox performed a Foucauldian social policing role. See Kevin P. Siena, "Pollution, Promiscuity and the Pox: English Venereology and the Early Modern Medical Discourse on Social and Sexual Danger," *Journal of the History of Sexuality* 8 (4) (1998): 553–74.

26. The only diagnosis that may have been more common than "venereal" was the equally poorly defined, amorphous early modern condition "fever." While

plague and smallpox were dreadful there is reason to believe that Pelling's claim for Norwich is applicable to London: "The [public] authorities were plainly more concerned about the 'French disease' than any other." "Healing the Sick Poor: Social Policy and Disability in Norwich, 1550–1640," reprinted in her *The Common Lot: Sickness, Medical Occupations and the Urban Poor in Early Modern England* (London: Longman, 1998), 95.

27. Donna Andrew, *Philanthropy and Police: London Charity in the Eighteenth Century* (Princeton: Princeton University Press, 1989), 54–73.

28. Jütte also shows that moral indignation did not prevent hospitals from extending charity to foul patients in the sixteenth century. However, he argues a step further to claim that entering a pox house carried no stigma at all. "Syphilis and Confinement," 114. Here our views diverge, or at least the English case seems very different. Elsewhere, Jütte discusses the role segregation played in early modern campaigns to mark and control deviants, usually the poor. On stigmatization and segregation see Robert Jütte, *Poverty and Deviance in Early Modern Europe* (Cambridge, Cambridge University Press, 1994), 158–77.

29. Jon Arrizabalaga, et al., *The Great Pox*, 18–9.

30. David Harley, "Rhetoric and the Social Construction of Sickness and Healing," *Social History of Medicine* 12 (3): 407–35. The point cited here appears on p. 419.

31. James Boswell, *London Journal 1762–1763*, ed. Frederick A. Pottle (London: Heinemann, 1950), 155–6. For a discussion of Boswell's illness see William B. Ober, *Boswell's Clap and Other Essays: Medical Analyses of Literary Men's Afflictions* (Carbondale, Illinois: Southern Illinois University Press, 1994.); and Roy Porter and Dorothy Porter, *Patient's Progress*, 108–10.

32. For example, in the rape trial of Julian Brown, a midwife, a surgeon, and an apothecary each testified that they inspected Susan Marshall's linen a fortnight after the attack, finding it "discoulr'd with a Running," and each testified that she was clapped. OBSP, October 15, 1735, 161–2.

33. Joseph Cam, for example, cited nine different types of "gonorrhoea," ranging in seriousness and virulence, including several non-sexually transmitted conditions. *A Practical Treatise: or Second Thoughts on the Consequences of the Venereal Disease in Three Parts. I. On the Simple Gonorrhoea, Gleets, and other Weaknesses whether from Venereal Embraces, Self-Pollution, improperly called Onanism, or Natural Imbecillity. II. On the Virulent Gonorrhoea, or Clap. III. on the Venereal Lues* . . . 3rd. ed. (London, 1729), 12–35.

34. RCS, "John Pearson Casebook (1798)" Lock MS 'Mr. Pearson.'

35. Linda Merians, "Introduction," in Merians, ed., *The Secret Malady*, 8.

36. Phillip K. Wilson, *Surgery, Skin and Syphilis: Daniel Turner's London (1667–1741)* (Amsterdam: Rodopi, 1999), 152–5.

37. Useful discussions of early modern symptomology can be found in Susan P. Connor, "The Pox in Eighteenth-Century France," in Merians, ed., *The Secret Malady*, 16–20; Arrizabalaga, et al., *The Great Pox*, 25–7.

38. Peter Charles, *Observations on the Venereal Disease, with the true way of curing the same* (London, 1693), 23; Wilson, *Surgery, Skin and Syphilis*.

39. Wilson, *Surgery, Skin and Syphilis*, 153.

40. Anon., *The Tomb of Venus: or, A Plain and Certain Method, by which All People That ever labour'd under any Venereal Distemper may infallibly know whether they are cured or not . . . by a Foreign Physician* (London, 1710), 66–70; Henry Wastell, *Observations on the Efficacy of a New Mercurial Preparation for the Cure of the Venereal Disease, in its most malignant state* (London, 1779), 60–1; Jesse Foot, *A complete treatise of the origin, theory, and cure of the Lues Venerea* (London, 1792), 28.

41. OBSP, May 14, 1766, 209–10.

42. On medical perceptions of skin and dermatological disorders with attention to the pox see Wilson, *Surgery, Skin and Syphilis*, 59–84.

43. For example, Claude Quétel, *The History of Syphilis*, trans. Judith Braddock and Brian Pike (Baltimore: Johns Hopkins University Press, 1990), 32, Arrizabalga, et al. *The Great Pox*, 53, 89; Fabricius, xv, 4, 11, and 12; Theodore Roseburg, *Microbes and Morals: The Strange History of V.D.* (New York: Viking Press, 1971), 133–9; W. A. Pusey, *The History and Epidemiology of Syphilis* (Springfield, Ill.: Thomas, 1933), 6–10; R. S. Morton, "Syphilis in Art: An Entertainment in Four Parts," *Genitourinary Medicine* 66 (3) 1990: 208; Sander Gilman, *Disease and Representation: Images of Illness from Madness to AIDS* (Ithaca: Cornell University Press, 1988), 246–54.

44. On the pox in Hogarth see N. F. Lowe, "The Meaning of Venereal Disease in Hogarth's Graphic Art," in Merians, ed., *The Secret Malady*, 168–82 and Peter Wagner "The Satire on Doctors in Hogarth's Graphic Works," in *Literature and Medicine During the Eighteenth Century*, eds. Marie Mulvey Roberts and Roy Porter (London: Routledge, 1993), 206–10.

45. RCS, "'Watercolour drawing of London Lock hospital patients' conditions' by E. Sewell and J. Holt, (1849–1851)." The collection contains 101 separate sheets with roughly 300 images. With permission of the Royal College of Surgeons. I consider these images useful despite the fact that they come some fifty years after the period under consideration. The lack of any significant development in treatment in that time means that patients in the Lock Hospital in earlier decades very likely resembled these depictions.

46. On the cultural constructions of the syphilitic nose see Sander Gilman, "The Phantom of the Opera's Nose," in his *Health and Illness: Images of Difference* (London: Reaktion Books, 1995), 67–92.

47. *The Diary of Samuel Pepys*, ed. Robert Latham and William Matthews (Berkeley: University of California Press, 1971), 5:242 and 4:12.

48. Wilson provides a useful description of various approaches to mercury treatment in *Surgery, Skin, and Syphilis*, 161–6.

49. Quétel, *History of Syphilis*, 58–63, 83–6.

50. As quoted in Temkin, "Therapeutic Trends and the Treatment of Syphilis," 521–2.

51. Jean Astruc outlines the care of dozens of specific patients in his *A Treatise of Venereal Diseases, in Nine Books*, trans. by William Barrowby (London, 1754), 183–92. On average these regimens lasted over six weeks.

52. For example, see Nicholas Robinson's description of patients in his *A Treatise of the Venereal Disease* (London, 1736), 256–7. For descriptions of mercury's effects see Quétel, *History of Syphilis*, 29–31.

53. Herman Boerhaave, *A Treatise on the Venereal Disease and its Cure in all its Stages and Circumstances* (London, 1729), 74.

54. Robinson, *A Treatise of the Venereal Disease*, 302–3.

55. OBSP, January 15–19, 1718, 1–2.

56. Robinson, *A Treatise of the Venereal Disease*, 259.

57. Anon., *The Tomb of Venus*, 66–70.

58. OBSP, March 1, 1721, 7.

59. WALM, Westminster Coroner's Inquests, 1760–1799; William Urin, December 27, 1771.

60. Other witnesses at Robertson's trial testified that Robertson "was weary of being in the world": OBSP, May 2, 1753, 150.

61. Ibid., Richard Miller, March 16, 1770.

62. Ibid., John Jackson, March 31, 1771.

63. Ibid., Mary Elson, April 25, 1780.

64. Ibid., James Sharp, May 28, 1789.

65. Michael MacDonald and Terrance R. Murphy, *Sleepless Souls: Suicide in Early Modern England* (Oxford: Clarendon press, 1990). See chap. 8, "Motives for Suicide," esp. 268–89.

66. MacDonald and Murphy make note of illness, though not in their discussion of motives for suicide. Rather they show how contemporaries believed that illness might impair one's thinking, render them out of their senses, and result in a suicide: *Sleepless Souls*, 233–8. Specifying such illnesses in inquests, especially fevers that brought delusions, often helped juries to declare the deceased a "lunatick" (*non compos mentis*), and avoid rendering the felony verdict *felo de se*. However, in many cases discussions of physical illnesses do not seem like mere devices allowing juries to circumvent issuing felony verdicts. On the impact of that verdict see Macdonald and Murphy, *Sleepless Souls*, 15–41, and Michael MacDonald, "The Secularization of Suicide in England, 1660–1800," *Past and Present* 111 (1986): 50–100, as well as the subsequent debate, between Macdonald and Donna Andrew, "Debate: The Secularization of Suicide in England 1660–1800," *Past and Present* 119 (1988): 158–70.

67. WALM, Westminster Coroner's Inquests 1760–99, Mary Simpson, Jan. 14, 1765; Mary Harris, April 4, 1765; Lydia Palmer, May 1, 1765; Thomas Cheeseman, June 15, 1765; Charels, Duke of Bolton, July 6, 1765; Rebecca Carpenter, August 13, 1765; Elizabeth Hornell, Nov. 27, 1765; Swain Jenkins, Dec. 21, 1765.

68. Boswell, *London Journal*, 155–205. Boswell pronounced himself infected on Jan. 20, 1763 and cured on Feb. 27, 1763. However, his health may not have

been as rosy as he thought. Stone is right to suspect that his seventeen seemingly separate infections may have simply been one infection flaring up periodically over a very long time. Lawrence Stone, *The Family, Sex and Marriage in England, 1500–1800* (London: Weidenfeld and Nicolson, 1977), 575.

Chapter 1

1. Some of the arguments in this chapter appeared in my article, "The 'Foul Disease' and Privacy: The Effects of Venereal Disease and Patient Demand on the Medical Marketplace in early modern London," *Bulletin of the History of Medicine* 75 (2) (2001): 199–224.

2. Owsei Temkin, "On the History of 'Morality and Syphilis'," in *The Double Face of Janus and Other Essays in the History of Medicine* (Baltimore: Johns Hopkins University Press, 1977), 477–9; Linda Merians, "Introduction," and Rose Zimbardo, "Satiric Representation of Venereal Disease: The Restoration Versus the Eighteenth-Century Model," in Linda E. Merians, ed., *The Secret Malady: Venereal Diseases in Eighteenth-Century Britain and France* (Lexington: University of Kentucky Press, 1996), 3–4, 183–94. By contrast, see Raymond A. Anselment, "The Seventeenth-Century Pox: The Medical and Literary Realities of Venereal Disease," *Seventeenth Century* 4 (2) (1989): 189–211; Claude Quétel, *The History of Syphilis*, trans. Judith Braddock and Brian Pike (Baltimore: Johns Hopkins University Press, 1990), 71–5, 94–5. Winfried Schleiner demonstrates the powerful moralizing that informed English medical constructions of the pox. "Moral Attitudes toward Syphilis and its Prevention in the Renaissance," *Bulletin of the History of Medicine* 68 (1994): 389–410.

3. John Wynell, *Lues Venerea, or, A Perfect Cure of the French Pox: Wherein The Names, Nature, Subject, Causes and Signes of the Disease are handled. Mistakes in these discovered, Doubts and Questions succinctly Resolved* (London, 1660), Preface (unnumbered).

4. B.P., *Pilulae Antipudendagriae: or Venus's Revenge. Whereby every one may Secretly Cure and Preserve themselves from all Venereal Evils* (London, 1669), "Epistle" (unnumbered); Charles Peter, *A description of the Venereal Disease: Declaring the Causes, Signs, Effects, And Cure thereof. With a Discourse of the most Wonderful Antivenereal Pill* (London, 1678), 8.

5. John Archer, *Every Man His Own Doctor* (London, 1671), 116.

6. Schleiner makes the case that English medical responses, when compared to those in Italy, demonstrate a much greater propensity to assume a moral high ground, linking the pox to promiscuity, whoring, and debauchery. This manifested itself in reluctance on the part of English doctors to advocate preventative measures then circulating in Italy, for fear that such advice was tantamount to promoting vice. "Moral Attitudes toward Syphilis and its Prevention," 389–97 and 404–10.

7. *The Diary of Samuel Pepys*, ed. Robert Latham and William Matthews (Berkeley: University of California Press, 1971), 5:84 (March 14, 1664).

8. Ibid., 86 (March 15, 1664).

9. This discussion of Pepys draws on Anselment, "The Seventeenth-Century Pox," 202.

10. Pepys, *Diary*, 5:86 (March 15, 1664).

11. Raymond Anselment, *The Realms of Apollo: Literature and Healing in Seventeenth-Century England* (Newark: University of Delaware Press, 1995), 134–5; Quétel, *History of Syphilis*, 94–5; Margaret Pelling, "Illness among the Poor in Early Modern English Towns," in her *The Common Lot: Sickness, Medical Occupations and the Urban Poor in Early Modern England* (London: Longman, 1998), 66–7; John Wynell noted the phenomenon as well in *Lues Venerea, or, A Perfect Cure of the French Pox* (London, 1660), unnumbered Preface.

12. Pepys, *Diary*, 8:423. For other instances when Pepys noted the infections, or rumors of infections, of others see 5:275, 6:12 and 60, 8:79, 114, 217, 334, and 337, and 9:413. Lawrence Stone notes how Pepys strove both to inform himself of other's gossip and conceal the details of his own sexual adventures, in *Family, Sex and Marriage, 1500–1800* (London: Weidenfeld and Nicolson, 1977), 560.

13. Anselment, *Realms of Apollo*, 134–5.

14. See Zimbardo, "Satiric Representation of Venereal Disease," and Leon Guilhamet, "Pox and Malice: Some Representations of Venereal Disease in Restoration and Eighteenth-Century Satire," in Merians, ed., *The Secret Malady*, 196–212; Anselment, "Seventeenth-Century Pox," 198–201.

15. "The Phlegm Pot" author and date unknown. BL. Add. MS, 27, 825, Place papers. Vol. 37: Manners and Morals, vol. I, A. Grossness. Books &c 1705–1795. B. Grossness Songs. C. Drunkenness. 1736–1826. D. Poor. Beggars. Lotteries. 1706–1826. Fol. 163. Thanks to Tim Hitchcock for alerting me to this text.

16. Joseph Cam, *A Practical Treatise: or Second Thoughts on the Consequences of the Venereal Disease* (London, 1729), 100.

17. Solomon Sawrey, *A Popular View of the Effects of the Venereal Disease upon the Constitution: Collected from the Best Writers* (Edinburgh, 1793), 8.

18. John Burrows, *A Dissertation on the Nature and Effects of a New Vegetable Remedy, Known by the Name of Velno's Vegetable Syrup, An Acknowledged Specific in all Venereal and Scorbutic Cases . . .* (London: 1772), 39.

19. Joseph Plenck, *A new and easy method of giving mercury, to those affected with the Venereal Disease*, trans. William Saunders, 2nd. ed. (London, 1768), 53–4.

20. Sawrey, *A Popular View of the Effects of the Venereal Disease*, 9–10.

21. Phillip K. Wilson, *Surgery, Skin and Syphilis: Daniel Turner's London (1667–1741)* (Amsterdam: Rodopi, 1999), 61.

22. Sara F. Matthews Grieco, "The Body, Appearance and Sexuality," in *A History of Women in the West*, vol. 3, *Renaissance and Enlightenment Paradoxes*, ed. Natalie Zemon Davis and Arlette Farge (Cambridge, Mass.: Belknap Press, 1993), 46–64; Roy Porter, Bodies *Politic: Disease, Death and Doctors in Britain, 1650–1900* (London: Reaktion Books, 2001), 35–88, and 96; Margaret Pelling, "Appearance and

Reality: Barber-Surgeons, the Body and Disease," in A. L. Beier and Roger Finlay, eds., *London 1500–1700: The Making of the Metropolis* (London: Longman, 1986), 89–102.

23. Possible comparisons might be drawn with madness, which families often tried to conceal. See Roy Porter, *Mind Forg'd Manacles: A History of Madness in England from the Restoration to the Regency* (Cambridge, Mass.: Harvard University Press, 1987), 137–8. Plague victims also concealed infection. Giulia Calvi has linked these actions on some levels to issues of shame, although the desire to avoid state-enforced quarantine serves to differentiate these cases in important ways: *Histories of a Plague Year: The Social and the Imaginary in Baroque Florence*, trans. Dario Biocca and Bryant T. Ragan (Berkeley: University of California Press, 1989), 104–17. Some medical discussions linked impotence to moral decay: see Judith C. Mueller, "Fallen Men: Representations of Male Impotence in Britain," *Studies in Eighteenth-Century Culture* 28 (1999): 85–102.

24. For recent articulations on the debate see Dario Castiglione and Lesley Sharpe, eds., *Shifting the Boundaries: Transformation of the Languages of Public and Private in the Eighteenth Century* (Exeter: University of Exeter Press, 1995). Especially useful for background is John Brewer's essay in that volume, "This, that and the other: Public, Social and Private in the Seventeenth and Eighteenth Centuries," 1–21. See also Dena Goodman, "Public Sphere and Private Life: Toward a Synthesis of Current Historiographical Approaches to the Old Régime," *History and Theory* 31 (1992): 1–20.

25. Jürgen Habermas, *The Structural Transformation of the Public Sphere: An Inquiry into a Category of Bourgeois Society*, trans. from the German by Thomas Burger (Cambridge, Mass.: MIT Press, 1991).

26. Phillipe Ariès, "Introduction" to *Histoire de la vie privée*, vol. 3, *De la Renaissance aux Lumiéres*, ed. Roger Cartier (Paris, 1986), trans. by Arthur Goldhammer as *A History of Private Life*, vol. 3, *Passions of the Renaissance* (Cambridge, Mass.: Belknap Press, 1989), 1–11.

27. Norbert Elias, *The Civilizing Process: The Development of Manners: Changes in the Code of Conduct and Feeling in Early Modern Times*, trans. from the German by Edmund Jephcott (New York: Urizen Books, 1978).

28. For a rich look at the issue through analysis of bodily functions and humoral medicine in drama see Gail Kern Paster, *The Body Embarrassed: Drama and the Disciplines of Shame in Early Modern England* (Ithaca: Cornell University Press, 1993).

29. Ariès "Introduction," 5. Medical historians have borne this out. See Roy Porter, "A Touch of Danger: The Man-Midwife as Sexual Predator," in *Sexual Underworlds of the Enlightenment*, eds. George Rousseau and Roy Porter (Manchester: Manchester University Press, 1987), 206–12 and "The Rise of the Physical Exam" in *Medicine and the Five Senses*, eds. Roy Porter and W. F. Bynum (Cambridge: Cambridge University Press, 1993), 191–4.

30. Elias, *Civilizing Process*, 72. That Elias quotes Erasmus here indicates that this was a feature of reactions to the pox from a very early date.

31. Mary Douglas, *Purity and Danger: An Analysis of the Concepts of Pollution and Taboo* (London; Routledge and Kegan Paul, 1966).

32. Richard Wunderli, *London Courts on the Eve of the Reformation* (Cambridge, Mass.: The Medieval Academy of America, 1981), 81; J. A. Sharpe, "Defamation and Sexual Slander in Early Modern England: The Church Courts at York," *Bothwick Papers* 58 (1980): 1–36; Martin Ingram, *Church Courts, Sex and Marriage in England*, (Cambridge: Cambridge University Press, 1987), 301–13. Most recently, the debate has been advanced by Laura Gowing, *Domestic Dangers: Women, Words, and Sex in Early Modern London* (Oxford University Press, 1998) and Robert Shoemaker, "The Decline of Public Insult in London 1660–1800," *Past and Present* 169 (2000): 97–131. On VD and slander see Johannes Fabricius, *Syphilis in Shakespeare's England* (London: Kingsley, 1994), 268–9.

33. OBSP, December 8, 1731, 22 and, December 10, 1735, 38.

34. See Anna Foa, "The Old and the New: The Spread of Syphilis (1494–1530)," in *Sex and Gender in Historical Perspective*, eds. Guido Ruggiero and Edward Muir (Baltimore: Johns Hopkins University Press, 1990), 26–45.

35. For a taste of the debate see Tim Meldrum, "Domestic Service, Privacy and the Eighteenth-Century Metropolitan Household," *Urban History* 26 (1) (1999): 27–39.

36. Linda A. Pollock, "Living on the Stage of the World: The Concept of Privacy Among the Elite of Early Modern England," in *Rethinking Social History: English Society 1570–1920 and its Interpretation*, ed. Adrian Wilson (Manchester: Manchester University Press, 1993), 83, 85–91.

37. Pepys's reaction here mirrors the sad story of Charles and Frances Williams told by Mary Margaret Stewart. "'And blights with plagues the Marriage hearse': Syphilis and Wives," in Merians, ed., *The Secret Malady*, 103–13.

38. Although as Vincent Brome points out, Samuel secretly wondered whether he had infected her thanks to his infidelity. *The Other Pepys* (London: Weidenfeld and Nicolson, 1992), 78.

39. Pepys, *Diary*, 5:385 (November 17, 1663).

40. One study that brings discussions further down the social scale is Lawrence E. Klein, "Politeness for Plebes: Consumption and Social Identity in Early Eighteenth-Century England," in *The Consumption of Culture 1660–1800: Image, Object, Text*, eds. Ann Bermingham and John Brewer (London: Routledge, 1995), 362–82.

41. Randolph Trumbach, *Sex and the Gender Revolution*, vol. 1, *Heterosexuality and the Third Gender in Enlightenment London* (Chicago: University of Chicago Press, 1998), 208–9.

42. OBSP, September 7, 1743, 237–8.

43. OBSP, May 22, 1735, 88.

44. On the experience and effects of sexual slander in eighteenth-century London see Tim Meldrum, "A Women's Court in London: Defamation at the Bishop of London's Consistory Court, 1700–1745," *London Journal* 1994 19(1): 1–20.

45. OBSP, October 16, 1723, 2.

46. SBH, HA/1/14, 90 On January 22, 1772, St. Bartholomew's board ruled: "it now appearing that her Disorder was the Foul Disease It is ordered that the said Elizabeth Starling be for the said Imposition Discharged as well as a Patient as a Nurse of this Hospital."

47. Anna Clark, *The Struggle for the Breeches: Gender and the Making of the British Working Class* (Berkeley, University of California Press, 1995), 42–62. John Gillis also suggests a cleavage between elite and plebeian sexuality in "Married but Not Churched: Plebeian Sexual relations and Marital Nonconformity in Eighteenth-Century Britain," in *Tis' Nature's Fault: Unauthorized Sexuality During the Enlightenment*, ed. Robert Maccubbin (Cambridge: Cambridge University Press, 1987), 31–42.

48. Trumbach, *Sex and the Gender Revolution*, chap. 9, "Shame and Rape," 276–301. See also Laura Gowing, "Secret Births and Infanticide in Seventeenth-Century England," *Past and Present* 156 (1997): 87–115.

49. Trumbach, *Sex and the Gender Revolution*, 289.

50. Laura Gowing, "Ordering the Body: Illegitimacy and Female Authority in Seventeenth-Century England," in *Negotiating Power in Early Modern Society: Order, Hierarchy and Subordination in Britain and Ireland*, eds. Michael J. Braddick and John Walter (Cambridge: Cambridge University Press, 2001), 58.

51. Roy Porter, *Health For Sale: Quackery in England 1660–1850* (Manchester: Manchester University Press, 1989), 153.

52. John Profily, *An Easy and Exact Method of Curing the Venereal Disease*, 2nd ed. (London, 1748),144–6.

53. Boues de Sigogne, *A New Method of Curing the Venereal Disease Much safer and easier than any hitherto used with a Confutation of the Old hypotheses concerning that Distemper* (London, 1724), v. John Clubbe reiterated that "It is often fear of Pain and Danger, or the Shame of a suspicious Retirement and the enormous prices Surgeons exact now a days to treat Patients at their Houses, which makes them continue in [their] miserable condition" and not seek help: *An Essay on the Virulent Gonorrhoea* (London, 1786), 62.

54. Nicholas Jewson, "Medical Knowledge and the Patronage System in Eighteenth-Century England," *Sociology* 8 (1974): 369–85. Some examples of how the effects of commercial forces on the medical scene have been more fully explored include Harold Cook, *The Decline of the Old Medical Regime in late Stuart London* (Ithaca: Cornell University Press, 1986) and *Trials of an Ordinary Doctor: Johannes Groenvelt in Seventeenth-Century London* (Baltimore: Johns Hopkins University Press, 1994); Mary Fissell's *Patients, Power and the Poor in Eighteenth-Century Bristol* (Cambridge: Cambridge University Press, 1991), 37–73, and Porter, *Health For Sale*, op. cit.

55. Margaret Pelling, "Medicine and the Environment in Shakespeare's England," in her *The Common Lot: Sickness, Medical Occupations and the Urban Poor in Early Modern England* (London: Longman, 1998), 28.

56. Stone, *The Family, Sex and Marriage*, 600; Roy Porter, "Laymen, Doctors and Medical Knowledge in the Eighteenth Century: The Evidence of the *Gentlemen's*

Magazine," in *Patients and Practitioners*, ed. Roy Porter, 283–314; Dorothy Porter and Roy Porter, *Patient's Progress: Doctors and Doctoring in Eighteenth-Century England* (Oxford: Polity Press, 1989), 108.

57. BL C.112.f.9; BL 551.a.32; and BL Harley 5931. These collections comprise 162, 195 and 155 medical advertisements respectively. I have omitted from these calculations a handful of nonmedical handbills in these collections. For work on early modern medical advertising in England, in addition to relevant sections of Porter, *Health for Sale*, see Patricia Crawford, "Printed Advertisements for Women Medical Practitioners in London 1670–170," *Society for the Social History of Medicine Bulletin* 35 (1984): 66–70; Francis Doherty, *A Study in Eighteenth-Century Advertising Methods: The Anodyne Necklace* (Lewiston, N.Y.: Edwin Mellen Press, 1992); and Elizabeth Lane Furdell, *Publishing and Medicine in Early Modern England* (Rochester, N.Y.: University of Rochester Press, 2002), 135–54.

58. Pierre Desault, *A Treatise on the Venereal Distemper*, trans. from the French by John Andree (London, 1738), 52.

59. Sigogne, *A New Method of Curing the Venereal Disease*, 36–42.

60. Wilson, *Surgery, Skin and Syphilis*, 157.

61. Owsei Temkin, "Therapeutic Trends and the Treatment of Syphilis before 1900," in his *The Double Face of Janus and Other Essays in the History of Medicine* (Baltimore: Johns Hopkins University Press, 1977), 522.

62. Francois Chicoyneau, *The Practice of Salivating Shewn to be of no Use or Efficacy in the Cure of Venereal Disease*, trans. from the French by C. Willoughby (London, 1723), 6; Wilson, *Surgery, Skin and Syphilis*, 174.

63. Nicholas Robinson, *A Treatise of the Venereal Disease* (London, 1736), 287; Franz Swediaur also noted the phenomenon of patients traveling for a cure, though he notes Montpellier's climate, which facilitated recovery, as its key to success: *Practical Observations on Venereal Complaints* (London, 1786), 151.

64. Craig Muldrew, *The Economy of Obligation: The Culture of Credit and Social Relations in Early Modern England* (New York: St. Martin's Press, 1998). See especially chap. 6, "The Cultural Currency of Credit and the Construction of Reputation," 148–72.

65. Sigogne, *A New Method of Curing the Venereal Disease*, v.

66. Desault, *A Treatise on the Venereal Distemper*, 52.

67. For an examination of Turner's anti-quack publishing battles see Philip K. Wilson, "Exposing the Secret Disease: Recognizing and Treating Syphilis in Daniel Turner's London," in Merians, ed., *The Secret Malady*, 68–84. A fuller study of Turner's career is available in Wilson's *Surgery, Skin and Syphilis*, op. cit.

68. Daniel Turner, *Syphilis, A Practical Dissertation on the Venereal Disease . . . With some Remarks, by way of Supplement, on Dr. Willoughby's Translation of Monsieur Chicoyneau's Method of Cure* (London, 1724), 374–6.

69. Joseph Cam, *The Practice of Salivating Vindicated: In answer to Dr. Willoughby's Translation of Mons. Chicoyneau's Pamphlet against mercurial salivations* (London, 1724), 12.

70. Ibid., 19.

71. Turner, *Syphilis*, 211, 255–6. As quoted in Wilson, *Surgery, Skin and Syphilis*, 159.

72. For another missive in the fight see Richard Brown, *A Letter from a Physician in London to his Friend in the Country; Giving an Account of the Montpellier Practice in curing the Venereal Disease* (London 1730).

73. Wilson, *Surgery, Skin and Syphilis*, 174–6.

74. Temkin, "On the History of 'Morality and Syphilis'," 472–84; Quétel, *History of Syphilis*, 29–32.

75. Marie E. McAllister, "John Burrows and the Vegetable War," in Merians, ed., *The Secret Malady*, 85–102.

76. John Leake, *A Dissertation on the Properties and Efficacy of the Lisbon Diet-Drink, and its Extract in the cure of the Venereal Disease and Scurvy, Rheumatic Gout, the Scrofula, Consumption, and other disorders* (London: 1790).

77. Porter, *Health for Sale*, 155.

78. McAllister, "John Burrows and the Vegetable Wars," 89; Robert Grubb, *A New Treatise on the Venereal Disease* (London, 1780?) (includes transcription of patent prior to p. 1); M Henriet, *The Antisiphylitic: or Public Health. A sure Method of guarding against all sorts of Venereal Disorders, and of curing our selves if afflicted by using, an anti-venereal water called water of Salubrity* (London, 1772) (includes patent prior to p. 1); Leake, *A Dissertation on the Properties . . .*, unnumbered postscript.

79. Francis Hauksbee, *A Further Account of the Effects of Mr. Hauksbee's Alternative Medicine, as applied in the Cure of the Venereal Disease* (London, 1743).

80. I. F. Nicholson, *The Modern Siphylis: or, The True Method of Curing every Stage and Symptom of the Venereal Disease* (London, 1718), unnumbered preface.

81. Fully 38 of the 117 handbills that mention venereal disease in BL 551.a.32 specify private treatment.

82. BL C.112.f.9, (117); BL Harley 5931, (77).

83. BL 551.a.32, (14).

84. BL 551.a.32, (33).

85. BL 551.a.32, (14).

86. BL 551.a.32, (90).

87. Porter, *Health for Sale* (n. 3), 154.

88. BL 551.a.32, (153).

89. BL C.112.f.9, (159).

90. Wilson, *Surgery, Skin, and Syphilis*, 150.

91. For a discussion of some of the new marketing experiments see Cook, *Trials of an Ordinary Doctor* (n. 14), 136–48. See also Porter, *Health for Sale*, 155, and Porter and Porter, *Patients Progress*, 76–8. On the form of medical correspondence see Wayne Wild, "Doctor-Patient Correspondence in Eighteenth-Century Britain: A Change in Rhetoric and Relationship," *Studies in Eighteenth-Century Culture* 29 (2000): 47–64.

92. Grubb, *A New Treatise on the Venereal Disease*, 87–8.

93. Anon., *The Practical Scheme of the Secret Disease, a Gleet and a Broken Constitution, caused by Fast-Living, Former Cares, Salivations, taking of Mercury, Self Abuses, &c. Shewing those*

who want a cure of either the Venereal Distemper, or a Gleet, what will best do It . . . 17th ed. (London, 1728), 4.

94. Ibid., 5.

95. Freeman Stephan, *A New Essay on the Venereal Disease* (London, 1780), 69–70.

96. Furdell gives a useful account of retail medicine in London bookshops in *Publishing and Medicine in Early Modern England*, 131–4. See also her article on bookshops, "'At the King's Arms in the Poultrey': The Bookshop Emporium of Dorman Newman, 1670–1694," *London Journal* 23 (2) (1998): 1–19.

97. T. C., *The Charitable Surgeon: or The Best Remedies for the Worst Maladies, reveal'd. Being a New and True way of Curing (without Mercury) the several degrees of the Venereal Disease in both Sexes* 2nd ed. (London, 1709), x.

98. BL 551.a.32, (33).

99. BL 551.a.32, (156).

100. See for example Raymond Villey, *Histoire du Secret Médical* (Paris: Seghers, 1986), which itself demonstrates that medical confidentiality became widely assumed only during the nineteenth and twentieth centuries, despite Villey's attempt to argue for a continuation of medical confidentiality from antiquity to modernity.

101. Robert Baker, "The History of Medical Ethics," in *Companion Encyclopaedia of the History of Medicine*, eds. W. F. Bynum and Roy Porter (London: Routledge, 1993), 852–87.

102. Mary E. Fissell, "Innocent and Honorable Bribes: Medical Manners in Eighteenth-Century Britain," in *The Codification of Medical Morality*, eds. Robert Baker, Dorothy Porter, and Roy Porter (Dordrecht: Kluwer Academic Publishers, 1993), 19–45.

103. Laurence McCullough has published widely on medical ethics and the career of John Gregory. This quotation comes from "John Gregory (1724–1773) and the Invention of Professional Relationships in Medicine," *Journal of Clinical Ethics* 8 (1997): 11. See also his *John Gregory and the Invention of Professional Ethics and the Profession of Medicine* (Dordecht: Kluwer Academic Publishers, 1998).

104. Porter, *Health for Sale*, 154.

105. Andrew Wear, "Medical Ethics in Early Modern England," in *Doctors and Ethics: The Earlier Historical Setting of Professional Ethics*, eds. Andrew Wear, Johanna Geyer-Kordesch, Roger K. French (Amsterdam: Rodopi, 1993), 98–130.

106. Winfried Schleiner, *Medical Ethics in the Renaissance* (Washington, D.C.: Georgetown University Press, 1995), 25. Schleiner has also shown that the central element of ethical debate for doctors who confronted the pox in the sixteenth century concerned not patient privacy, but whether discussing ways to prevent the disease may have promoted vice: "Moral Attitudes towards Syphilis and Its Prevention in the Renaissance."

107. Margaret Pelling calls the confidential consultation a "comparatively late development." Elsewhere she says that "private transactions may not have been

a major part of medicine in the early modern period" despite assumptions, and that "[t]he stress on privacy and confidentiality can be seen as an aspect of later phases of social differentiation." See "Occupational Diversity: Barber-Surgeons and Other Trades, 1550–1640," 223–4, and "Trade or Profession? Medical Practice in Early Modern England," 232, 244–6, both reprinted in her *The Common Lot: Sickness Medical Occupations and the Urban Poor in Early Modern England* (London: Longman, 1998).

108. Pepys, *Diary*, 9:413.

109. Porter, *Health for Sale*, 151.

110. BL C.112.f.9, (45).

111. BL Harley 5931, (147). Porter, *Health for Sale*, 154; Fissell notes this claim in Bristol advertisements as well: *Patients Power and the Poor*, 45.

112. BL 551.a.32, (28).

113. Jill Harsin has shown that this dynamic continued well into the late-nineteenth century. "Syphilis, Wives, and Physicians: Medical Ethics and the Family in Late Nineteenth-Century France," *French Historical Studies* 16 (1) (1989): 72–95.

114. John Marten, *A Treatise of all the Degrees and Symptoms of the Venereal Disease*, 6th ed. (London, 1708), 18–9.

115. On Marten's self promotion, including his use of titillating details in case studies, see Peter Wagner, "The Discourse on Sex—or Sex as Discourse: Eighteenth-Century Medical and Paramedical Erotica," in Rousseau and Porter, eds., *Sexual Underworlds of the Enlightenment*, 46–68, and Roy Porter, "'Laying aside any Private Advantage': John Marten and Venereal Disease," in Merians, ed., *The Secret Disease*, 51–67.

116. Henry Wastell, *Observations on the Efficacy of a New Mercurial Preparation for the Cure of the Venereal Disease* (London, 1779), 55–6.

117. Anonymous, *The Tomb of Venus: or, A Plain and Certain Method, by which All People That ever labour'd under any Venereal Distemper may infallibly know whether they are cured or not* (London, 1710), 59–61.

118. Joseph Plenck, *A new and easy method of giving mercury to those affected with the Venereal Disease*, trans. from the Latin by William Saunders (London, 1768), 148–9, 158. Saunders augments Plenck's treatise with his own commentary and case studies, 121–58.

119. Schleiner, *Medical Ethics in the Renaissance*, chap. 2, "Mentiamur sane: Lying for Health in Renaissance Medical Ethics," 5–48.

120. Consider as well the case detailed by Charles Peter of a woman who claimed to be infected with the pox despite his diagnosis to the contrary. Peter discussed the situation behind the woman's back with her husband and agreed to appease the woman by proclaiming her infected and administering worthless purges. By this treatment Charles mused that she was "cured of her imaginary pox": *Observations on the Venereal Disease, with the true way of curing the same* (London, 1693), 20–22.

121. Bernard Capp, "The Double Standard Revisited: Plebeian Women and Male Sexual Reputation in Early Modern England," *Past and Present* 162 (1999): 70–100, and Tim Hitchcock. "Sociability and Misogyny in the Life of John

Cannon, 1684–1743," in *English Masculinities, 1660–1800*, eds. Tim Hitchcok and Michele Cohen (London: Longman, 1999), 32.

122. Stewart, "And blights with plagues the Marriage hearse," 104.

123. On the tension inherent in the relationship between male doctors and female patients see Porter, "A Touch of Danger," and "The Rise of the Physical Exam."

124. Profily, *An Easy and Exact Method of Curing the Venereal Disease*, 144–5; Jesse Foot discusses the sensitivity of having to request permission to examine his female patients: *A complete treatise of the origin, theory, and cure of the Lues Venerea* (London, 1792), 242.

125. Sawrey, *A Popular View of the Effects of the Venereal Disease*, 15–18.

126. Alice Clark discussed women and medicine in her pioneering book on early modern working women: *Working Life of Women in the Seventeenth Century* (London: Routledge, 1919, reprinted 1982). More recently, see the work of Doreen Evenden, *Popular Medicine in Seventeenth-Century England* (Bowling Green, Ohio: Bowling Green State University Press, 1988), 54–78 and "Gender Differences in the Licensing and Practice of Male and Female Surgeons in Early Modern England," *Medical History* 42 (1998): 194–216; A. L. Wyman, "The Surgeoness: The Female Practitioner of Surgery 1400–1800," *Medical History* 28 (1984): 22–41. Margaret Pelling's work has also been crucial: Margaret Pelling and Charles Webster, "Medical Practitioners," in *Health, Medicine and Mortality in the Sixteenth Century*, ed. Charles Webster (Cambridge: Cambridge University Press, 1979), 165–236 and "Thoroughly Resented? Older Women and the Medical Role in Early Modern London," in *Women, Science and Medicine 1500–1700*, eds. Lynette Hunter and Sarah Hutton (Glouchestershire: Sutton Publishers, 1997), 63–88. It goes without saying that the extensive research on midwifery is also crucial here. See Doreen Evenden, *The Midwives of Seventeenth-Century London* (Cambridge: Cambridge University Press, 2000).

127. BL 551.a.32, (90).

128. BL 551.a.32, (199).

129. One study that looks at medicine, gender, and the family in the period is Margaret Pelling, "The Women of the Family? Speculations Around Early Modern British Physicians," *Social History of Medicine* 8 (3) (1995): 383–401.

130. On apprenticeship in medicine see Joan Lane, "The Role of Apprenticeship in Eighteenth-Century Medical Education in England," in *William Hunter and the Eighteenth-Century Medical World*, eds. W. F. Bynum and Roy Porter (Cambridge: Cambridge University Press, 1985), 57–104; and Margaret Pelling, "Medical Practice in Early Modern England: Trade or Profession?," in *The Professions in Early Modern England*, ed. Wilfrid Prest (London: Croom Helm, 1987), 90–128.

131. Laverenst proclaimed, "My Parents before me were so far skill'd in the Art and Knowledge of Physick. . . . From them I receiv'd the great Knowledge and Experience that I now profess," BL C.112.f.9, (26); BL 551.a.32, (31), (71). Soeburgh thanked her parents, "from whom I have receiv'd my greatest Experience

and Knowledge, with which I have cured many thousands of Women," BL C.112.f.9, (138).

132. BL C.112.f.9, (61); BL 551.a.32, (49), BL Harley 5931, (122).
133. BL Harley 5931, (68).
134. BL C.112.f.9, (78).
135. BL C.112.f.9, (93).
136. BL C.112.f.9, (128); BL 551.a.32, (51), 56; BL Harley 5931, (117), (125).
137. BL C. 112.f.9, (140).
138. BL Harley 5931, (76).
139. OBSP, May 28 1756, 93.
140. BL C.112.f.9, (28), (29).
141. BL 551.a.32, (49).
142. BL 551.a.32, (29).
143. Elizabeth Mears, called on to inspect and treat a child infected with the pox through a rape, and testifying in the subsequent trial, described herself thus: "I am a widow to a surgeon and apothecary." OBSP, January 13, 1749, 27–9.
144. BL 551.a.32, (49).
145. See for example BL C.112.f.9, (21) or BL 551.a.32, (201).
146. BL Harley 5931, (111).
147. BL 551.a.32, (187).
148. Porter, *Health for Sale*, 142; Fissell, *Patients, Power and the Poor*, 45.
149. Roy Porter, "Laymen, Doctors and Medical Knowledge in the Eighteenth Century: The Evidence of the *Gentlemen's Magazine*," Porter, ed., in *Patients and Practitioners*, 283–314; William F. Bynum, "Treating the Wages of Sin: Venereal Disease and Specialism in Eighteenth-Century Britain," in *Medical Fringe and Medical Orthodoxy, 1750–1850*, eds. W. F. Bynum and Roy Porter (London: Croom Helm, 1987), 11.
150. OBSP, September 11, 1754, 284; September 3, 1746, 225–6; December 4, 1717, 6; October 25, 1758, 323; October 11, 1732, 246–7.
151. Peter Clare, *A Practical Treatise on the Gonorrhoea* (London, 1784), iv. Clare lamented that only wealthy patients enjoyed the benefits of most alternative treatments, poignantly asking, "if one class of men are cured in his manner, why should not all enjoy the same benefit ?" (10).
152. John Andree, *Observations on the Theory and Cure of the Venereal Disease* (London, 1779), v. As quoted in Bynum, "Treating the Wages of Sin," 16.
153. Wilson, *Surgery, Skin, and Syphilis*, 175.

Chapter 2

1. Charles Peter, *Observations on the Venereal Disease, with the true way of curing the same* (London, 1693), 35–7.

2. The other royal hospitals, Bethlem and Christ's, specialized in treating the mad and children, respectively.

3. Anonymous, *The Tomb of Venus: or, A Plain and Certain Method, by which All People That ever labour'd under any Venereal Distemper may infallibly know whether they are cured or not* (London, 1710), 7–9.

4. W. Wall, *A New System of the French Disease. With An easy method of curing it, unknown to the Ancients or Moderns, with all its Common and Remote Symptoms, Obvious to the meanest Capacities* (London, 1696), unnumbered Preface.

5. LMA, HT/ST/A4/1, f. 7 "Yt is ordered that Eme Danyall being Diseased with the Pox she ys admyted." This was not an anomaly, for the governors also accepted an apprentice, Randall Brytton, "diseased with the poxe" the following year. As quoted in Parsons, Frederick Gymer. *The History of St. Thomas's Hospital* (London: Methuen, 1932), vol. 1:191–2.

6. SBH, HA 1/1, ff. 23–29. This was first recognised by Margaret Pelling in "Appearance and Reality: Barber-Surgeons, the Body and Disease," in A. L. Beier and Roger Finlay, eds., *London 1500–1700: The Making of the Metropolis* (London: Longman, 1986), 97. Johannes Fabricius draws on Pelling in his discussion of hospitals, *Syphilis in Shakespeare's England* (London: Kingsley Publishers, 1994), 71–2.

7. William Clowes, *A Short and profitable Treatise touching the cure of the disease called Morbus Gallicus by Unctions* (London, 1579), f. 3. On Clowes's career see Elizabeth Lane Furdell, *The Royal Doctors 1485–1714: Medical Personnel at the Tudor and Stuart Courts* (Rochester: University of Rochester Press, 2001), 87–9; Fabricius, *Syphilis in Shakespeare's England*, 34–7, 98–9, 106–12, and 191–5; Bruce T. Boehrer, "Early Modern Syphilis," *Journal of the History of Sexuality* 1 (1990): 11–7; Winifred Schleiner, "Moral Attitudes Towards Syphilis and its Prevention in the Renaissance," *Bulletin of the History of Medicine* 68 (1994): 394–7.

8. SBH, HA 1/1, ff. 23–9.

9. Jon Arrizabalaga, John Henderson, and Roger French, *The Great Pox: The French Disease in Renaissance Europe* (New Haven: Yale University Press, 1997), chaps. 7 and 8. Several southwestern German towns also had institutions called the *Blatternhaus* ("pox house"). One of the largest of these was run by the Fuggers in Augsberg: Guenter Risse in *Mending Bodies, Saving Souls: A History of Hospitals* (Oxford: Oxford University Press, 1999), 215. For a list of German pox houses see Robert Jütte, "Syphilis and Confinement: Hospitals in Early Modern Germany," in *Institutions of Confinement: Hospitals, Asylums, and Prisons in Western Europe and North America, 1500–1950*, eds. Norbert Finzsch and Robert Jütte (Cambridge: Cambridge University Press, 1996), 102.

10. See Claude Quétel, *The History of Syphilis*, trans. Judith Braddock and Brian Pike (Baltimore: Johns Hopkins University Press, 1990), 9–32.

11. On Germany see Jütte, "Syphilis and Confinement." On Norwich see Margaret Pelling, "Healing the Sick Poor: Social Policy and Disability in Norwich, 1550–1640." *Medical History* 29 (1985): 115–37.

12. Marjorie B. Honeybourne, "The Leper Hospitals of the London Area," *Transactions of the London and Middlesex Archaeological Society* 21 (1) (1963): 9.

13. It is not explained in Norman Moore, *The History of St. Bartholomew's Hospital* (London: C.A. Pearson, 1918), 267–8, Honeybourne, "The Leper Hospitals of the London Area," or Nicholas Orme and Margaret Webster, *The English Hospital, 1070–1570* (New Haven: Yale University Press, 1995).

14. Honeybourne, "Leper Hospitals of the London Area," 9.

15. Of the ten London leper hospitals that still operated in 1500 only five were still open by 1600, all of which were run by St. Bartholomew's. For example, the lazar house in St. Gile's-in-the-Fields closed in 1539, and Aldersgate Hospital closed in 1548. See Honeybourne's discussion of individual leper houses, "Leper Hospitals of the London Area," 13–61.

16. Quétel, *History of Syphilis*, 33. Arrizabalaga, et al., *The Great Pox*, 14, 52–4, and Nancy Siraisi, *The Clock and the Mirror: Girolamo Cardano and Renaissance Medicine* (Princeton; Princeton University Press, 1997), 136.

17. Steven Blanckaert, *A New Method of Curing the French-Pox. Written by an Eminent French Author. Together with the Practice an Method of Monsieur Blanchard. As also Dr. Sydenham's Judgement on the Same. To which is added Annotations and Observations by William Salmon* (London, 1690), 131–43.

18. These were originally published separately between 1718 and 1720. They are reprinted in William Beckett, *A Collection of Chirurgical Tracts . . . written and collected by William Beckett* (London, 1740), 91–119. They were still being discussed in the 1790s. Jesse Foot, *A complete treatise of the origin, theory, and cure of the Lues Venerea* (London, 1792), 10–48.

19. Saul N. Brody, *The Disease of the Soul: Leprosy in Medieval Literature* (Ithaca, N.Y.: Cornell University Press, 1974), 12–13.

20. Kevin P. Siena, "Pollution, Promiscuity, and the Pox: English Venereology and the Early Modern Discourse on Social and Sexual Danger," *Journal of the History of Sexuality* 8 (1998): 566–74. The most detailed collection of early origin theories is Jean Astruc, *A Treatise on the Venereal Disease* (London, 1754).

21. As quoted in Brody, *The Disease of the Soul*, 56–57.

22. Peter Lowe, for example, believed that the pox, if untreated "doth degenerate into leprosie": *An Easie, Certaine, and Perfect Method, to cure and prevent the Spanish Sicknes* (London, 1596), chap. 4.

23. Clowes, *A Short and profitable Treatise*, n.5. Siena, "Pollution, Promiscuity, and the Pox," 553–7; Fabricius, *Syphilis in Shakespeare's England*, 96–103 and 118–9.

24. Anna Foa, "The Old and the New: The Spread of Syphilis (1494–1530)," in *Sex and Gender in Historical Perspective*, eds. Guido Ruggiero and Edward Muir (Baltimore: Johns Hopkins University Press, 1990), 37–42.

25. Orme and Webster, *The English Hospital*, 41–8. For a map of London including the sites of medieval leper hospitals see Honeybourne, "Leper Hospitals of the London Area," 6; Margaret Pelling notes this term applied to surgeons in Norwich lazar houses as well in "Healing the Sick Poor," 91.

26. Pelling, "Appearance and Reality: Barber-Surgeons, the Body and Disease," 97; Moore, *History of St. Bartholomew's Hospital*, vol. 2: 280; W. Dunsmuir, "The First Wave of Syphilis: Bart's in the 16th Century," *St. Bartholomew's Hospital Journal* (Spring, 1995): 11.

27. SBH, HA 1/4, f. 122, f. 126,

28. As cited in James Paget, *The Records of Harvey in Extracts from the Journals of the Royal Hospital of St. Bartholomew's* (London, 1846), 16.

29. See Paget, *The Records of Harvey*, 16–7, n.1; Honeybourne, "Leper Hospitals of the London Area," 10; and Pelling, "Healing the Sick Poor," 91.

30. Arrizabalaga, et al., *The Great Pox*, 27–8, 145–230. It is also not likely that Bart's sent many patients to the outhouses who were beyond cure (i.e., *incurable*). Harvey's own rules indicate that the bulk of hospital resources should not be committed to those who could not be helped, a belief that constituted standard operating procedure in early modern hospitals. Just prior to his rule on outhouses Harvey ordered "that none be taken into the hospitall but such as be curable or but a c[er]taine number of such as are incurable." As quoted in Paget, *The Records of Harvey*, 16; Pelling, "Healing the Sick Poor," 93. On the general hospital policy of excluding those beyond cure see Risse, *Mending Bodies, Saving Souls*, 232.

31. SBH, HA 1/7, f. 199.

32. Plague victims were housed in the outhouses at several points in the sixteenth century as well as during the outbreak of 1665. Moore, *History of St. Bartholomew's Hospital*, 2:325. On the politics of segregation in London see Paul Slack, *The Impact of Plague in Tudor and Stuart England* (London: Routledge, 1985), 199–226.

33. Jessica A. Browner, "Wrong Side of the River: London's Disreputable South Bank in the Sixteenth and Seventeenth Century," *Essays in History* 36 (1994): 34–71.

34. For example, after 1667 a rather large gap appears in minutes of the Grand Committee, one of the main administrative bodies responsible for many day-to-day decisions. The minutes record virtually nothing up to 1689 (LMA, HT/ST/A6/2). Moreover, the ensuing volume of Grand Committee Minutes (1691–1701, LMA, HT/ST/A6/3) has been damaged and was classified "unfit for consultation" at the time of research.

35. LMA, H1/ST/A4/1, f. 33 (27 Jan 1560). "Yt is ordered that the sisters of the Sweate Ward shall not suffer the poore to go a brode in the streets before such tyme as they be perfectly hole and cured of their disease and yf they go out without License before they be hole they shall not be Received in to the house again."

36. LMA, HT/ST/A4/1. "For a poor woman put to the Locke: Yt is agreed that William Royse of the Locke shall receive in to the spitle there a poore woman whose name is Elizabeth, and the said Royse to have towards the Releafe of the said woman for one moneth VIs VIIId . . ." f. 5.

37. LMA, H1/ST/A4/5, f. 44; Parsons, *History of St. Thomas's Hospital*, 2:46.

38. Ibid.

39. These figures are given in pounds–shillings–pence. For calculations, all such figures were first converted to pence. Hence, this year Bart's paid 256,407 pence to feed patients in the main hospital, and 80,146 pence—23.8 percent of its total diet payments for the year—to feed patients in the outhouses.

40. SBH, HA 1/7, f. 96.

41. Were the twenty-patient limit adhered to by both outhouses the maximum diet payment for any year would be £243.6.8. (This assumes the hospital paid 4 pence per day for 40 patients over 365 days.) The actual payments exceed this figure every year throughout the first 45 years after 1622, more than doubling this figure in many years.

42. Ben Coates notes that wounded soldiers put a strain on the Royal hospitals, and that the hospitals complained to the Court of Alderman about the burden in 1644. "Poor Relief in London during the English Revolution Revisited," *London Journal* 25 (2) (2000): 50.

43. Parsons, *History of St. Thomas's Hospital*, 2:51.

44. LMA, H1/ST/B2/1. This admissions register covers April 1679–Dec. 1683.

45. Parsons, *History of St. Thomas's Hospital*, 2:100–1; Moore, *History of St. Bartholomew's Hospital*, 2:328–9.

46. SBH, HA 1/6, f. 26; Moore, *History of St. Bartholomew's Hospital*, 2:330.

47. LMA, H1/ST/A4/5, f. 160.

48. Ibid.

49. SBH, HA 1/6, f. 40.

50. SBH, HA 1/6, f. 40.

51. SBH, HA 1/6, f. 76. On February 6 the governors ordered "that with all such patients as the said two Guides [surgeons] or either of them shall agree with for their Victualls, Cures or Dyett Drink, the patients so agreeing shall pay the same themselves according to theire agreement with the said guides and this house for such shall pay nothing."

52. SBH, HA 1/7, f. 78 and f. 123.

53. For example, in 1673 soldiers and sailors injured in the war with the Dutch displaced venereal patients: LMA, H1/ST/A4/5, f. 168.

54. LMA, H1/ST/A4/6, f. 35.

55. Sir D'Arcy Power, *A Short History of St. Bartholomew's Hospital, 1123–1923* (London: Whittingham and Griigs, 1923), 83.

56. SBH, HA 1/8, f. 125.

57. SBH, HA 1/8, f. 127.

58. Ibid.

59. Peter Lewis Allen, *The Wages of Sin: Sex and Disease, Past and Present* (Chicago: University of Chicago Press, 2000), 57–8.

60. SBH, HA 1/5, f. 72.

61. SBH, HA 1/5, f. 137.

62. SBH, HA 1/8, f. 19.

63. LMA, H1/ST/A6/1, 15.
64. LMA, H1/ST/A4/5, f. 74; Parsons, *History of St. Thomas's Hospital*, 2:42.
65. SBH, HA 1/7, f. 199.
66. This runs counter to Jütte's claim that hospitalizsation in pox houses "involved neither social penalties nor stigmatisation," and may indicate that the seventeenth-century English experience differed from the sixteenth-century German one: "Syphilis and Confinement," 114.
67. This well-established aspect of the early modern medical exchange has been described by Nicholas Jewson, "Medical knowledge and the patronage system in eighteenth-century England," *Sociology* 8 (1974): 369–85, and Dorothy Porter and Roy Porter, *Patient's Progress: Doctors and Doctoring in Eighteenth-Century England* (Oxford: Polity Press, 1989), 78–9. On lay self-diagnosis see Andrew Wear, *Knowledge and Practice in English Medicine 1550–1680* (Cambridge: Cambridge University Press, 2000), 105–16.
68. LMA, H1/ST/A2, 81; H1/ST/A4/5, f. 160. Parsons, *History of St. Thomas's Hospital*, 2:104.
69. SBH, HA 1/7, f. 147.
70. SBH, HA 1/8, f. 19.
71. H1/ST/A6/1, 21; Parsons, *History of St. Thomas's Hospital*, 2:105.
72. LMA, H1/ST/A6/1, f. 161. Although the order specifies foul patients at the beginning, it switches near the end to claim that "no other p.sons coming from or inhabiting in or neere the City of London or Burrough of Southwark shalbe received into this hospital for cure of anie disease whatsoever without such certificate."
73. Philip K. Wilson, *Surgery, Skin and Syphilis: Daniel Turner's London (1667–1741)* (Amsterdam: Rodopi, 1999), 156; William F. Bynum, "Treating the Wages of Sin: Veneral Disease and Specialism in Eighteenth-Century Britain," in *Medical Fringe and Medical Orthodoxy 1750–1850*, ed. W. F. Bynum and Roy Porter (London: Croon Helm, 1987), 8–11. While Bynum shows that a range of practitioners treated the disease, it is hardly surprising, considering the money to be made, that surgeons clearly outnumbered physicians when it came to treating the pox.
74. Although physicians were supposed to attend the foul wards (see LMA, H1/ST/A4/5, f. 31), a 1687 petition by the surgeons complained that they failed to do so: LMA, H1/ST/A4/6, f. 48.
75. SBH, HA 1/5, f. 30, 32. When they hired his replacement, John Kent, they warned him that he would keep the position only "soe long as hee keeps no victualling house or sold or suffer to be sold in the said house any ale, beare wyne or cakes" in the hospital. Given the traditional connections between barber-surgeons and the food trades, perhaps Topliffe's actions should not surprise. See Margaret Pelling, "Occupational Diversity: Barber–Surgeons and Other Trades, 1550–1640," in her *The Common Lot: Sickness, Medical Occupations and the Urban Poor in Early Modern England* (London: Longman, 1998), 224–6.

76. The Governors eventually had to have the house rules affixed where the patients could read them in order that they knew what the surgeons were obliged to give them. SBH, HA 1/7, ff. 114–115.

77. LMA, H1/ST/A4/6, f. 48.

78. SBH, HB 1/4, (1629).

79. Usually the ledgers indicate pay for "cures by dyett." But in 1667 both guides received pay for "Cures by Dyett Drink": SBH, HB 1/7, Annual Account, 1667.

80. This seems likely since hospital ledgers show entries for medicines for the Kingsland outhouse that specify two substances, diet drink and mercury. See the Annual Account (1653), SBH, HB 1/6; and Quétel, *History of Syphilis*, 63.

81. Peter, *Observations on the Venereal Disease*, 34.

82. LMA, H1/ST/A4/6, f. 31.

83. LMA, H1/ST/A4/5, f. 135; Parsons, *History of St. Thomas's* II, 95.

84. LMA, H1/ST/A4/5, f. 137.

85. Three periods demonstrate clear patterns: 1628–42 and 1671–82, when surgeons were paid per head, and 1645–56, when surgeons received the identical (presumably salaried) payments year after year.

86. SBH, HB 1/9 (1682).

87. LMA, H1/ST/A4/5, f. 158.

88. Ibid., f. 160.

89. Wilson, *Surgery, Skin, and Syphilis*, 175. This is not surprising given that surgeons, who performed most of the work in the foul wards, were more prone to salivation than physicians: Owsei Temkin, "Therapeutic Trends and the Treatment of Syphilis before 1900," in his *The Double Face of Janus and Other Essays in the History of Medicine* (Baltimore: Johns Hopkins University Press, 1977), 523.

90. Anonymous, *The Tomb of Venus*, 8–9.

91. That year the outhouses had an average capacity of 79.5 patients. An estimated duration of treatment of 40.5 days (which emerges from hospital records the following century and will be discussed in chapter 6) suggests that the outhouses treated approximately 715 patients that year. The 33 guaiac cures would represent less than 5 percent of all patients.

92. Owsei Temkin, "On the History of 'Morality and Syphilis'," in *The Double Face of Janus*, 475–7, and "Thereapeutic Trends and the Treatment of Syphilis," 524.

93. LMA, H1/ST/A4/6, f. 27; Parsons, *History of St. Thomas's Hospital*, 2:116.

94. Parsons, *History of St. Thomas's Hospital*, 2:104.

95. See David Innes Williams, *The London Lock: A Charitable Hospital for Venereal Disease 1746–1952* (London: Royal Society of Medicine Press, 1996), 8; James Bettley, "Post Voluptatem Misercordia: The Rise and Fall of the London Lock Hospitals," *London Journal* 10 (1984): 167. The *Oxford English Dictionary* also supports this theory of the etymology of "lock."

96. SBH, HA 1/8, f. 78. This seems like standard practice, for in 1693 Nicholas Field, the Surgeon at the Kingsland outhouse, was reprimanded "for not

advancing weekely the allowance of 4d p[er] diem allowed the patients within that hospitall to buy meate and drink. . . ."

97. In December 1652, the governors complained that many patients "mispend it [their allowance] at Alehouses to the disordering of their bodyes, and after it is spent are forced to begg to the great dishonour of this house. . . ." SBH, HA 1/5, f. 89.

98. SBH, HA 1/7, f. 140 A sister at Kingsland was dismissed because she allowed patients to "go abroad" without the Guide's knowledge.

99. SBH, HA 1/5, f. 137. The discussion of women's work here clearly indicates that the purpose of their labor was not reform, but economics. Were this a drive to instill a work ethic, we would expect to see men being made to labor as well. For a discussion of women's various labor tasks in English Hospitals see Margaret Pelling, "Nurses and Nursekeepers: Problems of Identification in the Early Modern Period," in her *The Common Lot*, 179–202.

100. SBH, HA 1/5, f. 319; HA 1/7, ff. 107 and 171.

101. SBH, HA 1/8, f. 289.

102. SBH, HA 1/7, f. 118. The governors reiterated this order in August 1682. See HA 1/7 f. 123; Moore, *History of St. Bartholomew's Hospital*, 2:339.

103. Roy Porter, "The Gift Relation: Philanthropy and Provincial Hospitals in Eighteenth-Century England," in *The Hospital in History*, ed. Roy Porter and Lindsay Granshaw (London: Routledge, 1989), 149–78.

104. *Orders and Ordinances for the better Government of the Hospitall of Bartholomew the lesse* (London, 1652): "A Thanksgiving unto Almighty God to be said by the poor that are cured in the Hospitall, at the time of their delivery from thence" (32–3).

105. See, for example, Mary Fissell, "Charity Universal? Institutions and Moral Reform in Eighteenth-Century Bristol," in Lee Davison et al., eds., *Stilling the Grumbling Hive: The Response to Social and Economic Problems in England, 1689–1750* (New York: St. Martin's Press, 1992), 121–44.

106. LMA, H1/ST/A4/1, f. 33.

107. LMA, H1/ST/A6/2, 65.

108. LMA, H1/ST/A4/6, ff. 73–75.

109. For example, in 1673 an anonymous practitioner challenged the notion that the disease was "received from any common infection in the Air." L.S. Προφυλακτικον [Profulacticon]: *Some Considerations of a Notable Expedient To root out the French Pox From the English Nation* (London, 1673), 4–5. The issue remained debated. As late as 1678 we still find commentators like Charles Peter who continued to hold that infected bed linen might spread infection. Peter, *Description of the Venereal Disease*, 5–6.

110. Quétel, *History of Syphilis*, 73–5.

111. LMA, H1/ST/A4/2, f. 77.

112. As quoted in Parsons, *History of St. Thomas' Hospital*, 1:240–1.

113. Much scholarship followed the lead set by Natalie Zemon Davis in her work on charivaris, "The Reasons of Misrule" (1971), reprinted in her *Society and*

Culture in Early Modern France (Stanford: Stanford University Press, 1975), 97–123. For an example of some of the literature that links the work on defamation to corporeal punishment, see the work on scolds: Susan Dwyer Amussen, *An Ordered Society: Gender and Class in Early Modern England* (New York: Basil Blackwell, 1988), 117–33; David Underdown, "The Taming of the Scold': The Enforcement of Patriarchal Authority in Early Modern England," in Anthony Fletcher and John Stephenson, eds., *Order and Disorder in Early Modern England* (Cambridge: Cambridge University Press, 1985), 116–36.

114. Faramerz Dabhoiwala shows that such an approach to policing sexual offences remained common in London until the end of the seventeenth century, giving way to newer policing methods only in the eighteenth century: "Sex, Social Relations and the Law in Seventeenth- and Eighteenth-Century London," in Michael J. Braddick and John Walter, eds., *Negotiating Power in Early Modern Society: Order, Hierarchy and Subordination in Britain and Ireland* (Cambridge: Cambridge University Press, 2001), 85–101.

115. Susan Dwyer Amussen, "Punishment, Discipline, and Power: The Social Meanings of Violence in Early Modern England," *Journal of British Studies* 34 (1) (1995): 11.

116. LMA, H1/ST/A6/1, 81.

117. Craig Rose, "Politics and the London Royal Hospitals, 1683–92," in Lindsay Granshaw and Roy Porter, eds., *The Hospital in History* (London: Routledge, 1989), 123–48, and "Politics, Religion and Charity in Augustan London, c.1680–c.1720" (Ph.D. diss., Cambridge University, 1988).

118. Keith Thomas, "Puritans and the English Adultery Act of 1650 Reconsidered," in Keith Thomas and Donald Pennington, eds., *Puritans and Revolutionaries: Essays in Seventeenth–Century History Presented to Christopher Hill* (Oxford: Clarendon Press, 1978), 257–82.

119. William Clowes, "A Prayer," in *The Selected Writings of William Clowes, 1544–1604* ed. F. N. L. Poynter (London: Harvey & Blythe, 1948), 168; Schleiner, "Moral Attitudes toward Syphilis and its Prevention," 394–6.

Chapter 3

1. SBH, HA 1/12, 489–91.

2. Daniel Turner, *Syphilis: A Practical Dissertation on the Venereal Disease* (London, 1724), 374–6, and *Aphrodisiacus: Containing a Summary of the Ancient Writers on the Venereal Disease* (London, 1736), xxxv. On Turner's career, see the work of Phillip K. Wilson, especially *Surgery, Skin and Syphilis: Daniel Turner's London (1667–1741)* (Amsterdam: Rodopi, 1999). On Palmer's connection with Turner see pp. 158 and 168.

3. Calculated from the duration of time patients remained in the Lock Hospital (discussed in chapter 6).

4. This estimates stems from the outhouses's annual diet payments for the period 1622–96. The annual average over these seventy-five years was £365. Assuming patients stayed 40.5 days in the hospital this sum translates to 481 patients annually.

5. The £155.5.0 needed to feed all 230 outhouse patients represents just 5.3 percent of the £2878.13.10 expended to feed clean patients in the twelve months of 1719.

6. SBH, HA 1/12, 489–491.

7. With a forty-day treatment regimen, each bed could see nine different patients per year, or roughly 540 in the two outhouses combined. This is not to say that the outhouses functioned at maximum capacity at all times. But anything approaching that capacity would result in hundreds of patients treated each year.

8. LMA, H1/A41, *A True Report of the great number of Poor Children, and other Poor People, maintained in the several Hospitals, under the pious Care of the Lord Mayor, Commonality, and Citizens, of the City of London, the Year last past* (London, 1753, 1754, 1755, 1757, 1758, 1759, 1760).

9. Ibid. 1753 = 19.3%, 1754 = 19.0%, 1755 = 16.1%, 1757 = 16.8%, 1758 = 17.1%, 1759 = 16.3%, 1760 = 16.2%. For these calculations I have related the number reported in the outhouses each year to the number of inpatients reported remaining in the main house at the time of the report. In all cases I used the fluctuating figure of patients recorded remaining in the main house at Easter, rather than the figure of 420, which was also given each of these years as a figure for the capacity of the main house. I take the former to be a more accurate account of the number of patients on the wards.

10. LMA, H1/ST/A4/6, f. 172.

11. LMA, H1/ST/A7/1, 1–2; LMA, H1/ST/A4/6, f. 221.

12. LMA, H1/ST/A4/6, f. 221.

13. Ibid. For example, in 1731 the sisters of the foul wards received £40 per annum, while the sisters of the clean fluxing ward received £32, and the sisters in all other wards just £25. Similarly, the nurses of the foul wards commanded salaries of £20 per year, while the clean fluxing ward nurses received £18, and regular nurses just £16.

14. Two late-century surviving petitions from Guy's Hospital for foul patients sent by the parish of St. Margaret's, Westminster, include handwritten accounts itemizing the costs incurred. Both list a charge of six shillings, six pence for "Salivating Flannels" over and above the other admission fees and subsistence charges. WCA, St. Margaret's, Westminster, Hospital Petitions: MS 226.

15. LMA, H1/ST/A6/5, f. 172.

16. LMA, H1/ST/A4/7, f. 172.

17. Ibid., July 28, 1768.

18. Ibid., 356.

19. Ibid.

20. John Howard, *An Account of the Principal Lazarettos in Europe* (London, 1791), 132.

21. LMA, H2/WH/A1/5, Westminster Infirmary, Trustees Minutes (1738–54), 175–7.

22. SBH, HA 1/13, 297–8.

23. SBH, HA 1/12, 489–91.

24. Ibid.

25. Ibid.

26. Ibid., 510–11.

27. SBH, HA 1/12, 607 and HA 1/13, 8.

28. Norman Moore does not relate the incident, merely noting that Tuff "absconded" and that he was dismissed as Treasurer: *The History of St. Bartholomew's Hospital* (London: C.A. Pearson, 1918), 2:244.

29. SBH, HA 1/13, 122–33.

30. SBH, HA 1/13, 164–5.

31. Ibid.

32. Moore, *History of St. Bartholomew's Hospital*, 2:376.

33. Once the outhouses closed, the only significant data available are the amounts recorded collected from paying foul patients after 1760. Unfortunately, these still remain incomplete because there is no way to know what portion of all venereal patients this represents. The hospital Renter responsible for collecting these fees began reporting the figures annually in 1760. They are recorded in the hospital journals, with corresponding entries made in the annual ledgers. For a discussion and a chart of the figures see Kevin P. Siena, "Poverty and the Pox: Venereal Disease in London Hospitals, 1600–1800" (Ph.D. diss., University of Toronto, 2001), 169–72.

34. SBH, HA 1/13, 8.

35. Ibid., 164–5. Although, it is worth pointing out that this actually translated into a gain of three beds for men, but a loss of one bed for women.

36. Ibid., 186, 192–3, and 196.

37. Ibid., 253.

38. Howard, *Account of the Principal Lazarettos*, 132.

39. SBH, HA 1/15, 258.

40. The registers are held with the rest of St. Thomas's Records at the LMA. The surviving registers are: H1/ST/B3/1/1 (June 1773–Sept. 1775); H1/ST/B3/1/2 (March–April 1775); H1/ST/B3/2 (Feb.–Aug. 1776); H1/ST/B3/3 (Oct. 1781–Jan. 1782); H1/ST/B3/4 (Jan.–July 1782); H1/ST/B3/5 (Dec. 1783–June 1784); H1/ST/B3/6 (Jan.–Dec. 1787); H1/ST/B3/7 (Dec. 1787–Sept. 1788); H1/ST/B3/8 (March 1789–April 1790); H1/ST/B3/9 (April 1790–June 1791); H1/ST/B3/10 (Dec. 1793– Oct. 1795); H1/ST/B3/11 (May–Dec. 1796); H1/ST/B3/12 (Dec. 1796–June 1797); H1/ST/B3/13 (Dec. 1798–Nov. 1800). It must be noted that there are many missing weeks in these records. These dates indicate the range of time that each register

covers. The registers become more complete in the 1780s and 1790s, but gaps remain in many of them.

41. Some of these names are, of course, not coincidental. Because Job suffered from terrible skin sores he was, in some sense, the patron saint of lepers and then of venereal patients. Naples, to many, remained the European birthplace of the pox. And Magdalen, the most famous prostitute in history, was an obvious choice for one of the female foul wards.

42. LMA, H1/ST/B4/1. The weekly registers from which this index was compiled do not survive.

43. LMA, H1/ST/B4/1.

44. Tabulated from St. Thomas's surviving admission registers covering 1773–76, LMA, H1/ST/B3/1/1, H1/ST/B3/1/2, and H1/ST/B3/2, discussed in greater detail below. Foul women who paid the foul patient fees made up just 37 percent of all women admitted to the ward.

45. LMA, H1/ST/B4/1, "Bird, Sus.," February 2, 1770.

46. LMA, H1/ST/B3/6. The Grand Committee attempted to reverse this decision in 1790 and reinstate Susannah as a women's ward. However, the admissions records of the 1790s show that it did not remain a women's ward very long and that it never received another foul patient through the end of the century. LMA, H1/ST/A6/7, p. 49; H1/ST/B3/8–13. Howard, *Account of the Principal Lazarettos*, 134.

47. John D. Thompson and Grace Goldin, *The Hospital: A Social and Architectural History* (New Haven: Yale University Press, 1975), 84.

48. LMA, H1/ST/B3/1/1, H1/ST/B3/1/2, H1/ST/B3/2. 225 of the 2,342 patients came from London parishes.

49. Ibid. Seventy-four of the 684 foul patients came from parishes. Looked at another way, nearly one-third of all parish patients occupied beds in the foul wards, even though foul patients only comprised 28 percent of the general population: 74 of the 225 total parish patients entered the venereal wards.

50. Ibid. Of the 225 total 127 parish patients were women.

51. Ibid. Of the 74 foul parish patients, fully 54 were women.

52. Ibid. Of all 2,432 patients, 809 entered the hospital with the help of a governor's nomination.

53. Ibid. Of the 225 parish patients 48 also had a governor's nomination.

54. Ibid. Of all 1,136 clean men 394 had nominations (34.6 percent); 213 of all 612 clean women had nominations (34.8 percent).

55. Ibid. Of 1,748 clean patients 607 had nominations (34.7 percent), while only 202 of 684 foul patients had nominations (29.5 percent).

56. Ibid. Of all foul men 35.7 percent had a nomination (155/433) while only 34.6 percent of all clean men were nominated (394/1136).

57. Ibid. Merely 47 of the 251 women (18.7 percent) who entered the foul wards were able to secure a nomination.

58. Ibid. Of the 197 foul women (17.7 percent) who did not come from a parish, only 35 had nominations. Therefore, the variable of whether of not a woman came from a parish had little effect on her chances to win a nomination from a governor.

59. Franz Swediaur, *Practical Observations on Venereal Complaints*, 2nd ed. (London, 1786), 12–3.

60. Indeed, it is telling that by the 1770s it was extremely rare to find full charity cases represented in the admission registers, in which the governors waved all fees. Despite all the discussion of treating patients without "friends, family or a parish settlement" earlier in the century, the records off the 1770s leave little evidence of many cases of full charity.

61. LMA, H1/ST/B3/1/1, H1/ST/B3/1/2, H1/ST/B3/2. The pox was one of the most common ailments that early modern soldiers encountered. Virtually half all the soldiers and sailors in St. Thomas's in the period (61 of 124) entered the foul wards. Such high incidence should caution against drawing conclusions about rates of venereal infection among the general population based on evidence from military medical records. Randolph Trumbach, *Sex and the Gender Revolution*, vol. 1, *Heterosexuality and the Third Gender in Enlightenment London* (Chicago: University of Chicago Press, 1998), 198–201. On public perceptions of the pox and the military see Donna T. Andrew, *Philanthropy and Police: London Charity in the Eighteenth-Century* (Princeton: Princeton University Press, 1989), 57.

62. Ibid. Of the 1,993 patients who listed a surety (ie., nonparish patients and nonsoldiers) 1,845 secured their bond from a man. Only 7.4 percent (148) secured a bond from a woman. I have relied on first names to determine sex.

63. Ibid. Of the 539 clean female patients who listed sureties 52 secured their bond from a woman, while only 41 of the 922 clean men who arranged their bonds privately went to a woman.

64. Ibid. Only 9 of the 352 male foul patients who secured a private bond relied on a woman.

65. Ibid. Women provided the security bond for 46 of the 197 (nonparochial) foul women.

66. Harris posted the bonds for both William Fenning and Stephan Mier, both of whome entered the foul wards on June 27, 1776. Both entries merely list Harris's name and parish: "Nich. Harris, St. Thomas, Southwark." See LMA, H1/ST/B3/2.

67. LMA, Guy's Hospital, Admission and Discharge Registers (1770) H9/GY/B4/4.

68. RCS, Lock Hospital Board Book 4, 272–3.

69. RCS, Lock Board Book 5, 2.

70. William Beckett, *Practical Surgery Illustrated and Improved: being Chirurgical Observations with Remarks, upon the most extraordinary Cases, Cures, and Dissections made at St. Thomas's Hospital, Southwark* (London, 1740), 156–9.

71. These were printed to be posted in the wards of the hospital. LMA, H1/ST/A25; Frederick Gymer Parson, *The History of St. Thomas's Hospital* (London: Methuen, 1932), 2:207.

72. Ibid.

73. Ibid.

74. LMA, H1/ST/A4/7, 154–8.

75. For the 1699 rule see LMA, H1/ST/A4/6, f. 73–75.

76. LMA, H1/ST/A6/5, 146.

77. Owsei Temkin, "On the History of 'Morality and Syphilis'," in his *The Double Face of Janus and Other Essays in the History of Medicine* (Baltimore: Johns Hopkins University Press, 1977), 475.

78. SBH, HA 1/8, f. 300; Moore, *History of St. Bartholomew's Hospital* vol. II, 353–4.

79. Guenter Risse, *Hospital Life in Enlightenment Scotland: Care and Teaching at the Royal Infirmary of Edinburgh* (Cambridge: Cambridge University Press, 1986). Toby Gelfand, *Professionalizing Modern Medicine: Paris Surgeons and Medical Science and Institutions in the Eighteenth Century* (Westport, Conn.: Greenwood Press, 1980); Othmar Keel, "The Politics of Health and Institutionalisation of Clinical Practices in Europe in the Second Half of the Eighteenth Century," in W. F. Bynum and Roy Porter, eds., *William Hunter and the Eighteenth-Century Medical World* (Cambridge: Cambridge University Press, 1985).

80. Susan Lawrence, *Charitable Knowledge: Hospital Pupils and Practitioners in Eighteenth-Century London* (Cambridge: Cambridge University Press, 1996).

81. St. Thomas's Campus, UMDS, "Charles Oxley Notebook, Case notes of surgical patients admitted to St. Thomas's Hospital, 1725–26," MS S2.a.6. For a discussion of this source see Philip K. Wilson, "'Sacred Sanctuaries of the Sick': Surgery at St. Thomas's Hospital 1725–26," *London Journal* 17 (1992): 136–53.

82. St. Thomas's Campus, UMDS, "Charles Oxley Notebook," f. 6.

83. See for example Percival Pott's lectures delivered to medical students at St. Bartholomew's Hospital, St. Thomas's Campus UMDS: "Chirurgical Lectures by Percivall Pott F.R.S. and Senior Surgeon of St. Bartholomew's Hosp[ita]l London," (1785/86?), RCS, 42.d.32 (MSS Pot).

84. William Acton, *Prostitution, considered in its Moral, Social and Sanitary Aspects, in London and other large cities* (London, 1857), 144–5.

85. Michel Foucault, *The Birth of the Clinic: An Archaeology of Medical Perception*, trans. from the French by Alan Sheridan (New York: Pantheon, 1973). See also Edwin H Ackerknecht, *Medicine at the Paris Hospital, 1794–1848* (Baltimore: Johns Hopkins University Press, 1967); William Bynum, *Science and the Practice of Medicine in the Nineteenth century* (Cambridge: Cambridge University Press, 1994); Owsei Temkin, "The Role of Surgery in the Rise of Modern Medical Thought," *Bulletin of the History of Medicine* 25 (1951): 248–59.

86. Mary Fissell's work is crucial here. See her *Patients, Power and the Poor in Eighteenth-Century Bristol* (Cambridge: Cambridge University Press, 1991), 148–70,

and "The Disappearance of the Patient's Narrative and the Invention of Hospital Medicine," in Roger French and Andrew Wear, eds., *British Medicine in an Age of Reform* (London: Routledge, 1991), 92–109.

87. Fissell, *Patients, Power and the Poor*, 156–9.

88. It is notable that when describing the negotiated give and take of hospital exchanges predating the birth of the clinic, Fissell highlights examples from the venereal wards in which patients argued with surgeons who diagnosed them to be poxed. She rightly suggests that "[v]enereal disease probably represents the most telling incidence of the negotiations that characterized doctor-patient relationships." "The Disappearance of the Patient's Narrative," 94–95.

89. The figures in this paragraph come from an analysis of LMA, H1/ST/B4/1.

90. Ibid. "Grocer, Wm."; "Steers, James"; "Mills, James"; "Swallow, Jn."; "Wooding, Mary."

91. Ibid. "Stocker, Eliz."

Chapter 4

1. Hilary Marland was right to point out that most studies of Poor Law medicine have tended to focus on the post-1834 period of the New Poor Law: *Medicine and Society in Wakefield and Huddersfield 1780–1870* (Cambridge: Cambridge University Press, 1987), 52. For some of the work which has been done on medicine and the Old Poor Law see, for the provinces, pp. 57–70, and Margaret Pelling, "Healing the Sick Poor: Social Policy and Disability in Norwich, 1550–1640," *Medical History* 29 (1985): 115–37 and "Illness Among the Poor in an Early Modern Town: The Norwich Census of 1570," *Continuity and Change* 3 (1988): 153–75; and for London, Andrew Wear, "Caring for the Sick Poor in the parish of St. Bartholomew Exchange, 1580–1679," in *Living and Dying in London*, ed. W. F. Bynum and Roy Porter (*Medical History* Supplement 11, London, 1991): 41–60.

2. Paul Slack's overview of London's charitable institutions includes discussion of seventeenth-century workhouses, though not their medical features. His promise of a fuller exploration to come gives cause for optimism: "Hospitals, Workhouses and the Relief of the Poor in Early Modern London," in *Health Care and Poor Relief in Protestant Europe, 1500–1700*, ed. Ole Peter Grell and Andrew Cunningham (London: Routledge, 1997), 240–2, 248 n.2.

3. David Green, "Medical Relief and the New Poor Law in London," in *Health Care and Poor Relief in Eighteenth and Nineteenth-Century Northern Europe*, ed. Ole Peter Grell, Andrew Cunningham, and Robert Jütte (Aldershot: Ashgate, 2002), 226–7. In the same volume Anne Crowther suggests that in the eighteenth century "the workhouse was hardly ever used as a centre for the sick." But Crowther is

commenting on the provinces, where perhaps 1834 may represent a sharper divide: "Health Care and Poor Relief in Provincial England," 203–19.

4. Overviews of how settlement legislation functioned include Tim Hitchcock and John Black's Introduction to *Chelsea Settlement and Bastardy Examinations, 1733–1766* (London: London Record Society, 1999), vii–xx; Thomas Sokoll, *Essex Pauper Letters, 1731–1837* (Oxford: Oxford University Press, 2001), 10–17; or Lynn Hollen Lees, *The Solidarities of Strangers: The English Poor Laws and the People* (Cambridge University Press, 1998), 28–33. More detailed studies of the complex system include J. S. Taylor, "The Impact of Pauper Settlement 1691–1834," *Past and Present* 73 (1976): 42–74; Norma Landau, "The Eighteenth-Century Context of the Laws of Settlement," *Continuity and Change* 6 (1991): 417–39 and "Who was Subject to the Laws of Settlement? Procedure under the Settlement Laws in Eighteenth-Century England," *Agricultural History Review* 43 (1996): 139–59.

5. On the English Poor Law see especially the work of Paul Slack: *Poverty and Policy in Tudor and Stuart England* (London: Longman, 1988) and *From Reformation to Improvement: Public Welfare in Early Modern England* (Oxford: Oxford University Press, 1999). For studies of seventeenth-century parish relief (specific to London), which did not witness the scale of institutional relief that this book seeks to explore, see several articles by R. W. Herlan, "Poor Relief in the London Parish of Antholin's Bridge Row, 1638–1664," *Guildhall Studies in London History* 2 (4) (1977): 179–99, "Poor Relief in the London Parish of Dunstan in the West during the English Revolution," *Guildhall Studies in London History* 3 (1) (1977): 13–36, and "Poor Relief in London in the English Revolution," *Journal of British Studies* 18 (1979): 30–51. See also Jeremy Boulton, *Neighbourhood and Society: A London Suburb in the Seventeenth Century* (Cambridge: Cambridge University Press, 1987), 92–7.

6. Michael J. Braddick, *State Formation in Early Modern England c. 1550–1700* (Cambridge: Cambridge University Press, 2000), 103–35; Keith Wrightson, "The Politics of the Parish in Early Modern England," in *The Experience of Authority in Early Modern England*, ed. Paul Griffiths and Steve Hindle (London: Macmillan, 1996), 10–46; Vanessa Harding, "Controlling a Complex Metropolis, 1650–1750: Politics, Parishes and Powers," *London Journal* 26 (1) (2001): 29–37; Mark Goldie, "The Unacknowledged Republic: Officeholding in Early Modern England," in *The Politics of the Excluded, c.1500–1850*, ed. Tim Harris (New York: Palgrave, 2001), 153–94.

7. Timothy V. Hitchcock, "The English Workhouse: A Study in Institutional Poor Relief in Selected Counties, 1696–1750" (Ph.D. diss., Oxford University, 1985). These plans drew on earlier schemes to put the poor to work. On antecedents see Valerie Pearl, "Puritans and Poor Relief: The London Workhouse 1649–60," *Puritans and Revolutionaries*, ed. Donald Pennington and Keith Thomas, 206–32. This study will not be considering Bridewell. Though nominally a "hospital" this incarcerating institution functioned more as a prison than a center of medical care and thus does not provide such a useful corollary. On Bridewells see

Joanna Innes, "Prisons for the Poor: English Bridewells, 1555–1800," in *Labour, Law and Crime: An Historical Perspective*, ed. Francis Snyder and Douglas Hay (London: Tavistock, 1987), 42–122.

8. Tim Hitchcock, "Paupers and Preachers: The SPCK and the Parochial Workhouse Movement," in *Stilling the Grumbling Hive*, ed. Lee Davison et al., 145–66.

9. Ibid., 146.

10. Ibid., 160.

11. On parish payments to doctors see Irving Loudon, *Medical Care and the General Practitioner 1750–1850* (Oxford: Clarendon, 1986), 231–5; Anne Digby, *Making a Medical Living: Doctors and Patients in the English Medical Market for Medicine, 1720–1914* (Cambridge: Cambridge University Press, 1994), 224–33; Joan Lane, "The Provincial Practitioner and His Services to the Poor 1570–1800," *Bulletin of the Society for the Social History of Medicine* 28 (1981): 10–14; Marland, *Medicine and Society*, 57–70.

12. Hitchcock, "The English Workhouse," 158–160.

13. St. George Hanover Square, Workhouse Committee Minutes, WCA, C870, 181 and C877, 69.

14. Roy Porter, "The Gift Relation: Philanthropy and Provincial Hospitals in Eighteenth-Century England," in *The Hospital in History*, ed. Roy Porter and Lindsay Granshaw (London: Routledge, 1989), 161–2.

15. W. F. Bynum, "Physicians, Hospitals and Career Structures in Eighteenth-Century London," in *William Hunter and the Eighteenth-Century Medical World* (Cambridge University Press, 1985), 105–28.

16. For example, in April 1727 an apothecary named Melmoth underbid a competitor named Jones for the position as the workhouse apothecary in St. Sepulchre, London Division. Jones had asked for a meager salary of £15, but Melmoth offered to do the job for nothing. However, in a year's time Melmoth asked for a salary £15 per year. St. Sepulchre, London Division, Workhouse Committee Minutes, GL, 3137/1, 6–13.

17. For example, St. Sepulchre, London, hired Elizabeth Crouch as its midwife shortly after opening in May 1727, GL, 3137/1, 31.

18. On this tradition see Margaret Pelling, "Older Women: Household, Caring and Other Occupations in the Late Sixteenth-century Town" and "Nurses and Nursekeepers: Problems of Identification in the Early Modern Period," in her *The Common Lot: Sickness, Medical Occupations and the Urban Poor in Early Modern England* (London: Longman, 1998), 155–75 and 179–202 respectively.

19. GL, 3137/1, 16.

20. One of St. Margaret's nurses, Martha Pillow, was brought to trial for stealing goods from a woman under her care for the foul disease. OBSP, May 1, 1717, 1.

21. Ibid., 87, 164.

22. All the dates in this discussion of the establishment of various workhouses come from Hitchcock, "The English Workhouse" ("Appendix: Workhouse Foundation Listed by County"), 258–83.

23. WCA, C869, 65.
24. GL, 3137/1, 29–30.
25. Ibid., 93, 96, and 101.
26. St. Margaret's, Westminster, Workhouse Minute Book, WCA, E2632, August 23, 1726. The February 23, 1726, entry indicates that before the foundation of the workhouse, the parish arranged to house Itch patients with women termed "nurses."
27. Ibid., October 27, 1727.
28. Ibid., November 16, 1727.
29. Ibid., August 22, 1727 and October 3, 1727.
30. Hitchcock, "The English Workhouse," 194–7.
31. A database that comprises the complete surviving workhouse admissions registers for St. Luke's Chelsea is now available on the CD-ROM, *Economic Growth and Social Change in the Eighteenth-Century English Town* (History Courseware Consortium, 1998), designed by Tim Hitchcock and Robert Shoemaker. My thanks to Tim Hitchcock for allowing me to consult this database prior to the publication of the CD-ROM.
32. Ibid. These figures are based on Hitchcock's categorization, which are themselves based on the reasons given for entry in the St. Luke's Chelsea Workhouse Register. The breakdown follows: "Ill/Lunatic/Injured/Lame" = 1213, "Infirm/Lame" = 525, "Returned from the Hospital" = 69. Mary Fissell discusses illness as a factor in poor relief applications in Bristol. See "The 'Sick and Drooping Poor' in Eighteenth-Century Bristol and its Region," *Social History of Medicine* 2 (1989): 35–58. For a discussion of the causes of poverty that highlights physical impairment see Robert Jütte, *Poverty and Deviance in Early Modern Europe* (Cambridge: Cambridge University Press, 1994), 21–44.
33. Dates have been adjusted to have the year begin on January 1.
34. These patterns emerge in other parishes, and continued into the nineteenth century. Lynn MacKay, "A Culture of Poverty? The St. Martin's in the Fields Workhouse, 1817," *Journal of Interdisciplinary History* 26 (2) (1995): 224. On the causes driving the poor into workhouses see also MacKay's "Moral Paupers: The Poor Men of St. Martin's, 1815–1819," *Social History/Histoire Sociale* 34 (2001): 115–31.
35. Lees, *Solidarities of Strangers*, 36–7. Several contributions to *Chronicling Poverty: The Voices and Strategies of the English Poor 1640–1840*, ed. Tim Hitchcock, Peter King, and Pamela Sharpe (London: Macmillan, 1997) discuss early modern views on the entitlement to parish relief: see Tim Hitchcock, Peter King, and Pamela Sharpe "Introduction," 10–11, and Jeremy Boulton, "Going on the Parish: The Parish Pension and its Meaning in the London Suburbs, 1640–1724," 19–21. On the tension over the issue between poor law administrators and welfare recipients see Steve Hindle, "Exhortation and Entitlement: Negotiating Inequality in English Rural Communities, 1550–1650," in *Negotiating Power in Early Modern Society: Order, Hierarchy and Subordination in Britain and Ireland*, ed.

Michael J. Braddick and John Walter (Cambridge: Cambridge University Press, 2001), 102–22.

36. Mary Fissell suggests a similar picture for Bristol, where she claims that the ways in which patients used medical institutions in the eighteenth century helped to shape the services that infirmaries provided: "The 'Sick and Drooping Poor'," 56–7. She, too, suggests that the expectation of entitlement played a major role, and makes a convincing case that "[p]erhaps English society may have been characterized by a belief in entitlement to free health care for far longer than we have thought" (58). This does not mean that churchwardens did not wield discretionary latitude concerning whom to relieve. On parochial flexibility in defining the "deserving" poor see Claire S. Schen, "Constructing the Poor in Early Seventeenth-Century London," *Albion* 32 (3) (2000): 450–63.

37. WCA, E2634, 242–3 and 383–6.

38. Amanda Berry points out that the relationships between hospitals and workhouses varied from context to context. Generally, voluntary hospitals in communities with workhouse infirmaries were more likely to exclude parish paupers: "Community Sponsorship and the Hospital Patient in Late Eighteenth-Century England," in *The Locus of Care: Families, Communities, Institutions and the Provision of Welfare Since Antiquity*, ed. Peregrine Horden and Richard Smith (London: Routledge, 1998), 128–9 and 132–4.

39. WCA, D1711/11.

40. WCA, E2632, 7 March, 1727.

41. WCA, E2634, 84; GL, 3137/3, 254; WCA, C869, 84; CLA, Weekly Meetings of the Overseers of the Poor and Churchwardens of St. George the Martyr and St. Andrew Holborn, P/AH/PO/2, February 10, 1738.

42. GL, 3137/1, 248.

43. WCA, E2634, 273.

44. GL, MS 1613. St. John Zachary, Bethlem Hospital Bond.

45. Akihito Suzuki, "The Household and the Care of Lunatics in Eighteenth-Century London," in Horden and Smith, eds., *The Locus of Care*, 153–75.

46. For example, several of the applicants in St. Sepulchre during the year 1735 specifically sought admission to a hospital. See: WCA, E2634 parishioners Tibbet, 1/30/35; Elizabeth Steward, 3/6/35 and Elizabeth Toogood, 1/6/35.

47. On the effects of illness on families related to workhouse provision see MacKay, "Culture of Poverty?" 223–4.

48. Reprinted in Francis Hauksbee, *A Further Account of the Effects of Mr. Hauksbee's Alternative Medicine, as Applied in the Cure of the Venereal Disease* (London, 1743), 29–31.

49. GL, 3137/4, July 17, 1735.

50. WCA, E2634, June 16, 1735.

51. Tim Hitchcock, "The Publicity of Poverty in Eighteenth-Century London," in *Imagining Early Modern London: Perceptions and Portrayals of the City from Stow*

to *Strype 1598–1720*, ed. J. F. Merritt (Cambridge: Cambridge University Press, 2001), 174, 179, and 180–1.

52. *Rules, orders and regulations for the government of the workhouse belonging to that part of the parish of St. Andrew, Holborn* (London, 1791), 38.

53. *Hints and Cautions for the Information of the Churchwardens and Overseers of the Poor of the Parishes of St. Giles in the Fields and St. George Bloomsbury, in the County of Middlesex: and Rules, Orders, and Regulations for Maintaining, Governing, and Employing the said Poor; Made by the Vestry of the said Parishes* (London, 1797), 6.

54. This account is related in Hitchcock, "The English Workhouse," 148–9.

55. Ibid., 177–92.

56. Ibid., 213–14.

57. See for example GL, 3137/10, 319; MacKay, "A Culture of Poverty?" 216.

58. *Regulations which were agreed upon and established the twelfth day of July, 1726 by the Gentlemen of the Vestry then present for the Better Government and Management of the Work-House belonging to the parish of St. Giles in the Fields* (London, 1727), 4. The marking of paupers had a long history, dating back to the Middle Ages, forming a common element of poor relief throughout Europe: Jütte, *Poverty And Deviance*, 158–65.

59. Donna T. Andrew, "'To the Charitable and Humane': Appeals for Assistance in the Eighteenth-Century London Press," in *Charity, Philanthropy and Reform from the 1690's to 1850*, ed. Hugh Cunningham and Joanna Innes (London: MacMillan, 1998), 87–107.

60. WCA, C877, 265; GL, 3137/1, 61.

61. WCA, C881, 256.

62. Such instances were common. For example, St. Luke's sent eighteen-year-old Mary Smith to Bridewell on January 5, 1787, for undisclosed "bad" behavior: see LMA, Microfilm X/15/37, Mary Smith, 1/5/87; see also MacKay, "A Culture of Poverty?" 216–7.

63. *An Account of Several Work-Houses for Employing and Maintaining the Poor* (London, 1725), 35.

64. For a discussion of informal pauper relief strategies, including temporary stays in the workhouse, see Jeremy Boulton, "'It is extreme necessity that makes me do this': Some 'Survival Strategies' of Pauper Households in London's West End During the early Eighteenth Century," *International Review of Social History* 45 (2000): 47–69.

65. Randolph Trumbach has been one of the only scholars to note this: *Sex and the Gender Revolution*, vol. 1, *Heterosexuality and the Third Gender in Enlightenment London* (Chicago: University of Chicago Press, 1998), 210–11.

66. Dorothy M. George, *London Life in the Eighteenth Century* (London, 1925), 55.

67. WCA, E2634, 242–3 and 383–6.

68. The records of the Shoe Lane Workhouse are held at the Guildhall Library MS 9598 (1–4), "Shoe Lane Workhouse Admission Books 1776–1796." I have compiled a list of all sick inmates for the five-year period 1776–81. After

1781 large gaps appear in the records. During these five years 33 of 47 sick inmates suffered from venereal disease.

69. "The Covent Garden Ague: The venereal disease. Many brothels, under the denominations of bagnios, were formerly kept in that parish. Some, it is said, are still remaining." Francis Grose, *Provincial Glossary; with a Collection of local Proverbs and Popular Superstitions* (London 1790).

70. Several patients listed as cured of the "pox" on St. Margaret's Infirmary lists were not diagnosed as foul when they entered the workhouse. For example, Elizabeth Ruxstone was merely listed as "in ye sick ward" when she was admitted, while John Bennet's admission merely states "till cured of his Illness." See WCA, E2634, February 28, 1733, and March 28, 1724.

71. WCA, E2632, October 3, 1727.

72. Ibid., October 3, 1728. For example Samuel Golding was admitted on Jan 9, 1735, and described as "in a salivation."

73. The annual workhouse account listed the amount of supplies required for each ward of the house. The account of coals required for each ward shows that the two main foul wards each received extra supplies of coals to stoke the salivation stoves. *Rules, orders and regulations for the government of the workhouse belonging to that part of the parish of St. Andrew, Holborn*, 33 and 38.

74. GL, 3137/3, 243.

75. Ibid., 244.

76. Mary Matthews first appears in October 1729. She is diagnosed foul and put into St. Thomas's Hospital. GL, 3137/2, 87.

77. GL, 3137/3, 314.

78. Mary Matthews was sent back to the Hospital on September. 21, 1732: Ibid., 384; Susannah Pain was sent to a hospital on Aug 3, 1732: Ibid., 365; and the third patient, Sarah Howard, went under LeBarr's care On June 1, 1732. She later applied to for a hospital petition on December 28, 1735: GL, 3137/4, 186.

79. *Hints and Cautions for the Information of the Churchwardens and Overseers of the Poor of the Parishes of St. Giles in the Fields and St. George Bloomsbury*, 6–7.

80. GL, 3137/10, 91.

81. For example, Norma Landau has explored the use of settlement to regulate the labor pool the parishes: see "The Laws of Settlement and Surveillance of Immigration in Eighteenth-Century Kent," *Continuity and Change* 3 (1988): 391–420 and "The Regulation of Immigration, Economic Structures and Definitions of the Poor in Eighteenth-Century England," *Historical Journal* 33 (1990): 541–72.

82. "St. Luke's Chelsea, Workhouse Register," LMA, Microfilm, X/15/37, Ann Harris, January 11, 1758.

83. Ibid., Thomas Morris, November 29, 1744; Martha Jackson, May 12, 1749; Samuel Russell, October 17, 1783. This phenomenon was hardly limited to applicants with venereal disease, as parishes also tried to pass other applicants who posed fiscal burdens, including single mothers, bastards, and the chronically ill.

84. GL, 3137/3, 331–2.

85. GL, 3137/1, 185.
86. Ibid., 186.
87. GL, 3137/3, 26.
88. Ibid., 66.
89. Ibid., 109, 188.
90. Ibid., 404. "Anne Cox applied to be taken into the house she was ordered to go to the Church wardens of St. Johns Clerkenwell to be passed hither they formerly having given a man 40s. to marry her in order to make her settlement in this parish."
91. Ibid., 434.
92. GL, 3137/4, 483.
93. CLRO, Coroner's Inquests, London and Southwark 1788–1837, 400A Box 7 (1794), no. 116.
94. On unwed mothers and their encounters with relief agencies see Tim Hitchcock, "'Unlawfully Begotten on Her Body': Illegitimacy and the Parish Poor in St. Luke's Chelsea," in Hitchcock et al., eds., *Chronicling Poverty*, 70–86; R. B. Outhwaite, "'Objects of Charity': Petitions to the London Foundling Hospital, 1768–72," *Eighteenth-Century Studies* 32 (4) (1999): 497–510. On the broader European context see several contributions to John Henderson and Richard Wall, eds., *Poor Women and Children of the European Past* (London: Routledge, 1994). On deserted mothers see David A. Kent, "'Gone for a Soldier': Family Breakdown and the Demography of Desertion in a London Parish, 1750–91," *Local Population Studies* 1990 (45): 27–42.
95. On the role of women in policing illegitimacy see Laura Gowing, "Ordering the Body: Illegitimacy and Female Authority in Seventeenth-Century England," in Braddick and Walter, eds., *Negotiating Power in Early Modern Society*, 43–62.
96. GL, 3137/3, 209. The minutes of September 30, 1731, read: "Ordered that for the future as to such women as shall lye in this house of bastard children application be made to a justice of the peace that they may from time to time be punished according to law and that this resolution of the committee be published in Parker's Penny Post for one month."
97. WCA, E2633, December 18, 1729.
98. WCA, E2634, 1, 2, 6.
99. St. Sepulchre issued such a threat to Mary Green who had relied on the parish when infected and when she became pregnant out of wedlock. The committee ordered that she "promise to be no further troublesome And if she comes here again to be sent to Bridewell according to the Order of the 13th instant." GL, 3137/4, 172.
100. LMA, E2633, October 21, 1728.
101. On pauper apprenticeships see Joan Lane, *Apprenticeship in England*, 81–93.
102. GL, 3137/3, 331–2.

103. WCA, E2633, January 22, 1730. The committee minutes read: "Ord[ere]d that Mr. Thompson & p[ar]tner do relieve at their discretion Mary Lloyd and three children till further order, that her husband be re examined to his settlement & that Mr. Westbrook be desired to give her such medicines for her recovery as he shall think fitt."

104. WCA, E2633, November 28, 1728.

105. GL, 3137/10, 104. The minutes of October 8, 1761, read: "Sarah Law (whom the committee ordered to be pass'd to her place of settlement & for whom the Lord Mayor refused to grant a pass 'till she was cured she being foul) apply'd to be gott into the hospital which was ordered."

106. WCA, E2634, 227. "Ord[ere]d that Margaret Coleman and two children be admitted till cured and then be sent to the place of her husband's settlement which appears to be in the Parish of Martlesome in the County of Suffolk."

107. GL, 3137/4, 274, 276, 281.

108. GL, 3137/8, 17, 136, 141, 147.

109. Ibid., 182, 183, 547.

110. GL, 3137/3, 326, 332.

111. GL, 3137/4, 40, 75.

112. Ibid., 183.

113. Ibid., 186, 202, 212.

114. This quotation and the figures cited in this paragraph come from Hitchcock, "The English Workhouse," 194–5.

115. I have identified as "foul" all inmates and applicants said to be "foul," suffering from "the pox" or "venereal disease," and all those sent to the Lock or Kingsland hospitals. I did not include those only said to be "under a salivation" since salivation was sometimes employed for other maladies, although many of these were very likely in the foul wards. I used patients' first names to determine gender.

116. The gender breakdown of the four parishes follow: St Luke's Chelsea = 56 foul women / 19 foul men; St. Margaret's Westminster = 70 foul women / 19 foul men; St. Sepulchre, London Division = 32 foul women / 11 foul men; Shoe Lane Workhouse, St. Andrew's Holborn = 25 foul women / 7 foul men, 1 foul patient gender unknown.

117. Charles Wigg was forty-four and Bartholomew Low was forty-one when they entered the Shoe Lane Infirmary for the pox. See GL, MS 9598 (1–4) Shoe Lane Workhouse Admission Books, "Charles Wigg," February 22, 1779; "Bartholomew Low," October 1, 1778.

118. Tony Henderson, *Disorderly Women in Eighteenth-Century London: Prostitution and Control in the Metropolis, 1730–1830* (London: Longman, 1999), 20–7; Trumbach, *Sex and the Gender Revolution*, 1:117–119, 278–9.

119. LMA, Microfilm, X/15/37, David Davis, July 23, 1789; August 22, 1789.

120. Demographers suggest that the average age of marriage for English women in the eighteenth century fluctuated within a range between 25 and 27. Considering that 62.4 percent of adult female foul patients were already infected and admitted to the infirmary by the age of twenty-five, we can assume that the majority of them were single. On age of marriage see E. A. Wrigley and Roger Schofield, *The Population History of England, 1541–1871* (Cambridge: Cambridge University Press, 1981); E. A. Wrigley, "Marriage, Fertility and Population Growth in Eighteenth-Century England," in *Marriage and Society: Studies in the Social History of Marriage*, ed. R. B. Outhwaite (London: Europa Publications, 1981), 137–85; Peter Laslett, et al., *Bastardy and its Comparative History* (London: E. Arnold, 1980). For a critique of the interpretations of these demographic historians see Bridget Hill, "The Marriage Age of Women and the Demographers," *History Workshop Journal* 28 (1989): 129–47.

121. WCA, E2633, September 7, 1727 and September 14, 1727.

122. WCA, E2634, 2, 14.

123. Ibid., 70.

124. Ibid., 88.

125. See above pp. 159–61.

126. GL, 3137/3, 383, 476.

127. Ibid., 498.

128. GL, 3137/4, 52.

129. Ibid., 172.

130. LMA, Microfilm, X/15/37, Joan Rumbold, March 4, 1757; March 18, 1758; April 26, 1758; May 2, 1758; July 10, 1758; John Rumbold, March 20, 1757; March 18, 1758; May 2, 1758. Joan's story is related in Hitchcock, "'Unlawfully Begotten on Her Body,'" 73–4.

131. On the customs of premarital sex in early modern England, see John Gillis, *For Better, For Worse: British Marriages, 1600 to the Present* (Oxford: Oxford University Press, 1985), 38, 114–6, and 126–8.

132. GL, 3137/3, 152, 166.

133. GL, 3137/2, 142, 166; and 3137/3, 62, 92.

134. GL, 3137/3, 151, 152.

135. GL, 3137/3, February 6, 1732.

136. Ibid., 105, 108, 112, 115.

137. Ibid., 121.

138. *Hints and Cautions for the Information of the Churchwardens and Overseers of the Poor of the Parishes of St. Giles in the Fields and St. George Bloomsbury*, 29–30.

139. GL, 3137/4, 94.

140. Ibid., 105, 109, 111, 167, 169.

141. Ibid., 290.

142. Ibid., 495; GL, 3137/5, 1.

143. GL, 3137/5, 10, 32, 391, 513; 3137/6, 492; and 3137/8, 186.

144. We cannot assume that all applicants who had been previously diagnosed "foul" and who later applied for relief claiming other ailments were

consciously strategizing to conceal a sexual condition. Many people who had undergone salivation may well have believed that they were cured. However, the propensity for foul patients in all the hospitals studied to lie about their conditions makes it highly likely that workhouse patients followed suit. This seems especially likely given the evidence of patients like John Rawlins who applied for a hospital petition, only to "confess" their infection a few days later.

145. GL, 3137/3, 487.

146. LMA, Microfilm, X/15/37, Mary Lewin, November 4, 1782, December 4, 1782; James Lamityer, January 21, 1791, February 21, 1791.

147. WCA, E2334, 76, 87, 98, 152, 293, 310.

148. GL, 3137/8, 161, 165.

149. Ibid., 240. "John Ashton saying he had the foul disease applyed to be got into the hospital but was denied having runaway with clothes formerly given him when he had an order for admittance into St. Thomas's Hospital."

150. LMA, Microfilm, X/15/37, Margaret Cock, September 5, 1746, Ann Cock, November 6, 1746, Grace Cock, November 6, 1746, Hannah Cock, September 21, 1746.

151. Ibid., Margaret Cock, November 18, 1746, April 17, 1748.

152. Ibid., May 2, 1748, December 8, 1749, December 13, 1749.

153. Ibid., October 3, 1750.

154. Ibid., December 24, 1750, January 9, 1751.

155. Ibid., January 17, 1751, February 21, 1751.

156. Ibid., July, 19, 1751.

157. Ibid., October 7, 1751, April 22, 1752.

158. Ibid., October 31, 1752, November 2, 1752.

159. Ibid., December 7, 1752, February 10, 1753.

160. Ibid., Elizabeth Wyatt, June 16, 1750, July 6, 1750, July 30, 1750, November 10, 1750, December 24, 1750, May 15, 1751.

161. Ibid., October 26, 1757, November 8, 1757, November 21, 1757, December 13, 1757, May 2, 1758, May 9, 1758, January 24, 1759, February 24, 1759, October 26, 1759, November 8, 1759.

162. Ibid., December 4, 1759.

163. Ibid., April 9, 1760.

164. Ibid., June 12, 1760, June 28, 1760, June 30, 1760.

165. Ibid., July 5, 1760.

166. Ibid., December 3, 1760, January 14, 1761, February 4, 1761.

167. Ibid., April, 8, 1761.

168. Ibid., October 10, 1761, November 10, 1761.

169. WALM, Westminster Coroners Inquests, Frances Gardner, April 21, 1796.

170. Michael MacDonald, and Terrance R. Murphy, *Sleepless Souls: Suicide in Early Modern England* (Oxford: Clarendon Press, 1990), 268–89.

Chapter 5

1. W. F. Bynum asserted that from the outset the Lock was "charged with a double mission of caring for disease and reforming morals." See his "Foreword" to David Innes William's *The London Lock: A Charitable Hospital for Venereal Disease 1746–1952* (London, Royal Society of Medicine Press, 1996), vii. See also W. A. Waugh, "Attitudes of Hospitals in London to Venereal Disease in the Eighteenth and Nineteenth Centuries," *British Journal of Venereal Disease* 47 (1971): 147. More recently Randolph Trumbach asserted that "The organizers of the charity [the Lock] were, however, always concerned not only with the cure of the disease, but also with the reformation of the manners of those who entered the hospital." *Sex and the Gender Revolution*, vol. 1, *Heterosexuality and the Third Gender in Enlightenment London* (Chicago: University of Chicago Press, 1998), 221.

2. Judith Walkowitz, *Prostitution and Victorian Society: Women, Class, and the State* (Cambridge: Cambridge University Press, 1980), 57–63.

3. See Walkowitz, *Prostitution and Victorian Society*, as well as Paula Bartley, *Prostitution: Prevention and Reform in England, 1860–1914* (London: Routledge, 2000), 25–41, and Linda Mahood, *The Magdalenes: Prostitution in the Nineteenth Century* (London: Routledge, 1990), 29–39, 124–30, 145–53.

4. Donna T. Andrew, *Philanthropy and Police: London Charity in the Eighteenth Century* (Princeton: Princeton University Press, 1989), 69–72, 133, and 161–2, and "Two Medical Charities in Eighteenth-Century London: The Lock Hospital and the Lying-in Charity for Married Women," in *Medicine and Charity before the Welfare State*, eds. Jonathan Barry and Colin Jones (London: Routledge, 1991), 82–97.

5. Linda E. Merians, "The London Lock Hospital and the Lock Asylum for Women," in *The Secret Malady: Venereal Disease in Eighteenth-Century Britain and France*, ed. Linda Merians (Lexington: University of Kentucky Press, 1996), 128–48.

6. Williams, *The London Lock*.

7. Merians, "The London Lock Hospital and the Lock Asylum," 129; Williams, *The London Lock*, 11–18.

8. The following biographic sketch is largely dependent on Williams's *The London Lock*, 11–15, and on the *Dictionary of National Biography*.

9. Terry Gould and David Uttley, *A Short History of St. George's Hospital* (London: Atlantic Highlands, 1997), 5, shows that St. George's regularly accepted soldiers with the pox. Medical commentators who noted that some of their patients had been salivated in St. George's include Henry Wastell, *Observations on the Efficacy of a New Mercurial Preparation* (London, 1779), 60–1 and James Cowper, *A Narrative of the Effects of a Medicine, newly discovered, by Mr. Keyser, a German Chymist in Paris, that covers The Venereal Disease in its most inveterate and malignant state, without Salivation or strict regimen, as is now practiced in France, both in private cases and in the Military Hospitals* (London, 1760), 35.

10. Lindsay Granshaw, "'Fame by Means of Bricks and Mortar': Specialist Hospitals and the Medical Profession in Britain, 1800–1948," in *The Hospital in History*, eds. Lindsay Granshaw and Roy Porter (London: Routledge, 1989), 199–220. See also, W. F. Bynum, "Physicians, Hospitals and Career Structures in Eighteenth-Century London," in *William Hunter and the Eighteenth-Century Medical World*, ed. W.F. Bynum and Roy Porter (Cambridge: Cambridge University Press, 1985), 105–23.

11. Philanthropists like Richard Fitzgerald often supported both charities. Upon his death, Fitzgerald left the hefty sum of £2,000 to St. George's, the Lock, and the Middlesex hospitals. In addition to the first two surgeons at the Lock, Bromfeild and Thomas Williams, who each served both hospitals simultaneously for decades, the Lock also allowed surgeons in training at St. George's to serve as "House Pupils" in the Lock. Among them were Robert Mapletoft, appointed in 1749, and Joseph Clark, hired in March 1755. RCS, Lock Court Book I, 49 and 158.

12. For a map that demonstrates the proximity of the two hospitals, see Williams, *The London Lock*, 20.

13. *London Evening Post* (no. 2919) July 19–22, 1746.

14. Merians, "The London Lock Hospital and the Lock Asylum," 129.

15. On the make up of the Charity's benefactors see William, *The London Lock*, 22–4, and Andrew, "Two Medical Charities in Eighteenth-Century London," 91. According to the *Dictionary of National Biography*, Bromfeild's association with the court began when he accompanied the Princess of Mecklenburg on her journey to wed George III. Thereafter he was appointed to the queen's household.

16. On the voluntary hospital movement see Andrew, *Philanthropy and Police*, 44–9; Roy Porter, "The Gift Relation: Philanthropy and Provincial Hospitals in Eighteenth-Century England," in Porter and Granshaw, eds., *The Hospital in History*, 149–78; Anne Borsay "'Persons of Honour and Reputation': The Voluntary Hospital in the Age of Corruption," *Medical History* 35 (1991): 281–94, and "Cash and Conscience" Financing the General Hospital at Bath, c.1738–1750," *Social History of Medicine* 4 (1991): 207–29.

17. RCS, Lock Court Book I, 17–20.
18. Ibid.
19. Ibid.
20. Williams, *The London Lock*, 22.
21. RCS, Lock Court Book I, 17–20, 23.

22. In no year between 1752 and 1800 did subscriptions ever cover total annual expenses. In 37 of 46 documented years, subscriptions were less than half of the charity's expenses. In some years they were less than a third. For the figures see Kevin P. Siena, "Poverty and the Pox: Venereal Disease in London Hospitals, 1600–1800" (Ph.D. diss., University of Toronto, 2001), 215.

23. Andrew, *Philanthropy and Police*, 79–83; Sarah Lloyd, "Pleasing Spectacles and Elegant Dinners: Conviviality, Benevolence, and Charity Anniversaries in Eighteenth-Century London" *Journal of British Studies* 41 (2002): 23–57.

24. Ibid., 81.

25. All figures have been compiled from the annual financial account given each year at the Annual Court Meeting each March. See RCS, Lock Court Books 1–3.

26. RCS, Lock Hospital Court Book, I, 64.

27. For a comparison of the fundraising efficacy of foul patients vs. pregnant women, see Andrew, "Two Medical Charities in eighteenth-century London."

28. Ibid., 90.

29. On Madan see Andrew, *Philanthropy and Police*, 69–70.

30. RCS, Lock Court Book I, 270–1.

31. These figures have been compiled from the annual financial reports read each year at the meeting of the Annual Court each spring. See RCS, Lock Court Books, 1–3.

32. Williams, *The London Lock*, 35–9.

33. RCS, Lock Court Book I, 40.

34. Ibid., 59.

35. Ibid., 136–7.

36. Ibid.

37. John Reynolds, *Compassionate Address to the Christian World . . . revised, corrected and published for the use of the Patients in the Lock-Hospital, near Hyde-Park Corner* (London, 1767), iii–iv. Madan's original revision came out in 1760, and his preface is dated December 10, 1760.

38. *An Address to the President, Vice-Presidents, and the other Governors of the Lock Hospital near Hyde park Corner on Behalf of that Charity* (London, 1781), 20–1.

39. *Account of the Proceedings of the Lock Hospital near Hyde-Park Corner* (London, 1749).

40. See *Account of the Proceedings of the Lock Hospital near Hyde-Park Corner* (London, 1789).

41. See *Accounts of the Proceedings of the Lock Hospital near Hyde-Park Corner* (London, 1749, 1751, 1767, and 1789).

42. Martin Madan, *Every man our Neighbour. A Sermon preached at the Opening of the Chapel, of the Lock-Hospital near Hyde-Park-Corner, March 28, MDCCLXII* (London, 1762), 22.

43. Martin Madan, *A Sermon Preached at the Parish-Church of St. George, Hanover-Square, For the Benefit of the Lock-Hospital* (London, 1777), 22–3.

44. On the general inefficacy of this approach to the Lock's charity sermons see Andrew, *Philanthropy and Police*, 8.

45. *An Account of the Proceedings of the Lock Hospital near Hyde-Park Corner* (London, 1747, 1751, 1767, 1789).

46. See Winfried Schleiner, "Infection and Cure through Women: Renaissance Constructions of Syphilis," *Journal of Medieval and Renaissance Studies* 24 (1994): 499–517, and Kevin P. Siena, "Pollution, Promiscuity, and the Pox: English Venereology and the Early Modern Discourse on Social and Sexual Danger," *Journal of the History of Sexuality* 8 (1998): 553–74.

47. Madan, *Every man our Neighbour*, 16–9.

48. See Madan, *A Sermon Preached at the Parish-Church of St. George*, in which he asserts that "Many a worthy Woman has There to lament the diabolical Profligacy of an abandoned Husband—Many a poor and helpless Infant to deplore its being the Offspring of a distempered Parent—Many a Young Creature of tender Years, yea, even in Infancy itself, has to bewail the inhuman Violence of a diseased filthy, and loathsome Ravisher. Are not *such* Objects of Pity? of the tenderest Commiseration? Surely none can be greater!" (19).

49. Anonymous, *Some Reasons Tending to remove the Prejudices against the Lock-Hospital* (London, 1770).

50. Schleiner, "Infection and Cure through Women," 499–517.

51. Anna Clark, *Women's Silence, Men's Violence: Sexual Assault in England, 1770–1845* (London: Pandora Press, 1987), 42, n.75. On children's infection in the period see Barbara J. Dunlap, "The Problem of Syphilitic Children in Eighteenth-Century France and England," in Merians, ed., *The Secret Malady*, 116–25. On the Lock Hospital's campaign see Merians, "The London Lock Hospital and the Lock Asylum," 134.

52. Trumbach, *Sex and the Gender Revolution*, 1:210–18.

53. The figures in this paragraph stem from an analysis of *Old Bailey Sessions Papers* newly digitized for "The Proceedings of the Old Bailey" (www.oldbaileyonline.org) for the period December 1714 to December 1759 (the period digitized as of June, 2003). For background on the Old Bailey Sessions Papers see Simon Devereaux, "The City and the Sessions Paper: 'Public Justice' in London, 1770–1800," *Journal of British Studies* 35 (1996): 466–503, and Andrea McKenzie, "Making Crime Pay: Motives, Marketing Strategies, and the Printed Literature of Crime in England 1670–1770," in *Criminal Justice in the Old World and the New: Essays in Honour of J.M. Beattie*, ed. Greg T. Smith, Allyson N. May, and Simon Devereaux (Toronto: Centre of Criminology, University of Toronto, 1998).

54. For the period 1714–59 there are 152 rape trials available These concern 125 rape victims. (Multiple assailants were tried separately.) Of these, 72 victims were girls under sixteen, and just 53 were women sixteen and older. (When no age was given I assumed the victim to have been an adult.). On child sexual abuse in the period see Clark, *Women's Silence, Men's Violence*, 96–103.

55. Only 7 of the 53 adult victims claimed to have been infected by the rapist. But fully 39 of the 72 child victims (54 percent) were said to have been so infected. The completion of the Old Bailey database will greatly facilitate further, much needed research on this phenomenon.

56. Trumbach also relies on the *Old Bailey Sessions Papers*. However, he suggests that prior to 1760 venereal infection only comes up in 30 percent of cases of child rape. The calculations now made possible by the Old Bailey database may indicate that the practice was even more widespread in this period than he suggests. However, this may also serve to minimize the extent of change over time that he

proposes when he contrasts the allegedly higher (60 percent) incidence of such cases after 1760. *Sex and the Gender Revolution*, 1:212.

57. *An Account of the Proceedings of the Lock Hospital* (London, 1747, 1751, and 1767).

58. At least one surgeon shared the opinion that people too quickly presumed that children's genital symptoms signaled the foul disease. Testifying in a rape case that an infant was not poxed, surgeon Richard Guy stated "I have seen a great many children brought into the hospital; and it is usual for people to say it is venereal, when it only proceeds from an inflammation." OBSP, April 11, 1749, 94.

59. Anon. *Some Reasons Tending to remove the Prejudices against the Lock-Hospital* (London, 1770).

60. Madan, *Every man our Neighbour*, 24–25.

61. RCS Lock Board Book 3, 232, 273; Merians, "The London Lock Hospital and the Lock Asylum," 135.

62. For gender and age breakdown of the Lock's clientele, see below pp. 235–36.

63. *Annual Account of the Proceedings of the Lock Hospital near Hyde Park Corner* (London, 1747).

64. RCS, Lock Court Book 1, 142.

65. RCS, Lock Board Book 3, 176–7.

66. RCS, Lock Board Book 7, 280.

67. RCS, Lock Board Book 11, 10.

68. Williams gives a good account of the crisis in chapter 6 of *The London Lock*, "Times of Turmoil, 1780–1781," 43–8.

69. RCS, Lock Board Book 2, December 22, 1759. Charles Bromfeild was immediately discharged from the office of secretary "he having behaved himself very insolently and improperly to the Rev. Mr. Madan."

70. Williams, *The London Lock*, 43.

71. Ibid.

72. RCS, Lock Court Book 3, 69–74.

73. Ibid., 75–6.

74. Ibid., 79–80.

75. Ibid.

76. Ibid., 88–92. Original emphasis.

77. Williams, *The London Lock*, 45–8. Each governor could vote for twenty-one names.

78. RCS, Lock Court Book 3, 105–17.

79. Ibid., 117.

80. Ibid., 126. Bromfeild sent cards to governors that read: "Mr. Bromfeild presents his respects to Lord ——————— and begs the favour of his Lordship not to pay his subscription to the Collector for the Lock Hospital, till such time as Mr. B. can assure his Lordship that the money will be applied according to his Lordship's intention."

81. Ibid.

82. See Williams, *The London Lock*, chapter 7, "The Age of the Evangelicals: 1781–1837," 49–60.
83. RCS, Lock Select Committee Book, June 11, 1781.
84. Ibid., July 13, 1781.
85. Ibid.
86. Ibid.
87. Ibid., July 20, 1781; RCS, Lock Court Book 3, 99–100.
88. RCS, Lock Select Committee Book, August 10, 1781.
89. Hugh Cunningham, "Introduction," in *Charity, Philanthropy and Reform from the 1690's to 1850*, eds. Hugh Cunningham and Joanna Innes (London: MacMillan, 1998), 6.
90. RCS, Lock Board Book 10, 209; RCS, Lock Court Book 3, 139–40.
91. RCS, Lock Court Book 3, 115.
92. Ibid., 177–8. This decision came about in part to ensure that De Coetlogon could continue to work to sustain the congregation, which the governors feared was dropping off to the detriment of the charity's coffers.
93. Ibid., 169–70.
94. RCS, Lock Court Book 3, 198. An indication of this new attitude at the charity is the following Select Committee report of April 27, 1786:

> "It gives them great pleasure that they are able to congratulate the Governors on the salutary effects, likely to be produced by the zeal, piety and unremitted attention with which the patients are now visited in the ward by the clergyman appointed for that humane purpose; They find that a visible reformation has taken place since he visited them. That they are much more orderly and tractable than before, and that several of those who have been discharged have in very effecting terms expressed their gratitude, not only for the tender care of the surgeons, with respect to the restoration of their bodily health, but also for those solemn warnings & exhortations which they have received from their Minister, and which they have declared they trust & believe will be no less beneficial to their souls.
>
> Thus these unhappy people have not only been restored to society, but have given great ground of hope that they may become valuable members of it. Whilst the disease of the patients only is cured, and they remain insensible of, and hardened in their vices, short and temporary must be the relief afforded them, but if by the means adopted by the Governors they acquire a sense of religion & a detestation of their former courses, greatly must the utility & advantage of the charity be increased." (204–5).

95. John Howard, *An Account of the Principal Lazarettos in Europe* (London, 1791), 138.
96. Andrew, *Philanthropy and Police*,163–9, Williams, *The London Lock*, 49–53, Trumbach, *Sex and the Gender Revolution*, 1:221–2; Merians, "The London Lock Hospital and the Lock Asylum," 136.

97. Williams, *The London Lock*, 50.

98. Andrew, *Philanthropy and Police*, 169. On evangelicalism and moral reformation see 168–77.

99. Ibid., 200.

100. Susan Staves, "British Seduced Maidens," *Eighteenth-Century Studies* 14 (2) (1980/81): 109–34; Sarah Lloyd well describes the manufacturing of such narratives in the Magdalen Hospital's promotional literature. "'Pleasure's Golden Bait': Prostitution, Poverty and the Magdalen Hospital in Eighteenth-Century London," *History Workshop Journal*, 41 (1996): 57–62.

101. Tony Henderson, *Disorderly Women in Eighteenth-Century London: Prostitution and Control in the Metropolis, 1730–1830* (London: Longman, 1999), 184–7; W. A. Speck, "The Harlot's Progress in Eighteenth-Century England," *British Journal for Eighteenth-Century Studies* 3 (1980): 127–39; Vivien Jones, "Scandalous Femininity: Prostitution and Eighteenth-Century Narrative," in *Shifting the Boundaries: Transformation of the Languages of Public and Private in the Eighteenth Century*, ed. Dario Castiglione and Lesley Sharpe (Exeter: University of Exeter Press, 1995), 54–70. See also Trumbach, *Sex and the Gender Revolution*, 1:169–95 and "Modern Prostitution and Gender in *Fanny Hill*: Libertine and Domesticated Fantasy," in *Sexual Underworlds of the Enlightenment*, ed. G. S. Rousseau and Roy Porter (Manchester: Manchester University Press, 1987), 69–85.

102. Trumbach, *Sex and the Gender Revolution*, 1:188–90. Just a few of the many impassioned responses to Madan's suggestion of polygamy include: Thomas Wills, *Remarks on Polygamy &c. in answer to Mr. M–d–n's Thelypthora* (London, 1781); Anonymous, *A Letter to the Rev. Mr. Madan, concerning the chapter of polygamy in his late publication, entitled Thelypthora, by a layman* (London, 1780); and Thomas Haweis, *A Scriptural refutation of the arguments for polygamy advanced in a treatise entitled Thelypthora* (London, 1781).

103. For the history of the Magdalen Hospital see Stanley Nash, "Prostitution and Charity: The Magdalen Hospital, A Case Study," *Journal of Social History* 17 (1984): 617–28; Lloyd, "Pleasure's Golden Bait," 51–72; Andrew, *Philanthropy and Police*, 119–27; Miles Ogborn, *Spaces of Modernity: London geographies, 1680–1780* (London: Guilford, 1998), 39–74.

104. RCS, Lock Board Book 2, September 16 1758.

105. RCS, Lock Board Book 2, September 23, 1758, September 30, 1758, and October 7, 1758.

106. RCS, Lock Board Book 3, 241.

107. In 1767 Bart's governors ordered not to extend charity beds to Magdalen inmates any longer, demanding the charity to pay four pence per day like other patients. Interestingly, Bart's governors complained that many women treated for free never returned to the Magdalen House, speculating that they used the Magdalen only to arrange a hospital bed, never intending to reform at all. They complained "that great Abuses are made of the Indulgence of this Hospital by many such Patients not returning to the Magdalen House when cured who make Use of that Recommendation only as a means to get into this Hospital on the

House Account": SBH, HA 1/13, 492–3. St. Thomas's admission registers (1773–76) notes the Magdalen's secretary regularly paying admission fees for women at that hospital: LMA, H1/ST/B3/1/1 H1/ST/B3/1/2, and H1/ST/B3/2.

108. James Stephen Taylor, devotes a single page to the "ephemeral" hospital: *Jonas Hanway, Founder of the Marine Society: Charity and Policy in Eighteenth-Century Britain* (London: Scolar Press, 1985), 122; R. Everett Jayne, *Jonas Hanway: Philanthropist, Politician, and Author* (London: Epworth Press, 1929), makes no mention of the charity.

109. On the currency of this term at the Kingsland outhouse see Bettley, "Post Voluptatem Misericordia: The Rise and Fall of the London Lock Hospitals," *London Journal* 10 (1984): 167.

110. *The Medical Register for the year 1780* (London, 1780), 53; *The Medical Register for the year 1783* (London, 1783), 34.

111. Herbert Jones, *A Sermon Preached before the Governors of the Misericordia Hospital at St. Andrew's Church, Holborn, on Wednesday the 16th of November 1774; on occasion of Opening the said Hospital* (London, 1775); William Dodd, *A Sermon on John XIII, 35. Preached on Wednesday January 24, 1776 at the parish church of St Stephen, Walbrook before the president . . . of the Misericordia Hospital* (London: Printed by the charity, 1776); Jonas Hanway, *An Account of the Misericordia Hospital for the Cure of indigent Persons, involved in the Miseries occasioned by promiscuous Commerce. With Moral and religious Advice to the Patients* (London, 1780); Anonymous, *Hymns to be sung by Mrs. Kennedy . . . at St. Dunstan, Fleet Street . . . for the benefit of the Misericordia Hospital* (1781).

112. Hanway describes the Misericordia as "devoted to the relief of the eastern parts of these vast cities": *Account of the Misericordia Hospital*, 84.

113. Ibid., ii. It was Blizzard who approached Hanway to be treasurer.

114. Lloyd, "Pleasure's Golden Bait," 55.

115. Taylor, *Jonas Hanway*, 123.

116. Jones, *Sermon Preached before the Governors of the Misericordia Hospital*, 21–2.

117. Ogborn sees the Magdalen as a key early attempt to produce the modern subjective self through the technology of solitude, *Spaces of Modernity*, 60–70.

118. Michael Ignatieff, *A Just Measure of Pain: The Penitentiary in the Industrial Revolution, 1750–1850* (London: Macmillan, 1978). On Hanway's role in the penitentiary movement see 54, 78, 83–4, 90, and Ogborn, *Spaces of Modernity*, 43–5. On the penitentiary movement see also the seminal contribution of Michel Foucault, *Discipline and Punish: The Birth of the Prison*, trans from the French by Alan Sheridan (New York: Vintage, 1979), as well as John Bender *Imagining the Penitentiary: Fiction and the Architecture of Mind in Eighteenth Century England* (Chicago: University of Chicago Press, 1987).

119. Ignatieff, *A Just Measure of Pain*, 63, 83 and 84.

120. Hanway, *Account of the Misericordia Hospital*, ii and vi.

121. Ibid., 28.

122. Ogborn, *Spaces of Modernity*, 58–60.
123. Hanway, *Account of the Misericordia Hospital*, 26.
124. Ibid., 6.
125. Ibid., 10–11.
126. Ogborn, *Spaces of Modernity*, 44.
127. Hanway, *Account of the Misericordia Hospital*, 66, 69, and 77; Nash, "Prostitution and Charity," 621; Ogborn, *Spaces of Modernity*, 61–63.
128. Ogborn, *Spaces of Modernity*, 79.
129. Ibid., 4–5.
130. Ibid., 39–44.
131. Ibid., 81–3.
132. Ibid., 58, 76.
133. Ignatieff, *A Just Measure of Pain*, 55–9; Andrew, *Philanthropy and Police*, 163–9. Andrew links the influence of the penitentiary model and Evangelical zeal to the Lock Asylum (187–90).
134. Merians, "The London Lock Hospital and Lock Asylum," 136–7; Williams, *The London Lock*, 57.
135. On members of the Clapham Sect see Kenneth Hylson-Smith, *Evangelicals in the Church of England* (Edinburgh: T & T Clark, 1988), 79–93. See also Leonore Davidoff and Catherine Hall, *Family Fortunes: Men and Women of the English Middle Class* (Chicago: University of Chicago Press, 1987), 76–106. On the philanthropic activities of the Clapham Sect, see Andrew, *Philanthropy and Police*, 164–9.
136. Davidoff and Hall, *Family Fortunes*. See also Anthony Fletcher, *Gender, Sex, and Subordination in England, 1500–1800* (New Haven: Yale University Press, 1995).
137. These various lines of criticism are explored in Robert Shoemaker, *Gender in English Society, 1650–1850* (London: Longman, 1998). See also the important article by Amanda Vickery "Golden age to Separate Spheres? A Review of Categories and Chronology of English Women's History," *Historical Journal* 36 (1993): 383–414.
138. Linda Colley, *Britons: Forging the Nation, 1707–1837* (New Haven: Yale University press, 1992), 237–81.
139. On the Evangelicals' vision of gender and domesticity see Catherine Hall, "The Early Formation of Victorian Domestic Ideology," in her *White, Male, and Middle Class: Explorations in the history of feminism* (New York: Routledge, 1992), 75–93; Anna Clark, *The Struggle for the Breeches: Gender and the Making of the British Working Class* (Berkeley. University of California Press, 1995), 92–118; Hall and Davidoff, *Family Fortunes*, 149–92; Shoemaker, *Gender in English Society*, 217–25.
140. On this outlook in late Enlightenment religious charity see Hugh Cunningham, "Introduction," and Jeroen J. H. Dekker, "Transforming the Nation and the Child: Philanthropy in the Netherlands, Belgium, France and England, c. 1780–c.1850," both in Cunningham and Innes, eds., *Charity, Philanthropy and Reform*, 7–8, 130–47.

141. Clark, *Struggle for the Breeches*, 42–62.

142. Ruth Perry, "Colonizing the Breast: Sexuality and Maternity in Eighteenth-Century England," in *Forbidden History: The State, Society and the Regulation of Sexuality in Modern Europe*, ed. John Fout, (University of Chicago Press, 1992), 137. Originally published in *The Journal of the History of Sexuality* 2 (2) (1991).

143. RCS, Lock Asylum Book I, 3–4.

144. *Account of the Institution of the Lock Asylum* (London, 1788); Merians, "The London Lock Hospital and the Lock Asylum," 141–2.

145. *Account of the Institution of the Lock Asylum* (London, 1790), 12–14.

146. Ibid., 30–1.

147. Merians, "The London Lock Hospital and Lock Asylum," 137–8.

148. *An Account of the Institution of the Lock Asylum; for the reception of Penitent Female patients, when discharged from the Lock Hospital* (London, 1793), 3–8; Andrew discusses this portrayal of prostitutes in the late eighteenth and early nineteenth centuries, *Philanthropy and Police*, 190.

149. *An Account of the Institution of the Lock Asylum* (London, 1790), 3–8.

150. For a fuller analysis see Siena, "Pollution, Promiscuity and the Pox," 562–71; Schleiner, "Infection and Cure Through Women," 505.

151. Marie E. McAllister, "Stories about the Origins of Syphilis in Eighteenth-Century England: Science, Myth and Prejudice," *Eighteenth-Century Life* 24 (2000): 22–44.

152. Mary Spongberg, *Feminizing Venereal Disease: The Body of the Prostitute in the Nineteenth-Century* (London: Macmillan, 1997), 15–35.

153. Jesse Foot, *A complete treatise of the origin, theory, and cure of the Lues Venerea, and obstructions in the urethra, illustrated by a great variety of cases* (London, 1792), 243–4.

154. John Pearson, *Observations on various items of Materia Medica used in the treatment of Lues Venerea* (London, 1800).

155. RCS, Lock Asylum Book I,10–17; Merians, "The London Lock Hospital and Lock Asylum," 139–40.

156. RCS, Lock Asylum Book I,10–17.

157. *An Account of the Institution of the Lock Asylum* (London, 1793), 7–8.

158. RCS, Lock Asylum Book I, 15–17, and 26–31.

Chapter 6

1. Randolph Trumbach seems the only scholar to have noted this aspect of the Lock's mission: *Sex and the Gender Revolution* vol. I, *Heterosexuality and the Third Gender in Enlightenment London* (Chicago: University of Chicago Press, 1998), 218.

2. Guy's records are held at the London Metropolitan Archive. Virtually the entire set of administrative records was damaged at the time of research, classified as "Unfit for Consultation."

3. See for example, Hector Charles Cameron, *Mr. Guy's Hospital, 1727–1948* (London: Longmans, 1954), or Clive Handler, ed., *Guys Hospital 250 Years* (London: Guys Hospital Gazette, 1976).

4. These figures come from Guy's admissions registers for 1760, LMA, H9/GY/B1/6. In order to identify the foul patients in the registers it was first necessary to identify Guy's foul wards by name. Without majority of the administrative records, this had to be done by identifying foul patients from other sources and tracking them into Guy's. Elizabeth Wyatt was sent from the workhouse in St. Luke's Chelsea to Guy's for the foul disease, and she entered Patience Ward on May 8, 1760. Likewise, an unmarried couple, Richard Day and Dorthory Kirk, entered Guy's for treatment of the pox several weeks later. He entered Job's Ward and she entered Patience. Job and Patience have been identified as Guy's foul wards on the assumption that Guy's segregated its venereal patients just like all other hospitals in the period. Patience was also noted as a "fluxing" ward (LMA, H9/GY/A3/1/1, 84). Patients in Job and Patience account for almost one-quarter of Guy's total clientele. It is also worth noting that the gender breakdown of Guy's foul wards seems to resemble St. Thomas's. Nearly 60 percent of Guy's foul patients (239 of 409 or 58.4 percent) were men. These figures must remain speculative until Guy's records are repaired.

5. It is entirely possible that another ward may have housed foul patients. Lazarus ward, for example, could well have been a foul ward, named for the Lazarus, the patron saint of Lepers. But this remains speculation.

6. That Guy's accepted venereal patients from very early is shown by the publication in 1734 of a set of prayers for foul patients: *Directions and Prayers for the use of Patients in the Foul Wards of the Hospital in Southwark, Founded at the Sole Costs and Charges of Thomas Guy Esq.* (London, 1734).

7. John Howard, *An Account of the Principal Lazarettos in Europe* (London 1791), 134.

8. A. E. Clark-Kennedy, *The London: A Study in the Voluntary Hospital System*, 2 vols. (London: Pitmann Medical Publications, 1962). An abridged version of the story can be found in the single volume, *London Pride: The Story of a Voluntary Hospital* (London: Hutchinson Benham, 1979).

9. Clark-Kennedy, *The London*, 1:29.

10. London Hospital General Court Minute Book (1740–1747), RLH, LH/A/2/1, 12.

11. Harrison seemed to be the driving force behind the scheme, having called for the first meeting to discuss it. See RLH, House Committee Minutes, LH/A/5/1, 30–31.

12. RHL, LH/A/2/1, 42.

13. Patients from the "Lock" are noted among the inpatients discharged as cured beginning on October 5, 1742. RLH, LH/A/5/1, 41.

14. Ibid., 45–6.

15. Ibid., 56. "This Charity being Obliged to refuse great numbers of patients whose cases require Salivation, for want of a convenient House . . . and

finding the necessity of such an addition to the Charity, from the miserable Condition those poor Objects have been in, who have applied here for relief, the Governors have thought proper to take a House in the same Street, as a Lock to the Infirmary . . . and to give this publick Notice, that any person under the misfortune of the foul disease, or other Complaints that require such provision, may apply to this Infirmary and on paying a moderate Sum for their Subsistence, during the time of Cure, if they are recommended, or appear to the Governors present to be proper objects of Compassion, they will be received or advised every morning from the hour of 11 till one."

16. RLH, House Committee Minutes, LH/A/5/1, 15.

17. See RLH, House Committee Minutes 1748–1750 (LH/A/5/3), 9–10 and 50.

18. RLH, Report Book, LH/A/4/1, 8.

19. At times patients discharged are noted to have been in the Lock. However, the clerk was not consistent. For example, we know that a patient named James Punch was in the Lock in 1746, because he was reprimanded for bad behavior. However, when Punch was discharged, he was simply listed among the other In Patients. See RHL, House Committee Minutes, LH/A/5/2, 54 and 68.

20. Ibid., 112.

21. RLH, House Committee Minutes, LH/A/5/2, 243.

22. Ibid., 330.

23. Ibid., 254.

24. Ibid., 324. Hospital publications would continue to list VD patients along with pregnant women and infectious patients as barred from the hospital. The *Charter of Incorporation of the London Hospital* (London, 1759) states: "No Women big with Child, no Children under seven years of age (except in cases of compound fractures, Amputations, or cutting for the Stone); no persons disordered in their Senses, or that have ulcerated Cases of long standing, or are suspected to have the Venereal Distemper, Small-Pox, Itch, or other infectious Disorder, or who are judged to have a consumptive or Asthmathic Complaint, shall be admitted on any Account Whatsoever" (9–10). This continued into the nineteenth century, as shown by the *Bye-Laws of the Governors of the London Hospital* (London, 1810), 12–13.

25. Gallicote and Wheeler were both children and victims of rape. Members of the London Hospital's medical staff testified that they were brought to the hospital where they were diagnosed with the foul disease: OBSP, July 3, 1751, 216–7 and May 2, 1753, 148.

26. RLH, LH/M/1/1, covers January 1–December 30, 1760; LH M/1/2, covers January 4, 1791–Jan 2, 1792; and LH/M/1/3 records January 3–December 30, 1792. To give a sense of scope, 1,461 patients appear in the 1760 register, while 1,074 patients appear in the 1791 register.

27. RLH, LH/A/9, 184–5.

28. Ibid., 187 and 190.

29. Neither the Minutes of the General Court of Governors, nor the Report Book for this period record any discussion of the topic.

30. There is enough circumstantial evidence to allow for the speculation that this proposal may have been linked to the Misericordia. Given that it opened just a month later in the same neighborhood and given Misericordia surgeon William Blizzard's connection to the London Hospital, where he was a student and would soon be a surgeon, it is not out of the question that Hanway and Blizzard may have approached the London about the venture.

31. Donna T. Andrew certainly speculates that increased military activity may have resulted in rising incidence of the disease: *Philanthropy and Police: London Charity in the Eighteenth-Century* (Princeton: Princeton University Press, 1989), 70.

32. Tim Hitchcock, "Redefining Sex in Eighteenth-Century England," *History Workshop Journal* 41 (1996), 73–92. The argument he presents here also appears in his *English Sexualities, 1700–1800* (New York, St. Martin's Press, 1997).

33. Adrian Wilson, "Illegitimacy and Its Implications in Mid Eighteenth-Century London: the Evidence of the Foundling Hospital," *Continuity and Change* 4 (1) (1989): 103–64; Nicholas Rogers, "Carnal Knowledge: Illegitimacy in Eighteenth-Century Westminster," *Journal of Social History* 23 (2) 1989): 355–75.

34. Trumbach, *Sex and the Gender Revolution*, 1:196–9.

35. Ibid., 201.

36. Trumbach, *Sex and the Gender Revolution*, 1:119. Trumbach proposes that "In the twenty-four years between 1747 and 1774, the hospital treated 10,897 patients. If half of these were men, and some were wives and children, this leaves approximately four thousand young unmarried women, most of whom were prostitutes." This recurs later when he states of the Lock's patient population, "Of the women, 15 percent were wives who had been infected by their husbands. . . . Some of the other females (possibly five percent) were young girls who had been raped. But the remainder, or at least four thousand, were prostitutes, some quite young" (218).

37. John Gillis, "Married But Not Churched: Plebeian Sexual Relations and Marital Nonconformity in Eighteenth-Century Britain," in *Tis' Nature's Fault: Unauthorized Sexuality During the Enlightenment*, ed. Robert Maccubbin (Cambridge: Cambridge University Press, 1987), and *For Better, For Worse: British Marriages, 1600 to the Present* (Oxford: Oxford University Press,1985), chap. 4.

38. On the sexuality of domestic servants see Tim Meldrum, "London Domestic Servants from Depositional Evidence, 1660–1750: Servant-Employer Sexuality in the Patriarchal Household," in *Chronicling Poverty: The Voices and Strategies of the English Poor 1640–1840*, ed. Tim Hitchcock, Peter King, and Pamela Sharpe (London: Macmillan, 1997), 46–69. See also Nicholas Rogers, "Carnal Knowledge: Illegitimacy in Eighteenth-Century Westminster," *Journal of Social History* 23 (2) (1989):369. Rogers examines bastardy enquiries and shows that many sexual unions were between relative social equals and often constituted a step in the process of courtship. He concludes that most illegitimate births resulted from "failed courtship

or the breakdown of consensual unions." Trumbach himself discusses the illegitimacy figures (*Sex and the Gender Revolution*, 1:229–75) and in doing so discusses the difficulties of women who begot children by men who had promised marriage only to desert them. Since Trumbach does recognize wider, nonmercenary sex among single women, his assumption that almost all single women who entered the Lock Hospital were prostitutes remains curious.

39. Trumbach, *Sex and the Gender Revolution*, 1:225.

40. A succinct chart of London's population growth (from which these figures come) can be found in Francis Sheppard, *London: A History* (London: Oxford University Press, 1998), 362. The population model described above relies on pioneering demographic works such as E. A. Wrigley and Roger Schofield, *The Population History of England, 1541–1871* (Cambridge: Cambridge University Press, 1981); E. A. Wrigley, "A Simple Model of London's Importance in Changing English Society and Economy, 1650–1750," *Past and Present* 37 (1967): 44–70; and Roger Finlay and Beatrice Shearer, "Population Growth and Suburban Expansion," in *London 1500–1700: The Making of a Metropolis*, ed. A. L. Beier and Roger Finlay (London: Longman, 1986), 39–57. For a review of the demographic literature see Vanessa Harding, "The Population of London, 1550–1700: a Review of the Published Evidence," *London Journal*, 15 (1990): 111–28.

41. A. L. Beir and Roger Finlay, "Introduction: The Significance of the Metropolis," in Beier and Finlay, eds., *London 1500–1700*, 9. For discussion on the relation between mortality rates and migration on London's population see Sheppard, *London: A History*, 127–9, and Roy Porter, *London: A Social History* (Cambridge, Mass.: Harvard University Press, 1995), 131–3.

42. Lynn Hollen Lees, *The Solidarities of Strangers: The English Poor Laws and the People, 1700–1948* (Cambridge: Cambridge University Press, 1998), 106–7; Nicholas Rogers, "Policing the Poor in Eighteenth-Century London: The Vagrancy Laws and Their Administration," *Social History/Histoire Sociale* 24 (1991): 127–47.

43. Peter Clark and David Souden, eds., *Migration and Society in Early Modern England* (London: Hutchinson, 1987); Malcolm Kitch, "Capital and Kingdom: Migration to Stuart London," in Beir and Finlay, eds., *London 1500–1700*, 224–51; and John Wareing, "Changes in the Geographical Distribution of the Recruitment of Apprentices to the London Companies, 1486–1750," *Journal of Historical Geography* 6 (1980), 241–9. For a recent study of intrametropolitan mobility that accounts for both rank and gender, see Robert Shoemaker, "Gendered Spaces: Patterns of Mobility and Perceptions of London's Geography, 1660–1750," in *Imagining Early Modern London: Perceptions and Portrayals of the City from Stow to Strype 1598–1720*, ed. J. F. Merritt (Cambridge: Cambridge University Press, 2001), 144–65.

44. RCS, Lock Court Book 1, 20.

45. RCS, Lock Board Book 1, June 4, 1757.

46. RCS, Lock Board Book 1, May 28, 1757.

47. Ibid., May 26, 1759.

48. RCS, Lock Board Book 2A, January 24, 1761, and July 11, 1761.
49. RCS, Lock Board Book 3, 134–5.
50. Ibid., 158–60.
51. Ibid., 224–5, 262.
52. Ibid., 205–8.
53. Ibid.
54. RCS, Lock Board Book 4, 144.
55. RCS, Lock Board Book 3, 205–8.

56. At four pence per day, an average patient at St. Thomas's would cost parishes thirteen shillings, four pence in per diem costs (forty days times four pence per day). When added to the admissions fee this still totals £1.3.10, barely more than half the two guineas charged at the Lock Hospital.

57. This remained the case even for the Lock's home parish of St. George's Hanover Square, which petitioned in 1767 to strike a deal with the hospital to take its foul paupers. The Lock denied the offer: St. George's Hanover Square, Vestry Minutes, WCA, C894, 212. RCS, Lock Court Book 2, 110.

58. LMA, St. Luke's Chelsea, Workhouse Register, Microfilm, X/15/37, Mary Roberts, July 3, 1756, Mary Robinson, February 13, 1757, Joan Rumbold, February 6, 1758, Margaret Cooten, April 23, 1759, Sarah Lutarthur, February 7, 1761, Hannah Abbott, March 8, 1766, Ann Hill, March 21, 1768, William Green, February 18, 1783, Robert Casher, July 22, 1784, Mary Smith, November 2, 1786, Charles Castleman, July 24, 1788, Mary Cammel, January 8, 1789, Francis Sproull, March 3, 1791, Sarah Smith, May 2, 1795, Jane Bearcor, October 4, 1798.

59. LMA, H1/ST/B3/7.
60. SBH, HA/1/15, 95, 132, 176, 219, 257, 288, 343, and 410; HB/15/2.
61. LMA H1/ST/B3/8.
62. Tony Henderson, *Disorderly Women in Eighteenth-Century London: Prostitution and Control in the Metropolis, 1730–1830* (London: Longman, 1999), 18–20.

63. These figures come from a database based on the weekly admission and discharge records contained in the Lock's Board Books, RCS. I reconstructed the admissions and discharges for 1760, 1765, 1770, 1780, and 1790. I used first names to determine sex.

64. Three women entered the wards with children, Isabella Brooks on July 25, 1765, Sarah Bowley on September 19, 1765, and Martha Wingate on November 21, 1765. The only married women to enter the Lock that year were Mary Lincoln on October 2, 1765, and Hannah Harrison on November 21, 1765. See RCS, Lock Board Books 3 and 4.

65. RCS, "[John] Pearson Case Book." Pearson's notes detail 86 patients admitted to the Lock during the ten-month period July 1798 to April 1799. For discussion of the symptomology of these patients see pp. 16–18.

66. RCS, John Ritchie, "The Practice of the Lock Hospital, Grosvenor Place 1813–1814," (MS).

67. RCS, "Pearson Case Book." The average length of time between first infection and entry into the Lock for the 73 patients for whom such data is available is 26.1 weeks. Of course, these figures are rough estimates, likely based on patients' own assumptions about when they first contracted the disease.

68. RCS, Ritchie, "The Practice of the Lock Hospital."

69. Ibid. Thirteen out of the fourteen fully detailed male cases include references to the types of treatments the patients attempted since becoming infected. The majority administered mercury in pill form. One man, John Barret, reported obtaining his pills from a "French surgeon." James Knight reported using injections, while Edward Bunsiss used mercurial ointment. One man, John Robinson, also reported having been treated in St. Thomas's Hospital.

70. Here James C. Scott's notion of the performative aspects of the "public transcript" in which both the dominant and the subordinate assume the roles they perceive to be expected of them is useful: *Domination and the Arts of Resistance: Hidden Transcripts* (New Haven: Yale University Press, 1990), 1–16. Lees similarly suggests the importance of this aspect of the welfare exchange at the parish level: *Solidarity of Strangers*, 33–9.

71. G. R. Quaiffe, *Wanton Wenches and Wayward Wives: Peasants and Illicit Sex in Early Seventeenth-Century England* (New Brunswick, N.J.: Rutgers University Press, 1979), 186.

72. On the dispensary movement see Robert Kilpatrick, "'Living in the Light': Dispensaries, Philanthropy and Medical Reform in Eighteenth-Century London," in *Medical Enlightenment of the Eighteenth Century*, ed. Andrew Cunningham and Roger French (Cambridge: Cambridge University Press, 1990), 254–80; Bronwyn Coxson, "The Public and Private Faces of Eighteenth-Century London Dispensary Charity," *Medical History* 41 (1997): 127–49.

73. *Medical Register for the Year 1783* (London, 1783), 36.

74. Wellcome Library for the History and Understanding of Medicine, Misericordia General Dispensary, Blank Recommendation Form (London, 1780–1783 [?]).

75. On the economic crisis's effect in London see Leonard Schwarz, *London in the Age of Industrialisation: Entrepreneurs, Labour Force and Living Conditions* (Cambridge: Cambridge University Press, 1992), 157–78. Deborah Valenze links the economic crisis to rising poor relief demands: "Charity, Custom, and Humanity: Changing Attitudes Towards the Poor in Eighteenth-Century England" in *Revival and Religion Since 1700: Essays for John Walsh*, eds. Jane Garnett and Colin Matthew (London: Hambledon Press, 1993), 64.

76. RCS, Lock Board Book 4, November 21, 1765, and January 16, 1766.

77. RCS, Lock Board Book 3, February 21, 1765 and Lock Board Book 4, July 27, 1765.

78. RCS, Lock Board Book 4, November 21, 1765, January 30, 1766, May 1, 1766, and May 29, 1766. Bland's final discharge reads, "Richard Bland recommended by Lord Sondes some time ago & then postpon'd for want of room was

discharged from the postpon'd file for not attending to be admitted in his turn . . . he not having appear'd since the 15th instant."

79. They were John Finnegan, John Ramsey, and Rocca Balcini. See Lock Board Book 4, April 18, 1765, May 9, 1765, May 30, 1765, June 6, 1765, December 12, 1765, and January 9, 1766.

80. RCS, Lock Board Book 3, 106–7.

81. Common in the weekly admissions records were statements like the following from November 7, 1765: "John Kirk recommended by The Dk. of Athol; Terrence Kelly recommended by Lord Vere, & Robt. Armstrong recommended by Anthony Gifford Esqr. cou'd not be Admitted as their respective recommenders have Patients in the House." RCS, Lock Board Book 4, 84.

82. WALM, Westminster Coroner's Inquests, 1760–1799; Robert Munro, November 22, 1762. I am thankful to Tim Hitchcock for acquainting me with this case.

83. Ibid.

84. Of 383 total applicants 128 were postponed that year (33.4 percent), and 27 of those 128 (21 percent) gave up waiting before they gained admission.

85. For a table on the total number of patients treated in the Lock Hospital during the eighteenth century, see Linda E. Merians, "The London Lock Hospital and the Lock Asylum for Women," in *The Secret Malady: Venereal Disease in Eighteenth-Century Britain and France*, ed. Linda E. Merians (Lexington: University of Kentucky Press, 1996), 133. The figures come from the annual report given each March at the annual meeting of the General Court: RCS, Lock Court Books 1–3.

86. RCS, Lock Board Book 2A, January 23, 1762.

87. There is only one case even similar in either of these two years. A patient named Benjamin Carpenter was discharged in 1760, "being not a clear case": RCS, Lock Board Book 2, February 23, 1760.

88. Pearson's casebook has several torn pages, leaving researchers biographical information on eighty-six patients, but complete clinical data on only seventy-five of them.

89. Ritchie administered mercury to virtually all of his patients.

90. RCS, "Pearson Case Book." This figure was calculated by counting all the days on which Pearson noted new information about each patient.

91. Ibid., This can only be calculated for the patients for whom precise dates of the start and end of salivation were noted. The average number of days Pearson's patients were not salivated was 8.4.

92. The twenty-four patients for whom we can calculate the exact duration of their salivation, salivated for an average of 40.4 days.

93. Ibid., Of the twenty-four patients, Elizabeth Field remained in the house fifteen days after her salivation ended, and Sarah Gardmen stayed twelve days.

94. John Pearson, *Observations on various items of Materia Medica used in the treatment of Lues Venerea* (London, 1800).

95. Bougies were one of the supplies that Bromfeild was accused of stealing during the 1780 enquiry: RCS, Lock Court Book 3, 69–74. Bougies seem to have been a French import in London. One prominent proponent whose treatises were published in English was Francis Lallier. See his *Instructions necessary for using the anti-venereal bougie in curing the retention of urine, and all maladies in the urethra* (London, 1753).

96. RCS, Lock Board Book 6, 19.

97. RCS, Lock Court Book 3, 116.

98. RCS, Lock Board Book 10, 239.

99. RCS, Lock Court Book 3, 334.

100. RCS, Lock Board Book 7, September 13, 1770. Similarly, 1765 witnessed fifteen such incidents. Nine patients ran away and six patients were discharged for failing to obey regulations.

101. Of 698 total applicants, six hundred actually entered the house when one accounts for outpatients and unsuccessful applicants. The twenty patients who ran away or were discharged represent just 3.33 percent of all patients.

102. RCS, Lock Asylum Book 1, 10–17.

103. The Lock Asylum Book that contains this information is extremely sparce from mid-1789 through the end of the century. Only a handful of sporadic entries exist from the 1790s and no information about the inmates is available from this period.

104. It is possible to reconstruct partially the admissions and discharges for the period July 1787 to December 1788, because both admissions and discharge comments survive for many of the inmates. Information exists for many women admitted in 1789. However, discharge comments exist for only two patients admitted after 1788. The Lock Asylum Book gives the names of ninety-five total inmates. Sixty-four inmates were admitted between July 1787 and December 1788. Eleven of those 64 (15.7 percent) ran away.

105. *An Account of the Institution of the Lock Asylum* (London, 1796).

106. Merians, "The London Lock Hospital and the Lock Asylum," 142.

107. The governors published histories of some of the "success" stories from the asylum. These tended to be stories of women placed in domestic service positions after discharge. But by 1796 the governors had to address their poor success rates. They asserted, "if amidst reiterated disappointments, we be successful only in a few instances, and a very small number be brought to true repentance and a Christian conversion: this will be an abundant compensation." *An Account of the Institution of the Lock Asylum* (London, 1796), 15. See also Merians's discussion of the topic, "The London Lock Hospital and the Lock Asylum," 139–42.

Conclusion

1. OBSP, March 27, 1734, 77.
2. Leonore Davidoff and Catherine Hall, *Family Fortunes: Men and Women of the English Middle Class* (Chicago: University of Chicago Press, 1987).
3. For instance, in Mary E. Fissell, "The Disappearance of the Patient's Narrative and the Invention of Hospital Medicine," in *British Medicine in an Age of Reform*, ed. Roger French and Andrew Wear (London: Routledge, 1991).

BIBLIOGRAPHY

I. Primary Sources

Selected Manuscript Sources

British Library
St. Martin's in the Fields, Disbursements for the Poor of St. Martins in the Fields. 1688–1689. MS 44934.

Camden Local Studies and Archive Centre
St. Andrew Holborn and St. George the Martyr, Combined Workhouse Registers, 1750–55, 1758–59. P/AH/PO/3–5.
———, Weekly Meetings of the Overseers of the Poor and Churchwardens, 1737–1738. P/AH/PO/2.

Corporation of London Records Office
London and Southwark Coroner's Inquests (1794). 400A Box 7.

Guildhall Library
St. Andrew's Holborn, Shoe Lane Workhouse Admission Books 1776–1796. MS 9598/1–4.
St. Sepulchre, London Division, Minutes of the Committee for Rebuilding the Workhouse, 1796–1838. MS 3226/1–3.
———, Workhouse Committee Minutes, 1727–1800, MS 3137/1–3137/16.
———, Workhouse Inspection Committee Minutes, 1730–1829, MS 3220/1–12.
———, Workhouse Maintenance Account, 1728–1746, MS 3227.
———, Workhouse Master's Receipts, 1708–1745, MS 3270.

330 Bibliography

St. Sepulchre, Workhouse Receipts, 1704–1729. MS 3269.
———, Workhouse Tradesmen Accounts, 1727–1825, MS 3177/1–6.
———, Workhouse Treasurer's Book, 1744–1746, MS 3237 1/2.
———, Vestry Minutes, 1653–1662, MS 3149/1.
———, Middlesex, Settlement Examinations, 1773–1780. MS 9095/2.

Royal London Hospital Archives
———, General Court Minute Books, 1740–1786, LH/A2/1–4.
———, House Committee Minutes, 1741–1794, LH/A/5/1–12.
———, House Committee Report Books, 1742–1775, LH/A/4/1–5.
———, House Visitors Book, 1749–1756, LH/A/16/1.
———, Patient Registers, 1760, 1791 & 1792, LH/M/1/1–3.
———, Steward's Cash Ledger, 1752–1760, LH/F/7/1–3.
———, Treasurer's Cash Books, 1740–1748. LH/F/1/1–2.

London Metropolitan Archives
Foundling Hospital, Account of Foundlings Treated in the Lock, 1760. A/FH/A6/7/8/1.
Guy's Hospital, Admission and Discharge Registers, 1760. H9/GY/B4/1.
———, Admission and Discharge Registers, 1770. H9/GY/B4/4.
———, Admission and Discharge Registers, 1780. H9/GY/B4/6.
———, Court of Committees Minutes, 1725 –1733. H9/GY/A3/1/1.
———, Stationers Bills. H9/GY/D81/1–15.
St. Luke's Chelsea, Workhouse Committee Book, 1735–1750. P74/LUK/3.
———, Workhouse Admissions and Discharges, 1743–1769, 1782–1799. Microfilm x/15/37.
St. Mary's Newington, Workhouse Register Book, 1783–1786. P85/MRY1/279.
St. Thomas's Hospital, Admission and Discharge Registers, 1773–1800. H1/ST/B3/1/1 H1/ST/B3/1/2, H1/ST/B3/2 –13.
———, General Court of Governors Minutes, 1556–1800. H1/ST/A4/1–8.
———, Grand Committee Minutes, 1634–1800. H1/ST/A6/1–7.
———, Hospital Rules, ca.1752. H1/ST/A25.
———, Patient Indexes, 1768–1772, 1786–1792 and 1796–1802. H1/ST/B4/1–3.
———, Sub–Committee Minutes, 1730–1765. H1/ST/A7/1.
Surrey Dispensary, Governors Minutes, 1777–1806, A/SD/2–4.
———, Patient Registers. 1782–1785. LMA A/SD/38.
Westminster Infirmary. "Resolutions and Orders of the Westminster Hospital," 1719–1774. H2/WH/A1/64.
———, Trustees' Minutes, 1738–1754. H2/WH/A1/5.

Royal College of Physicians
Drury Lane Dispensary, Account of Cash Received and Disbursed, 1782–1793, MS 2508 D.
———, Minute Book, 1796–1828, MS 2498 D.
Westminster Dispensary, Governors Minutes, 1774–1793, MS 629.

Royal College of Surgeons
Lock Asylum, Lock Asylum Cash Book, 1787–1806. Lock Asylum Cash Book 1.
———, Lock Asylum Committee Minute Book, 1787–1814. Lock Asylum Book 1.
Lock Hospital, Board of Governors' Minutes, 1755–1800. Lock Board Books 1–15.
———, General Court of Governors' Minutes, 1746–1816. Lock Court Books 1–4.
———, Lock Hospital Cash Book, 1765–1784 Lock Hospital Cash Books 1–2.
———, Select Committee Minutes, 1781–1842, Lock Hospital Select Committee Book 1.
Pearson, John, "Pearson Case Book." 1798. Lock MS, "Mr. Pearson."
Pott, Percivall. "Chirurgical Lectures by Percivall Pott F.R.S. and Senior Surgeon of St. Bartholomew's Hospl. London." 1785/86? 42.d.32 (MSS Pot).
Ritchie, John. "The Practice of the Lock Hospital, Grosvenor Place 1813–1814." Lock MS, "Drug book & case notes of J. Ritchie."

St. Bartholomew's Hospital Archives
———, Annual Accounts, 1615–1814. Ledger, HB 1/4–1/13, HB 2/1–2/4.
———, Governors' Minutes, 1549–1556, 1607–1800. Journal, HA 1/1, HA 1/4–1/15.
St. Bartholomew's the Less, Hospital Petitions, 1711–1773. SBL 49 (1–7).
———, Letter to Churchwarden from a Sister of the Lock Hospital (Southwark) Seeking Payment for a Discharged Patient, 1747. SBL 50.

UMDS, Medical Library, St. Thomas's Hospital Campus
Oxley, Charles. "Charles Oxley Notebook, Case Notes of Surgical Patients Admitted to St. Thomas's Hospital, 1725–26." MS. 31.d.2.

Westminster City Archives Centre
St. George Hanover Square, Committee of Governors and Directors of the Poor Minutes, 1765–1768. C. 893–5.
———, Workhouse Committee Minutes, 1726–1746 C. 869–79.
St. James, Westminster, "Money laid out att ye Hospitall," 1691–1695. D1711/11.

St. Margaret's, Westminster, Committee for the Workhouse Orders to Overseers, 1728–1729. E3382.
———, "Mr. Roger bill for ye bleeding & fisicking of ye Severall poore," 1706. E Unlisted <> ["Diamond"] 1035.
———, "Orders of Sessions re paupers" (1727). E3385/17.
———, Overseer's Vouchers, 1728–9. E 3382.
———, Overseer's Personal Pay Books (1740). E2038.
———, Overseer's Receipts, 1692–1748. E Unlisted <> ["Diamond"] 1035.
———, "Petty Sessions Orders Concerning Workhouse, 1728–1729. E3382.
———, "Reports of the state of the poor", 1752 and 1763. E Unlisted <> ["Diamond"] 1177.
———, Rules and Orders of the Workhouse, 1742. E Unlisted <> ["Diamond"] 931.
———, "St. Margaret's Westminster Hospital Petitions." MS 226. (1–3).
———, Vestry Minutes 1674–1790, E2416–E2422.
———, Workhouse Account Book, 1740–1746. E2152.
———, Workhouse Committee Minutes, 1726–1748, E2632–E2638.
———, Workhouse Orders 1727–1728, E Unlisted <> ["Diamond"] 1066.
———, Workhouse Rules, Orders and Correspondence, 1746–1766. E573.
St. Marylebone General Dispensary, Governors' Minutes, 1785–1821. M.403/38.

Westminster Abbey Library and Muniments Room.
Westminster Coroner's Inquests, 1760–1799.

Selected Published Sources

Anon., *A General Account of the Dispensary for the relief of the Infant Poor*. London, 1769.
Anon., *A Gonorrhea, And the effects of the Application of Venereal Matter, To the Parts of Generation of both Sexes, Explain'd on rational Principles; And short Account of A new invented Remedy, for the safe, easy and speedy Cure, and effectual Prevention of this Disease. With all its dreadful Consequences. By an eminent Surgeon in London*. London, 1780.
Anon., *A Letter to the Rev. Mr. Madan, concerning the chapter of polygamy in his late publication, entitled Thelypthora, by a layman*. London, 1780.
Anon., *A Plan of the Surrey Dispensary in Montague-Close, near St. Saviour's Church for the Relief of the Poor Inhabitants of the Borough of Southwark and places Adjacent, at their own habitations. Instituted in the year 1777*. London, 1777 (?).
Anon., *An Account of the General Dispensary for the Relief of the Poor instituted 1770 in Aldersgate-Street*. London, 1776.
Anon., *An Account of the Middlesex Dispensary. Instituted in the Year 1778*. London, 1783.

Anon., *An Account of Several Work-Houses for Employing and Maintaining the Poor*. London, 1725.

Anon., *Directions and Prayers for the Use of the Patients in the Foul Wards of the Hospital in Southwark, Founded at the sole costs and charges of Thomas Guy, Esq.*. London, 1734.

Anon., *Eronania: On the Misusings of the Marriage-Bed by Er and Onan, Judah's two sons: Genesis 38. Or the Hainous Crime of Self-Defilement . . . With another Treatise on the Cure of the Gout, a Gleet, and a Gonorrhoea, without taking any Thing at the Mouth*. London, 1724.

Anon., *General Sketch of the Plan of the New Finsbury Dispensary, in St. John's-Street, Clerkenwell, for administering Advice and Medicines to the Poor at the Dispensary, or at their own Habitations within certain districts. Instituted September 20, 1786*. London, 1786.

Anon., *Medical Register for the Year 1779*. London, 1779.

Anon., *Medical Register for the Year 1780*. London, 1780.

Anon., *Medical Register for the Year 1783*. London, 1783.

Anon., *Pharmacopaeia Venerea: Or A Compleat Venereal Dispensatory*. London, 1725.

Anon., *Plan of the London Dispensary, Primrose Street, Bishopsgate Without; for administering Advice and Medicines to the poor at the Dispensary or their own habitations. Instituted, 1777*. London, 1780.

Anon., *Plan of the Publick Dispensary, in Carey-Street, Lincoln's-Inn-Fields, Instituted 1783*. London, 1788.

Anon., *Plan of the Western Dispensary in Charles-Street, Westminster, Instituted in the year 1789, for administering Advice and Medicines to the Poor Inhabitants of the City of Westminster, and places adjacent, at the Dispensary, or at their own Habitations*. London, 1789 & 1801.

Anon., *Regulations which were agreed upon and established the twelfth day of July, 1726. By the Gentlemen of the vestry then present, for the better government and management of the workhouse belonging to the parish of St. Giles's in the fields, humbly offered to the consideration of the House of commons of Ireland*. Dublin, 1727.

Anon., *Rules and Orders for the Regulation of the Workhouse, of the Parish of Enfielf in the County of Middlesex*. London, 1787.

Anon., *Some Reasons of a Member of the Committee, &c. of the Trustees of the Infirmary in James Street Westminster, near St. James Park, for his dividing against the Admission of Venereal Patients. In a Letter to a Lady*. London, 1738.

Anon., *Some Reasons Tending to remove the Prejudices against the Lock-Hospital*. London, 1770.

Anon., *The Modern Practice of the London Hospitals viz. St. Bartholomew's, St. Thomas's, Guy's, St. Georges's, The Portuguese, and The Lock at Hyde-park Corner*. London, 1770.

Anon., *The Practical Scheme of the Secret Disease*. 17th ed. London, 1728.

Anon., *The Tomb of Venus: or, A Plain and Certain Method, by which All People That ever labour'd under any Venereal Distemper may infallibly know whether they are cured or not*. London, 1710.

Anon., *Thoughts on the Means of Alleviating the Miseries Attendant upon Common Prostitution*. London, 1799.

Andree, John. *Observations on the Theory and Cure of the Venereal Disease.* London, 1779.

Archer, John. *Every Man His Own Doctor.* London,1671.

———. *Secrets Disclosed, of Consumptions shewing how to distinguish between Scurvy and Venereal Disease.* London,1684.

Armstrong, Charles. *An Essay on the Symptoms and Cure of the Virulent Gonorrhea in Females.* London, 1783.

Armstrong, John. *A Synopsis of the History and Cure of Venereal Diseases.* London, 1737.

Astruc, Jean. *A Treatise of the Venereal Disease.* Translated from the French by William Barrowby. London, 1754.

Atkins, John. *The Navy Surgeon: or a Practical System of Surgery . . . To which is added a Treatise on the Venereal Disease.* London, 1734.

Beckett, William. *A Collection of Chirurgical Tracts . . . written and collected by William Beckett.* London, 1740.

Blanckaert, Steven. *A New Method of Curing the French-Pox.* London, 1690.

Blegny, Nicholas de. *New and Curious Observations concerning the art of Curing the Venereal Disease, and the accidents it produces . . . Explained by Natural and Mechanical principles with Motions, Actions, and Effects of Mercury.* Translated from the French by Walter Harris. London, 1676.

Boerhaave, Herman. *A Treatise on the Venereal Disease and its Cure in all its Stages and Circumstances.* London, 1729.

Boulton, Richard. *Phisico-Chirurgical Treatise of the Gout, the Kings Evil, and the Lues Venerea.* London, 1714.

B.P. *Pilulae Antipudendagriae: or Venus's Revenge. Whereby every one may Secretly Cure and Preserve themselves from all Venereal Evils.* London,1669.

Brown, Richard. *A Letter from a Physician in London to his Friend in the Country; Giving an Account of the Montpellier Practice in curing the Venereal Disease.* London, 1730.

Burroughs, Henry. *An Address to the Public; But more especially to the Inhabitants of the Parish of St. Leonard, Shoreditch, Relative to the master of their Workhouse. By Mr. Burroughs of Hoxton.* London, 1784.

Burrows, John. *A Dissertation on the Nature and Effects of a New Vegetable Remedy, Known by the Name of Velno's Vegetable Syrup, An Acknowledged Specific in all Venereal and Scorbutic Cases.* London, 1772.

Cam, Joseph. *A Practical Treatise: or Second Thoughts on the Consequences of the Venereal Disease in Three Parts* 3rd ed.. London, 1729.

———. *The Practice of Salivating Vindicated: In answer to Dr. Willoghby's Translation of Mons. Chicoyneau's Pamphlet against mercurial salivations.* London, 1724.

Chapman, Samuel. *An Essay on the Venereal Gleet; in which the Different Species of this Disorder are distinguished, and Their Causes assigned.* London, 1751.

Chicoyneau, Francois. *The Practice of Salivating Shewn to be of no Use or Efficacy in the Cure of Venereal Disease.* Translated from the French by C. Willoughby. London, 1723.

City of London. *A true Report of the great number of Poor Children, and other Poor People, maintained in the several Hospitals, under the pious Care of the Lord Mayor, Commonality,* and Citizens, of the City of London, the Year last past. London, 1689, 1721, 1727, and 1730–1800.

Clare, Peter. *A Practical Treatise on the Gonorrhoea; Recommending the use of Injection, as the most Speedy and effeaciacious Method of Cure.* 4th ed. London, 1784.

Clowes, William. *A Brief and Necessary Treatise, Touching on the Cure of the Disease now called Lues Venerea, by unctions and other approved waies of curing: newly corrected and augmented in the yeare of our Lord 1596.* London, 1596.

———. *A short and profitable Treatise touching on the cure of a disease now called Morbus Gallicus.* London, 1579.

———. *The Selected Writings of William Clowes, 1544–1604.* F. N. L. Poynter, ed. London: Harvey & Blythe, 1948.

Clubbe, John. *An Essay on the Virulent Gonorrhea.* London, 1786.

Cole, Abdiah, Felix Plater, and Nicholas Culpeper. *A Golden Practice of Pysick in Five Books and Three Tomes . . . Unto which is added two excellent Treatises. 1. Of the French Pox. 2. Of the Gout.* London, 1662.

Cowper, James. *A Narrative of the Effects of a Medicine, newly discovered, by Mr. Keyser, a German Chymist in Paris, that covers The Venereal Disease in its most inverterate and malignant state, without Salivation or strict regimen, as is now practiced in France, both in private cases and in the Military Hospitals.* London, 1760.

Daran, Jaques. *A Complete Treatise of the Virulent Gonorrhoea both in Men and Women,* Translated from the French by Andrew Blake. London, 1767.

De Coetlogon, Charles. *National Prosperity and National Religion Inseparably Connected: A Sermon Preached on Friday, December 13, 1776.* London, 1777.

Desault, Pierre. *A Treatise on the Venereal Distemper,* Translated from the French by John Andree. London, 1738.

Dodd, William. *A Sermon on John XIII, 35. Preached on Wednesday January 24, 1776 at the parish church of St. Stephen, Walbrook before the president . . . of the Misericordia Hospital.* London, 1776.

Dufour, William. *A Treatise on Diseases in the Urinary Passages . . . To which are added some New Observations on the Venereal Disease, with remarks on A Method of Cure, that requires no restraint, either with respect to diet or confinement, Subjects the patient to no exposure, nor his constitution to the least degree of danger.* London, 1795.

Dunbar, James Innes. *A Treatise of the Venereal Disease.* London, 1783.

Duncan, Andrew. *Observations on the Operation and Use of Mercury in the Venereal Disease.* London, 1772.

Dunn, Edward. *A compendius and new method of performing chirurgical operations, fit for young surgeons. To which are added, short and easy directions how to manage the venereal disease.* London, 1724.

Ellis, W. *An Essay on the cure of the Gonorrhoea, in a new method with some observations on Gleets.* London, 1771.

Falck, Nikolai. *A Treatise on the Venereal Disease.* London, 1772.

Foot, Jesse. *A complete treatise of the origin, theory, and cure of the Lues Venerea, and obstructions in the urethra, illustrated by a great variety of cases.* London, 1792.

Fordyce, George. *Elements of the Practice of Physick*. 5th ed. London, 1784.
Fordyce, William. *A Review of the Venereal Disease, and its Remedies*. London, 1767.
Freeman, Stephan. *A New Essay on the Venereal Disease: Or, every Person afflicted with that Disorder Their Own Physician*. 2nd ed. London, 1780.
Garlick, Thomas. *A Mechanical account of the Cause and Cure of a Virulent Gonorrhoea in both sexes*. 2nd ed. London, 1727.
Good, John Mason. *Dissertation on the Diseases of Prisons and Poor houses, Published at the request of the Medical Society of London, Having obtained the Premium offered by the Society for the best Essay on the Subject*. London, 1795.
Grubb, Robert. *A New Treatise on the Venereal Disease; or, every person afflicted with the disorder their own physician*. London, 1780 (?).
Hanway, Jonas. *An Account of the Misericordia Hospital for the Cure of indigent Persons, involved in the Miseries occasioned by promiscuous Commerce. With Moral and religious Advice to the Patients*. London, 1780.
Harris, Walter. *A Treatise of the Acute Diseases of Infants. To which are added, Medical Observations on Several grievous Diseases*. London, 1742.
Harvey, Gideon. *Great Venus Unmasked: Or a more exact Discovery of the Venereal Evil or French Disease*. 2nd ed. London, 1672.
Hauksbee, Francis. *A Further Account of the Effects of Mr. Hauksbee's Alternative Medicine, as applied in the Cure of the Venereal Disease*. London, 1743.
Haweis, Thomas. *A Scriptural refutation of the arguments for polygamy advanced in a treatise entitled Thelyphthora*. London, 1781.
Henriet, M. *The Antisiphylitic: or Public Health. A sure Method of guarding against all sorts of Venereal Disorders, and of curing our selves if afflicted by using, an anti-venereal water called water of Salubrity*. London, 1772.
Horne, Charles. *Serious Thoughts on the Miseries of Seduction and Prostitution, with a Full Account of the Evils that Produce them; plainly shewing, Seduction and Prostitution to be contrary to the Laws of Nature*. London, 1783.
Houlston, William. *Sketches of Facts and Opinions respecting the Venereal Disease*. London, 1794.
Howard, John. *An Account of the Principal Lazarettos in Europe*. London, 1791.
J. S. *A Short Compendium of Chirurgery . . . With The Several Methods of Treating the French Pox: The Cure of Baldness, Inflammation of the Eyes, and Tooth-ach: And an Account of Blood-letting, Cup-setting, and Blooding with Leeches*. London, 1678.
Jones, Herbert. *A Sermon Preached before the Governors of the Misericordia Hospital at St. Andrew's Church, Holborn, on Wednesday the 16th of November 1774; on occasion of Opening the said Hospital*. London, 1775.
Lallier, Francis. *A Dissertation on the Disorders which affect the Neck of the Bladder, the urinary Passage, and the neighboring parts*. 10th ed. London, 1795.
———. *Instructions Necessary for Using the Anti-venereal Bougie, in curing he Retention of Urine, and All Maladies in the Urethra*. London, 1753.
Leake, John. *A Dissertation on the Properties and Efficacy of the Lisbon Diet-Drink . . . To which is added an Appendix; with a succinct account of such remarkable Venereal and*

Scorbutic Cases as have been successfully treated by a Course of the Lisbon Diet Drink. 3rd ed. London, 1762.

———. *A Dissertation on the Properties and Efficacy of the Lisbon Diet-Drink . . . To which is now added A Supplement, containing Plain Rules for distinguishing Venereal Symptoms from those often mistaken for such by Patients.* London, 1790.

Lettsom, John Coackley. *Medical Memoirs of the General Dispensary in London, for part of the years 1773 and 1774.* London, 1774.

———. *Of the improvement of Medicine in London on the Basis of Public Good.* 2nd ed. London, 1770.

Lock Asylum. *An Account of the Institution of the Lock Asylum; for the reception of Penitent Female patients, when discharged from the Lock Hospital.* London, 1793.

———. *An Account of the Institution of the Lock Asylum; for the reception of Penitent Female patients, when discharged from the Lock Hospital.* London, 1796.

Lock Hospital. *Account of the proceedings of the Lock Hospital near Hyde-Park Corner.* London, 1749.

———. *Account of the proceedings of the Lock Hospital near Hyde-Park Corner.* London, 1751.

———. *Account of the proceedings of the Lock Hospital near Hyde-Park Corner.* London, 1767.

———. *Account of the proceedings of the Lock Hospital near Hyde-Park Corner.* London, 1789.

———. *An Account of the nature and Intention of the Lock Hospital near Hyde Park Corner; the Proceedings of the Governors, and the improvements lately adopted; with an Abstract of its income and expenditure, the state of its finances to Lady-Day, 1810, and a list of the Governors and Subscribers.* London, 1810.

———. *An Address to the President, Vice-Presidents, and the other Governors of the Lock Hospital near Hyde park Corner on Behalf of that Charity.* London, 1781.

———. *Some Reasons Tending to remove the Prejudices against the Lock-Hospital* London, 1770 (?).

London Hospital. *Bye-Laws of the Governors of the London Hospital, Made and Ordained Pursuant to their Charter, Bearing the date the Ninth of December, in the Thirty-second Year of the reign of his Majesty King George the Second. To which are annexed Rules and Orders for the Better Government of the Said Hospital.* London, 1810.

———. *Charter of Incorporation of the London Hospital.* London, 1759.

Lowe, Peter. *An Easie, certaine, and perfect method, to cure and prevent the Spanish Sicknes.* London, 1596.

L. S. Προφυλακτικον [Profulacticon]*: Or Some Considerations of a Notable Expedient To root out the French Pox From the English Nation. With Excellent Defensive Remedies to Preserve Mankind from the Infection of Pocky Women.* London, 1673.

Madan, Martin. *A collection of psalm and hymn tunes, never published before. To be had at the Lock Hospital, near Hyde Park Corner.* London, 1769.

———. *A new, improved edition of the collection of psalm and hymn tunes sung at the chapel of the Lock Hospital with considerable additions.* London, 1792.

———. *A Sermon Preached at the Parish-Church of St. George, Hanover-Square, For the Benefit of the Lock-Hospital, on Tuesday, February 25th, 1777.* London, 1777.

———. *An account of the triumphant death of F. S. A converted prostitute, who died April, 1763, aged twenty-six years.* London, 1764 (?).

———. *The Magdalen: or Dying Penitent. Exemplified in the Death of F. S.. Who died April, 1763, aged Twenty-six Years. To which is added A short Account of Poor Joseph.* Dublin, 1789.

———. *Thelyphthora, or a Treatise on Female Ruin.* London, 1780.

———. *Every man our Neighbour. A Sermon preached at the Opening of the Chapel, of the Lock-Hospital near Hyde-Park-Corner, March 28, MDCCLXII.* London, 1762.

Marriot, John. *A prepresentation of some Mismanagements by Parish-Officers In the Method at present followed for Maintaining the Poor; with a Proposal Humbly offered for Rectifying the same.* London, 1726.

Marten, John, *A Treatise of all the Degrees and Symptoms of the Venereal Disease.* 6th ed. London, 1708.

Maynwaring, Everad. *The History and Mystery of the Venereal Lues.* London, 1673.

Millar, John. *Observations on the Practice in the Medical Department of the Westminster General Dispensary: together with an Arithmetical calculation of the comparative success of various establishments for the Relief of the Sick.* London, 1777.

Mooney, M. *A Dissertation on the Nature and Cure of the Venereal Disease.* London, 1756.

Needham, Thomas. *A Treatise of the Consumption and the Venereal Disease: the signs or symptoms of the venereal Infection with various methods of cure.* London, 1700.

Nicholson. I. F. *The Modern Siphylis: or, The True Method of Curing every Stage and Symptom of the Venereal Disease.* London, 1718.

Parfect, Caleb. *Proposals made in the year 1720. To the Parishioners of Stroud, near Rochester in Kent, for building a Work-house there; With an account of the good success thereof.* London, 1725.

Pearson, John. *Observations on various items of Materia Medica used in the treatment of Lues Venerea.* London, 1800.

Peter, Charles. *A Description of the Venereal Disease: Declaring the Causes, Signs, Effects, And Cure thereof. With a Discourse of the most Wonderful Antivenereal Pill.* London, 1678.

———. *Observations on the Venereal Disease, with the true way of curing the same.* London, 1693.

Plenck, Joseph James. *A new and easy method of giving mercury, to those affected with the Venereal Disease.* Translated from the Latin by William Saunders. 2nd ed. London, 1768.

Profily, John. *An Easy and Exact method of Curing the Venereal Disease.* 2nd ed. London, 1748.

Reynolds, John. *Compassionate Address to the Christian World . . . revised, corrected and published for the use of the Patients in the Lock-Hospital, near Hyde-Park Corner.* London, 1767.

Robinson, Nicholas. *A Treatise of the Venereal Disease.* London, 1736.

St. Andrew's Holborn and St. George the Martyr. *Rules, orders and regulations for the government of the workhouse belonging to that part of the parish of St. Andrew, Holborn,* which lies above the bars, in the County of Middlesex, and the parish of St. George the Martyr, in the said county. Made by the governors and directors of the poor. London, 1791.

St. Bartholomew the Less. *Orders and Ordinances for the better Government of the Hospitall of Bartholomew the lesse. As also Orders enacted for Orphans and their Portions*. London, 1652.

St. Giles in the Fields. *Regulations agreed upon and established this twelfth day of July 1726 by the gentlemen of the vestry then present, for the better government and management of the work-house belonging to the parish of St. Giles's in the Fields*. London, 1726.

St. Giles in the Fields and St. George Bloomsbury. *Hints and Cautions for the Information of the Churchwardens and Overseers of the Poor of the Parishes of St. Giles in the Fields and St. George Bloomsbury, in the County of Middlesex: and Rules, Orders, and Regulations for Maintaining, Governing, and Employing the said Poor; Made by the Vestry of the said Parishes*. London, 1797.

St. James, Westminster. *The case of the Parish of St. James's Westminster; As to their Poor, and A Work-House designed to be built for Employing them*. London, 1730.

Salmon, William. *Praxis Medica: The Practice of Physick*. London, 1707.

Sanches, Antoni'o Nunes Ribeiro. *An historical investigation in to the first appearance of the Venereal Disease in Europe, with remarks on its particular nature*. London, 1790.

Sawrey, Solomon. *A Popular View of the Effects of the Venereal Disease upon the Constitution: Collected from the Best Writers. To which are Prefixed Miscellaneous Observations by a Physician*. Edinburgh, 1793.

Scott, Thomas. *A Discourse on Repentance*. London, 1783.

———. *Hints for the consideration of patients in hospitals*. London, 1797.

Sennert, Daniel, Nicholas Culpeper, and Abdiah Cole. *Two Treatise. The First of the Venereal Pocks. The Second Treatise of the Gout*. London, 1660.

Sennert, Daniel. *Two Treatises: the first on the Venereal Pox*, Translated from the Latin by Nicholas Culpepper. London, 1673.

Sigogne, Bouez de. *A New Method of Curing the Venereal Disease*. London, 1724.

Sintelaer, Johannes. *The Scourge of Venus and Mercury: Represented in a treatise of the Venereal Disease*. London, 1709.

Simmons, Samuel Foart. *Observations on the Cure of the Gonorrhea*. London, 1780.

Smith, Hugh. *Formulae Medicamentorum Concinnatae: or Elegant Medical Prescriptions for Various Disorders. Translated from the Latin of the late Dr. Hugh Smith. To which is prefixed, a Sketch of his Life*. London, 1791.

Smyth, John Hamilton. *A New Treatise on the Venereal Disease, Gleets, Seminal Weaknesses; The Dreadful Effects of Self-Pollution, and the Cause of Impotency*. 5th ed. London, 1771.

Sparrow, John. *A Mechanical Dissertation upon the Lues Venerea*. 4th ed. London, 1739.

Spinke, John. *Quackery Unmask'd: or Reflections on the Sixth Edition of Mr. Martin's Treatise of the Venereal Disease and its Appendix*. London, 1709.

Swediaur, Franz. *Practical Observations on Venereal Complaints*. 2nd ed. London, 1786.

Swift, Charles. *Salivation Exploded: or A Practical Essay on the Venereal Disease, Fully Demonstrating the Inefficacy of Salivation*. 4th ed. London, 1782.

T. C. *The Charitable Surgeon: or The Best Remedies for the Worst Maladies, reveal'd. Being a New and True way of Curing (without Mercury) the several degrees of the Venereal Disease in both Sexes.* 2nd ed. London, 1709.

Tongue, James. *An Inaugural Dissertation upon the three folowing subjects: an attempt to prove that the Lues venerea was not introduced into Europe from America. II An Experimental Inquiry into the Modus Operendi of Mercury in Curing the Lues Venerea. III. Experimental Proofs That the Lues Venerea and Gonorrhoea, are Two Distinct Forms of Disease.* Philadelphia, 1801.

Turner, Daniel. *Aphrodisiacus. Containing a Summary of the Ancient Writers on the Venereal Disease.* London, 1736.

———. *Syphilis. A Practical Dissertation on the Venereal Disease.* London, 1724.

Wall, W. *A New System of the French Disease. With An easy method of curing it, unknown to the Ancients or Moderns, with all its Common and Remote Symptoms, Obvious to the meanest Capacities.* London, 1696(?).

Wastell, Henry. *Observations on the Efficacy of a New Mercurial Preparation for the Cure of the Venereal Disease.* London, 1779.

Westminster, J. T. *The Hunting of the Pox: A Pleasant Discourse between the Author and Pild Garlicke. Wherin is declared the nature of the Disease, how it came, and how it may be cured.* London, 1619.

Wills, Thomas. *Remarks on Polygamy &c. in answer to Mr. M-d-n's Thelypthora.* London, 1781.

Wynell, John. *Lues Venerea, or, A Perfect Cure of the French Pox: Wherein The Names, Nature, Subject, Causes and Signes of the Disease are handled. Mistakes in these discovered, Doubts and Questions succinctly Resolved.* London, 1660.

II. Secondary Sources

Acton, William. *Prostitution, considered in its moral, social, and sanitary aspects in London and other large cities.* London: J. Churchill, 1857.

Ackerknecht, Edwin H. *Medicine at the Paris Hospital, 1794–1848.* Baltimore: Johns Hopkins University Press, 1967.

Adler, M. W. "History of the Development of a Service for the Venereal Diseases." *Journal of the Royal Society of Medicine* 75 (1982): 124–8.

Allen, Peter Lewis, *The Wages of Sin: Sex and Disease, Past and Present.* Chicago: University of Chicago Press, 2000.

Amussen, Susan Dwyer. *An Ordered Society: Gender and Class in Early Modern England.* New York: Basil Blackwell, 1988.

———. "Punishment, Discipline, and Power: The Social Meanings of Violence in Early Modern England." *Journal of British Studies* 34 (1) (1995): 1–34.

Anselment, Raymond. "Seventeenth-Century Pox: The Medical and Literary Realities of Venereal Disease." *Seventeenth Century* 4 (1989): 189–211.

———. *The Realms of Apollo: Literature and Healing in Seventeenth-Century England.* Newark: University of Delaware Press, 1995.

Ariès, Phillipe. Introduction to *A History of Private Life*. Edited by Roger Cartier, translated from the French by Arthur Goldhammer. Vol. 3, *Passions of the Renaissance*, 1–11. Cambridge, Mass.: Belknap Press, 1989.

Arrizabalaga, Jon. "Medicina Universitaria y *Morbus Gallicus* en La Italia de Finales del siglo XV: El Arquatra pontifico Gaspar Torella (c. 1452–c.1520)." *Asclepio* 60 (1) (1988): 3–33.

Arrizabalaga, Jon, John Henderson, and Roger French. *The Great Pox: The French Disease in Renaissance Europe*. New Haven: Yale University Press, 1997.

Andrew, Donna T. "Debate: The Secularization of Suicide in England 1660–1800." *Past and Present* 119 (1988): 158–70.

———. *Philanthropy and Police: London Charity in the Eighteenth Century*. Princeton: Princeton University Press, 1989.

———. "'To the Charitable and Humane': Appeals for Assistance in the Eighteenth- Century London Press." In *Charity, Philanthropy and Reform from the 1690's to 1850*, edited by Hugh Cunningham and Joanna Innes. London: MacMillan, 1998.

———. "Two Medical Charities in Eighteenth-Century London: The Lock Hospital and the Lying-in Charity for Married Women." In *Medicine and Charity before the Welfare State*, edited by Jonathan Barry and Colin Jones. London: Routledge, 1991.

Baker, Robert. "The History of Medical Ethics." In *Companion Encyclopaedia of the History of Medicine*, edited by W. F. Bynum and Roy Porter, 852–87. London: Routledge, 1993.

Baker, Robert, Dorothy Porter, and Roy Porter. *The Codification of Medical Morality: Historical and Philosophical Studies of the Formalization of Western Medical Morality in the Eighteenth and Nineteenth Centuries*. Dordrecht: Kluwer Academic Publishers, 1993.

Bakhtin, Mikhail. *Rabelais and his World*. Translated from the Russian by Héléne Iswolsky. Cambridge, Mass.: Harvard University Press, 1968.

Baldwin, Peter. *Contagion and the State in Europe, 1830–1930*. Cambridge: Cambridge University Press, 1999.

Barry, Jonathan, and Colin Jones, eds. *Medicine and Charity Before the Welfare State*. London: Routledge, 1991.

Bartley, Paula. *Prostitution: Prevention and Reform in England, 1860–1914*. London: Routledge, 2000.

Beir, A. L. "Social Problems in Elizabethan London." *Journal of Interdisciplinary History* 9 (1978–79): 203–21.

Beir, A. L. and Roger Finlay, eds. *London 1500–1700: The Making of the Metropolis*. London: Longman, 1986.

Bender, John. *Imagining the Penitentiary: Fiction and the Architecture of Mind in Eighteenth-Century England*. Chicago: University of Chicago Press, 1987.

Berry, Amanda. "Community Sponsorship and the Hospital Patient in Late Eighteenth-Century England." In *The Locus of Care: Families, Communities, Institutions and the Provision of Welfare Since Antiquity*, edited by Peregrine Horden and Richard Smith. London: Routledge, 1998.

Bettley, James. "'Post Voluptatem Misercordia': The Rise and Fall of the London Lock Hospitals." *London Journal*, 10 (1984): 167–75.

Boeherer, Bruce T. "Early Modern Syphilis." *Journal of the History of Sexuality* 1 (1990): 197–214.

Boose, Lynda. "Scolding Brides and Bridling Scolds: Taming the Woman's Unruly Member." *Shakespeare Quarterly* 42 (2) (1991): 179–213.

Boulton, Jeremey. "Food Prices and the Standard of Living in London in the 'Century of Revolution' 1580–1700." *Economic History Review* 53 (3) (2000): 455–92.

———. "Going on the Parish: The Parish Pension and its Meaning in the London Suburbs, 1640–1724." In *Chronicling Poverty: The Voices and Strategies of the English Poor 1640–1840*, edited by Tim Hitchcock, Peter King, and Pamela Sharpe. London: Macmillan, 1997.

———. "'It is extreme necessity that makes me do this': Some 'Survival Strategies' of Pauper Households in London's West End during the early Eighteenth Century." *International Review of Social History* 45 (2000): 47–69.

———. *Neighbourhood and Society: A London Suburb in the Seventeenth Century*. Cambridge: Cambridge University Press, 1987.

———. "The Poor Among the Rich: Paupers and the Parish in the West End, 1600–1724." In *Londinopolis: Essays in the Cultural and Social History of Early Modern London*, edited by Mark S. R., Jenner and Paul Griffiths. Manchester University Press, 2000.

Braddick, Michael J. *State Formation in Early Modern England c. 1550–1700*. Cambridge: Cambridge University Press, 2000.

Braddick Michael J. and John Walter, eds. *Negotiating Power in Early Modern Society: Order, Hierarchy and Subordination in Britain and Ireland*. Cambridge: Cambridge University Press, 2001.

Brewer, John. "This, That and the Other: Public, Social and Private in the Seventeenth and Eighteenth Centuries." In *Shifting the Boundaries: Transformation of the Languages of Public and Private in the Eighteenth Century*, edited by Dario Castiglione and Lesley Sharpe. Exeter: University of Exeter Press, 1995.

Borsay, Anne. "Cash and Conscience" Financing the General Hospital at Bath, c.1738–1750." *Social History of Medicine* 4 (1991): 207–29.

———. "'Persons of Honour and Reputation': The Voluntary Hospital in the Age of Corruption." *Medical History* 35 (1991): 281–94.

Brody, Saul N. *The Disease of the Soul: Leprosy in Medieval Literature*. Ithaca, N.Y.: Cornell University Press, 1974.

Brome, Vincent. *The Other Pepys*. London: Weidenfeld and Nicolson, 1992.

Browner, Jessica A. "Wrong Side of the River: London's Disreputable South Bank in the Sixteenth and Seventeenth Century." *Essays in History* 36 (1994): 34–71.

Bynum, William F. "Physicians, Hospitals and Career Structures in Eighteenth-Century London." In *William Hunter and the Eighteenth-Century Medical World*, edited by W. F. Bynum and Roy Porter. Cambridge: Cambridge University Press, 1985.

———. *Science and the Practice of Medicine in the Nineteenth Century*. Cambridge: Cambridge University Press, 1994.

———. "Treating the Wages of Sin: Venereal Disease and Specialism in Eighteenth-Century Britain." In *Medical Fringe and Medical Orthodoxy, 1750–1850*, edited by W. F. Bynum and Roy Porter. London: Croom Helm, 1987.

Bynum, William F. and Roy Porter, eds. *Medical Fringe & Medical Orthodoxy: 1750–1850*. London: Croom Helm, 1987.

Calvi, Giulia. *Histories of a Plague Year: The Social and the Imaginary in Baroque Florence*. Translated from the Italian by Dario Biocca and Bryant T. Ragan. Berkeley: University of California Press, 1989.

Capp, Bernard. "The Double Standard Revisited: Plebeian Women and Male Sexual Reputation in Early Modern England." *Past and Present* 162 (1999): 70–100.

Castiglione, Dario and Sharpe, Lesley, eds. *Shifting the Boundaries: Transformation of the Languages of Public and Private in the Eighteenth Century*. Exeter: University of Exeter Press, 1995.

Cavallo, Sandra. *Charity and Power in Early Modern Italy: Benefactors and Their Motives in Turin, 1541–1789*. Cambridge: Cambridge University Press, 1995.

———. "Charity as Boundary Making: Social Stratification, Gender and the Family in the Italian States (Seventeenth–Nineteenth Centuries)." In *Charity, Philanthropy and Reform from the 1690's to 1850*, edited by Hugh Cunningham and Joanna Innes. London: MacMillan, 1998.

———. "Charity, Power, and Patronage in Eighteenth-Century Italian Hospitals." In *The Hospital in History*, edited by Lindsay Granshaw and Roy Porter. London: Routledge, 1989.

———. "The Motivations of Benefactors: An Overview of Approaches to the Study of Charity." In *Medicine and Charity Before the Welfare State*, edited by Jonathan Barry and Colin Jones. London: Routledge, 1991.

Cameron, Hector Charles. *Mr. Guy's Hospital, 1727–1948*. London: Longmans, 1954.

Clark, Alice. *Working Life of Women in the Seventeenth Century*. 1919. Reprint, London: Routledge, 1982.

Clark, Anna. *The Struggle for the Breeches: Gender and the Making of the British Working Class*. Berkeley: University of California Press, 1995.

———. *Women's Silence, Men's Violence: Sexual Assault in England 1770–1845*. New York: Pandora's Press, 1987.

Clark, Peter, and David Souden, eds. *Migration and Society in Early Modern England*. London: Hutchinson, 1987.
Clark-Kennedy, A. E. *London Pride: The Story of a Voluntary Hospital*. London: Hutchinson Benham, 1979.
———. *The London: A Study in the Voluntary Hospital System*. 2 vols. London: Pitman Medical Publications, 1962.
Coates, Ben. "Poor Relief in London During the English Revolution Revisited." *London Journal* 25 (2) (2000): 40–58.
Colley, Linda. *Britons: Forging the Nation, 1707–1837*. New Haven: Yale University Press, 1992.
Connor, Susan P. "The Pox in Eighteenth-Century France." In *The Secret Malady: Venereal Disease in Eighteenth Century Britain and France*, edited by Linda E. Merians. Lexington: University of Kentucky Press, 1996.
Cook, Harold. *The Decline of the Old Medical Regime in Late Stuart London*. Ithaca, N.Y.: Cornell University Press, 1986.
———. *Trials of an Ordinary Doctor: Joaanes Groenvelt in Seventeenth-Century London*. Baltimore: Johns Hopkins University Press, 1994.
Coxson, Bronwyn. "The Public and Private Faces of Eighteenth-Century London Dispensary Charity." *Medical History* 41 (1997): 127–49.
Crawford, Patricia. "Attitudes to Menstruation in Seventeenth-Century England." *Past and Present* 91 (1981): 47–73.
———. "Printed Advertisements for Women Medical Practitioners in London, 1670–1710." *Society for the Social History of Medicine Bulletin* 35 (1984): 66–70.
Crowther, Anne. "Health Care and Poor Relief in Provincial England." In *Health Care and Poor Relief in Eighteenth and Nineteenth-Century Northern Europe*, edited by Ole Peter Grell, Andrew Cunningham, and Robert Jütte. Aldershot: Ashgate, 2002.
Cunningham, Hugh, and Joanna Innes, eds. *Charity, Philanthropy and Reform from the 1690's to 1850*. London: MacMillan, 1998.
Dabhoiwala, Faramerz. "Sex, Social Relations and the Law in Seventeenth- and Eighteenth-Century London." In *Negotiating Power in Early Modern Society: Order, Hierarchy and Subordination in Britain and Ireland*, edited by Michael J. Braddick and John Walter Cambridge: Cambridge University Press, 2001.
Davidoff, Leonore, and Catherine Hall. *Family Fortunes: Men and Women of the English Middle Class*. Chicago: University of Chicago Press, 1987.
Davidson, Roger. *Dangerous Liaisons: A Social History of Venereal Disease in Twentieth-Century Scotland*. Amsterdam: Rodopi, 2000.
Davidson, Roger, and Lesley Hall, eds. *Sex, Sin, and Suffering: Venereal Disease and European Society Since 1870*. London: Routledge, 2001.
Debus, Allen. *The Chemical Philosophy: Paracelsian Science and Medicine in the Sixteenth and Seventeenth Centuries*. New York: Science History Publications, 1977.
———. *The English Paracelsians*. London: Oldbourne, 1965.

Dekker, Jeroen J. H. "Transforming the Nation and the Child: Philanthropy in the Netherlands, Belgium, France and England, c.1780–c.1850." In *Charity, Philanthropy and Reform from the 1690's to 1850*. Edited by Hugh Cunningham and Joanna Innes. London: MacMillan, 1998.

Devereaux, Simon. "The City and the Sessions Paper: 'Public Justice' in London, 1770–1800." *Journal of British Studies* 35 (1996): 466–503.

———. "The Fall of the Sessions Paper: Criminal Trial and the Popular Press in Late Eighteenth-Century London." *Criminal Justice History* 18 (2002).

Digby, Anne. *Making a Medical Living: Doctors and Patients in the English Medical Market for Medicine, 1720–1914*. Cambridge: Cambridge University Press, 1994.

Doherty, Francis Cecil. *A Study in Eighteenth-Century Advertising Methods: The Anodyne Necklace*. Lewiston, N.Y.: Edwin Mellen Press, 1992.

Douglas, Mary. *Purity and Danger: An Analysis of the Concepts of Pollution and Taboo*. London; Routledge and Kegan Paul, 1966.

Duden, Barbara. *The Woman Beneath the Skin: A Doctor's Patients in Eighteenth-Century Germany*. Translated from the German by Thomas Dunlap. Cambridge, Mass.: Harvard University Press, 1991.

Dunlap, Barbara J. "The Problem of Syphilitic Children in Eighteenth-Century France and England." In *The Secret Malady: Venereal Disease in Eighteenth-Century Britain and France*, edited by Linda E. Merians. Lexington: University of Kentucky Press, 1996.

Dunsmuir, W. "The First Wave of Syphilis: Bart's in the 16th Century." *St. Bartholomew's Hospital Journal* (Spring, 1995): 11.

Eamon, William. "Cannibalism and Contagion: Framing Syphilis in Counter-Reformation Italy." *Early Science and Medicine* 3 (I) (1998): 1–31.

Elias, Norbert. *The Civilizing Process: The Development of Manners. Changes in the Code of Conduct and Feeling in Early Modern Times*. Translated from the German by Edmund Jephcott. New York: Urizen Books, 1978.

Ell, Stephan R. "Blood and Sexuality in Medieval Leprosy." *Janus* 71 (1984): 153–64.

Evenden, Doreen. "Gender Differences in the Licensing and Practice of Female and Male Surgeons in Early Modern England." *Medical History* 42 (1998): 194–216.

———. *Popular Medicine in Seventeenth-Century England*. Bowling Green, Ohio: Bowling Green State University Press, 1988.

———. *The Midwives of Seventeenth-Century London*. Cambridge: Cambridge University Press, 2000.

Fabricius, Johannes. *Syphilis in Shakespeare's England*. London: Kingsley Publishers, 1994.

Finlay, Roger, and Beatrice Shearer. "Population Growth and Suburban Expansion." In *London 1500–1700: The Making of a Metropolis*, edited by A. L. Beier and Roger Finlay. London: Longman, 1986.

Finzsch, Norbert, and Robert Jütte, eds. *Institutions of Confinement: Hospitals, Asylums, and Prisons in Western Europe and North America, 1500–1950*. Cambridge: Cambridge University Press, 1996.

Fissell, Mary E. "Charity Universal? Institutions and Moral Reform in Eighteenth-Century Bristol." In *Stilling the Grumbling Hive: The Response to Social and Economic Problems in England, 1689–1750*, edited by Lee Davison, Tim Hitchcock, Tim Keirn, and Robert B. Shoemaker. New York: St. Martin's Press, 1992.

———. "Innocent and Honorable Bribes: Medical Manners in Eighteenth-Century Britain." In *The Codification of Medical Morality*, edited by Robert Baker, Dorothy Porter, and Roy Porter. Dordrecht: Kluwer Academic Publishers, 1993.

———. *Patients, Power and the Poor in Eighteenth-Century Bristol*. Cambridge: Cambridge University Press, 1991.

———. "The Disappearance of the Patient's Narrative and the Invention of Hospital Medicine." In *British Medicine in an Age of Reform*, edited by Roger French and Andrew Wear. London: Routledge, 1991.

———. "The 'Sick and Drooping Poor' in Eighteenth-Century Bristol and its Region." *Social History of Medicine* 2 (1989): 35–58.

Flegel, Kenneth M. "Changing Concepts of the Nosology of Gonorrhea and Syphilis." *Bulletin of the History of Medicine* 48 (4) (1974): 571–88.

Fletcher, Anthony. *Sex, Gender and Subordination in England 1500–1800*. New Haven: Yale University Press, 1995.

Foa, Anna. "The Old and the New: The Spread of Syphilis (1494–1530)" In *Sex and Gender in Historical Perspective*, edited by Guido Ruggiero and Edward Muir. Baltimore: Johns Hopkins University Press, 1990.

Foucault, Michel. *The Birth of the Clinic: An Archaeology of Medical Perception*, translated from the French by Alan Sheridan. New York: Pantheon, 1973.

———. *Discipline and Punish: The Birth of the Prison*, translated from the French by Alan Sheridan. New York: Vintage, 1979.

Fout, John, ed. *Forbidden History: The State, Society and the Regulation of Sexuality in Modern Europe*. Chicago: University of Chicago Press, 1992.

Furdell, Elizabeth Lane. "'At the King's Arms in the Poultrey': The Bookshop Emporium of Dorman Newman, 1670–1694." *London Journal* 23 (2) (1998): 1–19.

———. *Publishing and Medicine in Early Modern England*. Rochester, N.Y.: University of Rochester Press, 2002.

———. *The Royal Doctors 1485–1714: Medical Personnel at the Tudor and Stuart Courts*. Rochester, N.Y.: University of Rochester Press, 2001.

Gelfand, Toby. *Professionalizing Modern Medicine: Paris Surgeons and Medical Science and Institutions in the Eighteenth Century*. Westport, Conn.: Greenwood Press, 1980.

George, Dorothy M. *London Life in the Eighteenth Century*. New York: Harper and Row, 1964.

Gillis, John. *For Better, For Worse: British Marriages, 1600 to the Present*. Oxford: Oxford University Press, 1985.

———. "Married but Not Churched: Plebeian Sexual Relations and Marital Nonconformity in Eighteenth-Century Britain." In *Tis' Nature's Fault:*

Unauthorized Sexuality during the Enlightenment, edited by Robert Maccubbin. Cambridge: Cambridge University Press, 1987.

Gilman, Sander. *Disease and Representation: Images of Illness from Madness to AIDS*. Ithaca, N.Y.: Cornell University Press, 1988.

———. *Health and Illness: Images of Difference*. London: Reaktion Books, 1995.

Goldie, Mark. "The Unacknowledged Republic: Officeholding in Early Modern England." In *The Politics of the Excluded, c.1500–1850*, edited by Tim Harris. New York: Palgrave, 2001.

Goodman, Dena. "Public Sphere and Private Life: Toward a Synthesis of Current Historiographical Approaches to the Old Régime." *History and Theory* 31 (1992): 1–20.

Gould, Terry, and David Uttley. *A Short History of St. George's Hospital and the Origins of its Wards Names*. London: Atlantic Highlands, 1997.

Gowing, Laura. *Domestic Dangers: Women, Words, and Sex in Early Modern London*. Oxford University Press, 1998.

———. "Ordering the Body: Illegitimacy and Female Authority in Seventeenth-Century England." In *Negotiating Power in Early Modern Society: Order, Hierarchy and Subordination in Britain and Ireland*, edited by Michael J. Braddick and John Walter. Cambridge: Cambridge University Press, 2001.

———. "Secret Births and Infanticide in Seventeenth-Century England." *Past and Present* 156 (1997): 87–115.

Granshaw, Lindsay. "'Fame by Means of Bricks and Mortar': Specialist Hospitals and the Medical Profession in Britain, 1800–1948." In *The Hospital in History*, edited by Lindsay Granshaw and Roy Porter, 199–220. New York: Routledge, 1989.

Granshaw, Lindsay, and Roy Porter, eds. *The Hospital in History*. London: Routledge, 1989.

Green, David. "Medical Relief and the New Poor Law in London." In *Health Care and Poor Relief in Eighteenth and Nineteenth-Century Northern Europe*, edited by Ole Peter Grell, Andrew Cunningham, and Robert Jütte. Aldershot: Ashgate, 2002.

Grell, Ole Peter, and Andrew Cunningham, eds. *Health Care and Poor Relief in Protestant Europe, 1500–1700*. London: Routledge, 1997.

Grell, Ole Peter, Andrew Cunningham, and Robert Jütte, eds. *Health Care and Poor Relief in Eighteenth- and Nineteenth-Century Northern Europe*. Aldershot: Ashgate, 2002.

Grieco, Sara F. Matthews. "The Body, Appearance and Sexuality." In *A History of Women in the West*, general editors, Georges Dub and Michelle Perrot. Vol. 3, *Renaissance and Enlightenment Paradoxes*, edited by Natalie Zemon Davis and Arlette Farge. Cambridge, Mass.: Belknap Press, 1993.

Guilhamet, Leon. "Pox and Malice: Some Representations of Venereal Disease in Restoration and Eighteenth-Century Satire." In *The Secret Malady: Venereal Disease in Eighteenth Century Britain and France*, edited by Linda E. Merians. Lexington: University of Kentucky Press, 1994.

Habermas, Jürgen. *The Structural Transformation of the Public Sphere: An Inquiry into a Category of Bourgeois Society*, translated from the German by Thomas Burger. Cambridge, Mass.: MIT Press, 1991.

Hall, Catherine. *White, Male, and Middle Class: Explorations in the History of Feminism*. New York: Routledge, 1992.

Handler, Clive, ed. *Guy's Hospital 250 Years*. London: Guy's Hospital Gazette, 1976.

Harding, Vanessa. "Controlling a Complex Metropolis, 1650–1750: Politics, Parishes and Powers." *London Journal* 26 (1) (2001): 29–37.

———. "The Population of London, 1550–1700: A Review of the Published Evidence." *London Journal*, 15 (1990): 111–28.

Harley, David. "Rhetoric and the Social Construction of Sickness and Healing." *Social History of Medicine* 12 (3) (1999): 407–35.

Harsin, Jill. "Syphilis, Wives, and Physicians: Medical Ethics and the Family in Late Nineteenth-Century France." *French Historical Studies* 16 (1989): 72–95.

Henderson, John, and Richard Wall, eds. *Poor Women and Children of the European Past*. London: Routledge, 1994.

Henderson, Tony. *Disorderly Women in Eighteenth-Century London: Prostitution and Control in the Metropolis, 1730–1830*. London: Longman, 1999.

Herlan, R. W. "Poor Relief in London in the English Revolution." *Journal of British Studies* 18 (1979): 30–51.

———. "Poor Relief in the London Parish of Antholin's Bridge Row, 1638–1664." *Guildhall Studies in London History* 2 (4) (1977): 179–99.

———. "Poor Relief in the London Parish of Dunstan in the West During the English Revolution." *Guildhall Studies in London History* 3 (1) (1977): 13–36.

———. "Social Articulation and the Configuration of Parochial Poverty in London on the Eve of the Restoration." *Guildhall Studies in London History* 2 (2) (1976): 43–53.

Hill, Bridget. "The Marriage Age of Women and the Demographers." *History Workshop Journal* 28 (1989): 129–47.

Hindle, Steve. "Exhortation and Entitlement: Negotiating Inequality in English REural Communities, 1550–1650." In *Negotiating Power in Early Modern Society: Order, Hierarchy and Subordination in Britain and Ireland*, edited by Michael J. Braddick and John Walter. Cambridge: Cambridge University Press, 2001.

Hitchcock, Tim. *English Sexualities, 1700–1800*. New York, St. Martin's Press, 1997.

———. "Paupers and Preachers: The SPCK and the Parochial Workhouse Movement." In *Stilling the Grumbling Hive: The Response to Social and Economic Problems in England, 1689–1750*, edited by Lee Davison, Tim Hitchcock, Tim Keirn, and Robert B. Shoemaker. New York: St. Martin's Press, 1992.

———. "Redefining Sex in Eighteenth-Century England." *History Workshop Journal* 41 (1996): 73–92.

———. *Richard Hutton's Complaint Book: The Notebook of the Steward of the Quaker Workhouse at Clerkenwell*. London: London Records Society, 1987.

———. "Sociability and Misogyny in the Life of John Cannon, 1684–1743." In *English Masculinities, 1660–1800*, edited by Tim Hitchcock and Michele Cohen. London: Longman, 1999.

———. "The English Workhouse: A Study in Institutional Poor Relief in Selected Counties, 1696–1750." Ph.D. diss., Oxford University, 1985.

———. "The Publicity of Poverty in Eighteenth-Century London." In *Imagining Early Modern London: Perceptions and Portrayals of the City from Stow to Strype 1598–1720*, edited by J. F. Merritt. Cambridge: Cambridge University Press, 2001.

———. "'Unlawfully Begotten on Her Body': Illegitimacy and the Parish Poor in St. Luke's Chelsea." In *Chronicling Poverty: The Voices and Strategies of the English Poor 1640–1840*, edited by Tim Hitchcock, Peter King, and Pamela Sharpe. London: Macmillan, 1997.

Hitchcock, Tim, and John Black, eds. *Chelsea Settlement and Bastardy Examinations, 1733–1766*. London: London Record Society, 1999.

Hitchcock, Tim, and Michelle Cohen, eds. *English Masculinities, 1660–1800*. London: Longman, 1999.

Hitchcock, Tim, Peter King, and Pamela Sharpe, eds. *Chronicling Poverty: The Voices and Strategies of the English Poor 1640–1840*. London: Macmillan, 1997.

Honeybourne, Marjorie B. "The Leper Hospitals of the London Area." *Transactions of the London and Middlesex Archaeological Society* 21 (1963): 1–61.

Hunt, Margaret R. "Wives and Marital 'Rights' in the Court of Exchequer in the Early Eighteenth Century." In *Londinopolis: Essays in the Cultural and Social History of Early Modern London*, edited by Mark S. R. Jenner, and Paul Griffiths. Manchester University Press, 2000.

Hylson-Smith, Kenneth. *Evangelicals in the Church of England*. Edinburgh: T & T Clark, 1988.

Ignatieff, Michael, *A Just Measure of Pain: The Penitentiary in the Industrial Revolution, 1750–1850*. London: Macmillan, 1978.

Ingram, Martin. *Church Courts, Sex and Marriage in England*. Cambridge: Cambridge University Press, 1987.

———. "Scolding Women Ducked or Washed." In *Women, Crime and the Courts in Early Modern England*, edited by Jennifer Kermode and Garthine Walker. Chapel Hill: University of North Carolina Press, 1994.

Innes, Joanna. "Prisons for the Poor: English Bridewells, 1555–1800." In *Labour, Law and Crime: An Historical Perspective*, edited by Francis Snyder and Douglas Hay London: Tavistock, 1987.

Jayne, R. Everett. *Jonas Hanway: Philanthropist, Politician, and Author*. London: Epworth Press, 1929.

Jenner Mark S. R., and Paul Griffiths, eds. *Londinopolis: Essays in the Cultural and Social History of Early Modern London*. Manchester University Press, 2000.

Jewson, Nicholas. "Medical Knowledge and the Patronage System in Eighteenth-Century England." *Sociology* 8 (1974): 369–85.

Jones, Colin. "The Construction of the Hospital Patient in Early Modern France." In *Institutions of Confinement: Hospitals, Asylums, and Prisons in Western Europe and North America, 1500–1950*, edited by Norbert Finzsch and Robert Jütte. Cambridge: Cambridge University Press, 1996.

Jones, Vivien. "Scandalous Femininity: Prostitution and Eighteenth-Century Narrative." In *Shifting the Boundaries: Transformations of the Languages of Public and Private in the Eighteenth Centuries*, edited by Dario Castiglione and Lesley Sharpe. Exeter: University of Exeter Press, 1995.

Jütte, Robert. *Poverty and Deviance in Early Modern Europe*. Cambridge: Cambridge University Press, 1994.

———. "Syphilis and Confinement: Hospitals in Early Modern Germany." In *Institutions of Confinement: Hospitals, Asylums, and Prisons in Western Europe and North America, 1500–1950*, edited by Norbert Finzsch and Robert Jütte. Cambridge: Cambridge University Press, 1996.

Keel, Othmar. "The Politics of Health and Institutionalisation of Clinical Practices in Europe in the Second Half of the Eighteenth Century." In *William Hunter and the Eighteenth-Century Medical World*, edited by W. F. Bynum and Roy Porter Cambridge: Cambridge University Press, 1985.

Kent, David A. "'Gone for a Soldier': Family Breakdown and the Demography of Desertion in a London Parish, 1750–91." *Local Population Studies* 1990 (45): 27–42.

Kilpatrick, Robert. "'Living in the Light': Dispensaries, Philanthropy and Medical Reform in Eighteenth-Century London." In *Medical Enlightenment of the Eighteenth Century*, edited by Andrew Cunningham and Roger French. Cambridge: Cambridge University Press, 1990.

King, Lester. *The Philosophy of Medicine: The Early Eighteenth Century*. Cambridge, Mass.: Harvard University Press, 1978.

Kitch, Malcolm. "Capital and Kingdom: Migration to Stuart London." In *London 1500–1700: The Making of the Metropolis*, edited by A. L. Beier and Roger Finlay. London: Longman, 1986.

Klein, Lawrence E. "Politeness for Plebes: Consumption and Social Identity in early Eighteenth-Century England." In *The Consumption of Culture 1660–1800: Image, Object, Text*, edited by Ann Bermingham and John Brewer. London: Routledge, 1995.

Lake, Peter. *The Boxmaker's Revenge: "Orthodoxy," "Heterodoxy" and the Politics of the Parish in Early Stuart London*. Manchester: Manchester University Press, 2001.

Landau, Norma. "The Eighteenth-Century Context of the Laws of Settlement." *Continuity and Change* 6 (1991): 417–39.

———. "The Laws of Settlement and Surveillance of Immigration in Eighteenth-Century Kent." *Continuity and Change* 3 (1988): 391–420.

———. "The Regulation of Immigration, Economic Structures and Definitions of the Poor in Eighteenth-Century England." *Historical Journal* 33 (1990): 541–72.

———. "Who was Subject to the Laws of Settlement? Procedure Under the Settlement Laws in Eighteenth-Century England." *Agricultural History Review* 43 (1996): 139–59.

Lane Joan. *Apprenticeship in England, 1600–1914*. London, UCL Press, 1996.

———. "The Provincial Practitioner and His Services to the Poor 1570–1800." *Bulletin of the Society for the Social History of Medicine* 28 (1981): 10–14.

———. "The Role of Apprenticeship in Eighteenth-Century Medical Education in England." In *William Hunter and the Eighteenth-Century Medical World*, edited by William F. Bynum and Roy Porter. Cambridge: Cambridge University Press, 1985.

Lawrence, Susan. *Charitable Knowledge: Hospital Pupils and Practitioners in Eighteenth-Century London*. Cambridge: Cambridge University Press, 1996.

Lees, Lynn Hollen. *The Solidarities of Strangers: The English Poor Laws and the People, 1700–1948*. Cambridge: Cambridge University Press, 1998.

Lees, Robert. "The 'Lock Wards' of the Edinburgh Royal Infirmary." *British Journal of Venereal Disease* 37 (1961): 187–9.

Levine, Phillipa. "Rough Usage: Prostitution, Law and the Hhistorian." In *Rethinking Social History: English society 1570–1920 and Its Interpretation*, edited by Adrian Wilson. Manchester: Manchester University Press, 1993.

Lloyd, Sarah. "Pleasing Spectacles and Elegant Dinners: Conviviality, Benevolence, and Charity Anniversaries in Eighteenth-Century London." *Journal of British Studies* 41 (2002): 23–57.

———. "'Pleasure's Golden Bait': Prostitution, Poverty and the Magdalen Hospital in Eighteenth-Century London." *History Workshop Journal*, 41 (1996): 51–72.

Loudon, Irvine. *Medical Care and the General Practitioner, 1750–1850*. Oxford: Clarendon Press, 1986.

Lowe, N. F. "The Meaning of Venereal Disease in Hogarth's Graphic Art." In *The Secret Malady: Venereal Disease in Eighteenth-Century Britain and France*, edited by Linda E. Merians. Lexington: University of Kentucky Press, 1996.

Maccubbin, Robert, ed. *Tis' Nature's Fault: Unauthorized Sexuality During the Enlightenment*. Cambridge: Cambridge University Press, 1987.

MacDonald, Michael. "Debate: The Secularization of Suicide in England 1660–1800." *Past and Present* 119 (1988): 158–70.

———. "The Secularization of Suicide in England, 1660–1800." *Past and Present* 111 (1986): 50–100.

MacDonald, Michael, and Terrance R. Murphy. *Sleepless Souls: Suicide in Early Modern England*. Oxford: Clarendon Press, 1990.

MacFarlane, Stephan. "Social Policy and the Poor in the Later Seventeenth Century." In *London 1500–1700: The Making of the Metropolis*, edited by A. L. Beier and Roger Finlay. London: Longman, 1986.

MacKay, Lynn. "A Culture of Poverty? The St. Martin's in the Fields Workhouse, 1817." *Journal of Interdisciplinary History* 26 (2) (1995): 209–31.

———. "Moral Paupers: The Poor Men of St. Martin's, 1815–1819." *Social History/Histoire Sociale* 34 (2001): 115–31.
Marland, Hilary. *Medicine and Society in Makefield and Huddrersfield 1780–1870*. Cambridge; Cambridge University Press, 1987.
Marshall, Dorothy. *The English Poor in the Eighteenth Century*. London: George Routledge and Sons, 1926.
Mahood, Linda. *The Magdalenes: Prostitution in the Nineteenth Century*. London: Routledge, 1990.
McAllister, Marie E. "John Burrows and the Vegetable War." In *The Secret Malady: Venereal Disease in Eighteenth-Century Britain and France*, edited by Linda E. Merians. Lexington: University of Kentucky Press, 1996.
———. "Stories of the Origin of Syphilis in Eighteenth-Century England: Science, Myth, and Prejudice." *Eighteenth-Century Life* 24 (2000): 22–44.
McCullough, Laurence. "John Gregory (1724–1773) and the Invention of Professional Relationships in Medicine." *Journal of Clinical Ethics* 8 (1997): 11–21.
———. *John Gregory and the Invention of Professional Ethics and the Profession of Medicine*. Dordecht: Kluwer Academic Publishers, 1998.
McKenzie, Andrea. "Making Crime Pay: Motives, Marketing Strategies, and the Printed Literature of Crime in England 1670–1770." In *Criminal Justice in the Old World and the New: Essays in Honour of J. M. Beattie*, edited by Greg T. Smith, Allyson N. May, and Simon Devereaux. Toronto: Centre of Criminology, University of Toronto, 1998.
McLaren, Angus. "Privileged Communications: Medical Confidentiality in Late Victorian Britain." *Medical History* 37 (1993): 129–47.
Meldrum, Tim. "A Women's Court in London: Defamation at the Bishop of London's Consistory Court, 1700–1745." *London Journal* 1994 19(1): 1–20.
———. *Domestic Service and Gender, 1660–1750*. London: Longman, 2000.
———. "Domestic Service, Privacy and the Eighteenth-Century Metropolitan Household." *Urban History* 26 (1) (1999): 27–39.
———. "London Domestic Servants from Depositional Evidence, 1660–1750: Servant-Employer Sexuality in the Patriarchal Household." In *Chronicling Poverty: The Voices and Strategies of the English Poor*, edited by Tim Hitchcock, Peter King, and Pamela Sharpe. New York: St. Martin's, 1997.
Merians, Linda E. "The London Lock Hospital and the Lock Asylum for Women." In *The Secret Malady: Venereal Disease in Eighteenth-Century Britain and France*, edited by Linda E. Merians. Lexington: University of Kentucky Press, 1996.
Morton, R. S. "Syphilis in Art: An Entertainment in Four Parts." *Genitourinary Medicine* 66 (1–4) (1990): 33–44, 112–23, 208–21, 280–94.
Moore, Norman. *The History of St. Bartholomew's Hospital*. 2 vols. London: C.A. Pearson, 1918.

Mueller, Judith C. "Fallen Men: Representations of Male Impotence in Britain." *Studies in Eighteenth-Century Culture* 28 (1999): 85–102.
Muldrew, Craig. *The Economy of Obligation: The Culture of Credit and Social Relations in Early Modern England.* New York: St. Martin's Press, 1998.
Nash, Stanley. "Prostitution and Charity: The Magdalen Hospital—a Case Study." *Journal of Social History* 17 (1984): 617–28.
Niccolli, Ottavia. "'Menstruum Quai Monstruum'" Monstrous Births and Menstrual Taboo in the Sixteenth Century." In *Sex and Gender in Historical Perspectives*, edited by Edward Muir and Guido Ruggiero. Baltimore: Johns Hopkins University Press, 1990.
Norberg, Kathryn. "From Courtesan to Prostitute: Mercenary Sex and Venereal Disease, 1730–1802." In *The Secret Malady: Venereal Disease in Eighteenth-Century Britain and France*, edited by Linda Merians. Lexington: University of Kentucky Press, 1996.
Ober, William B. *Boswell's Clap and Other Essays: Medical Analyses of Literary Men's Afflictions.* Carbondale, Illinois: Southern Illinois University Press, 1994.
Ogbourn, Miles. *Spaces of Modernity: London Geographies, 1680–1780.* London: Guilford, 1998.
O'Shea, J. G. "'Two Minutes with Venus, Two Years with Mercury': Mercury as an Antisyphilitic Chemotherapeutic Agent." *Journal of the Royal Society of Medicine* 83 (1990): 392–5.
Orme, Nicholas, and Margaret Webster. *The English Hospital, 1070–1570.* New Haven: Yale University Press, 1995.
Oriel, J. D. *The Scars of Venus: A History of Venereology.* London: Springer-Verlag, 1994.
Outhwaite, R. B. "'Objects of Charity': Petitions to the London Foundling Hospital, 1768–72." *Eighteenth-Century Studies* 32 (4) (1999): 497–510.
Oxley, G. W. *Poor Relief in England and Wales 1601–1834.* London; North Pomfreet, Vt.: David & Charles, 1974).
Pagel, Walter. *Paracelsus: An Introduction to Philosophical Medicine in the Era of the Renaissance.* 2nd ed. New York: Karger, 1982.
Paget, James. *The Records of Harvey in Extracts from the Journals of the Royal Hospital of St. Bartholomew's.* London, 1846.
Parsons, Frederick Gymer. *The History of St. Thomas's Hospital.* 3 vols. London: Methuen, 1932.
Paster, Gail Kern. *The Body Embarrassed: Drama and the Disciplines of Shame in Early Modern England.* Ithaca, N.Y.: Cornell University Press, 1993.
Pearl, Valerie. "Puritans and Poor Relief: The London Workhouse 1649–60." In *Puritans and Revolutionaries: Essays in Seventeenth-Century History Presented to Christopher Hill*, edited by Donald Pennington and Keith Thomas. Oxford, Clarendon Press, 1978.
———. "Social Policy in Early Modern London." In *History and Imagination: Essays in Honour of H. R. Trevor-Roper*, edited by Hugh Lloyd-Jones, Valerie Pearl, and Blair Worden. London: Duckworth, 1981.

Pelling, Margaret. "Appearance and Reality: Barber-Surgeons, the Body and Disease." In *London 1500–1700: The Making of the Metropolis*, edited by A. L. Beier and Roger Finlay. London: Longman, 1986.

———. "Healing the Sick Poor: Social Policy and Disability in Norwich, 1550–1640." *Medical History* 29 (1985): 115–37.

———. "Illness among the Poor in an Early Modern Town: The Norwich Census of 1570." *Continuity and Change* 3 (1988): 153–75.

———. "Medical Practice in Early Modern England: Trade or Profession?" In *The Professions in Early Modern England*, edited by Wilfrid Prest. London: Croom Helm, 1987.

———. *The Common Lot: Sickness, Medical Occupations and the Urban Poor in Early Modern England*. London: Longman, 1998.

———. "The Women of the Family?: Speculations Around Early Modern British Physicians." *Social History of Medicine* 8 (1995): 383–401.

———. "Thoroughly Resented? Older Women and the Medical Role in Early Modern London." In *Women, Science and Medicine 1500–1700*, edited by Lynette Hunter and Sarah Hutton. Glouchestershire: Sutton Publishers, 1997.

Pelling, Margaret, and Charles Webster. "Medical Practitioners." In *Health, Medicine and Mortality in the Sixteenth Century*, edited by Charles Webster. Cambridge: Cambridge University Press, 1979.

Perry, Ruth. "Colonizing the Breast: Sexuality and Maternity in Eighteenth-Century England." *The Journal of the History of Sexuality* 2 (2) (1991): 204–34.

Pollock, Linda A. "Living on the Stage of the World: The Concept of Privacy Among the Elite of Early Modern England." In *Rethinking Social History: English Society 1570–1920 and Its Interpretation*, edited by Adrian Wilson. Manchester: Manchester University Press, 1993.

Porter, Dorothy, and Roy Porter. *Patient's Progress: Doctors and Doctoring in Eighteenth-Century England*. Oxford: Polity Press, 1989.

Porter, Roy. "A Touch of Danger: The Man-Midwife as Sexual Predator." In *Sexual Underworlds of the Enlightenment*, edited by George Rousseau and Roy Porter. Manchester: Manchester University Press, 1987.

———. *Bodies Politic: Disease, Death and Doctors in Britain, 1650–1900*. London: Reaktion Books, 2001.

———. *Health For Sale: Quackery in England 1660–1850*. Manchester: Manchester University Press, 1989.

———. "'Laying Aside and Private Advantage': John Marten and Venereal Disease." In *The Secret Malady: Venereal Disease in Eighteenth-Century Britain and France*, edited by Linda Merians. Lexington: University of Kentucky Press, 1996.

———. "Laymen, Doctors and Medical Knowledge in the Eighteenth century: The Evidence of the *Gentlemen's Magazine*." In *Patients and Practitioners: Lay Perceptions of Medicine in Pre-Industrial Society*, edited by Roy Porter. Cambridge: Cambridge University Press, 1985.

———. *London: A Social History*. Cambridge, Mass.: Harvard University Press, 1995.
———. *Mind Forg'd Manacles: A History of Madness in England from the Restoration to the Regency*. Cambridge, Mass.: Harvard University Press, 1987.
———. "The Gift Relation: Philanthropy and Provincial Hospitals in Eighteenth-Century England." In *The Hospital in History*, edited by Roy Porter and Lindsay Granshaw. London: Routledge, 1989.
———. *The Greatest Benefit to Mankind: A Medical History of Humanity from Antiquity to the Present*. London: Harper Collins, 1997.
———. "The Patient's View: Doing Medical History from Below." *Theory and Society* 14 (1985): 175–98.
———. "The Rise of the Physical Exam." In *Medicine and the Five Senses*, edited by Roy Porter and W. F. Bynum. Cambridge: Cambridge University Press, 1993.
Power, D'Arcy. *A Short History of St. Bartholomew's Hospital, 1123–1923*. London: Printed for the Hospital, Whittingham and Griggs, 1923.
Pusey, W. A. *The History and Epidemiology of Syphilis*. Springfield, Ill.: Thomas, 1933.
Quaiffe, G. R. *Wanton Wenches and Wayward Wives: Peasants and Illicit Sex in Early Seventeenth-Century England*. New Brunswick: Rutgers University Press, 1979.
Quétel, Claude. *The History of Syphilis*. Translated from the French by Judith Braddock and Brian Pike. Baltimore: Johns Hopkins University Press, 1990.
Rogers, Nicholas. "Carnal Knowledge: Illegitimacy in Eighteenth-Century Westminster." *Journal of Social History* 23 (2) (1989): 355–76.
———. "Policing the Poor in Eighteenth-Century London: The Vagrancy Laws and Their Administration." *Social History/Histoire Sociale* 24 (1991): 127–47.
Rodgers, Nicholas A. M. *The Wooden World: An Anatomy of the Georgian Navy*. London: Collins, 1986.
Rolleston, J. D. "Venereal Disease in Pepys's Diary." *British Journal of Venereal Disease* 19 (1943): 169–73.
Rose, Craig. "Politics and the London Royal Hospitals, 1683–92." In *The Hospital in History*, edited by Lindsay Granshaw and Roy Porter. London: Routledge, 1989.
———. "Politics, Religion and Charity in Augustan London, c.1680–c.1720." Ph.d. diss., University of Cambridge, 1988.
Rosebury, Theodore. *Microbes and Morals: The Strange Story of Venereal Disease*. New York: Viking Press, 1971.
Rousseau, G. S., and Roy Porter, eds. *Sexual Underworlds of the Enlightenment*. Manchester: Manchester University Press, 1987.
Risse, Guenter. *Hospital Life in Enlightenment Scotland: Care and Teaching at the Royal Infirmary of Edinburgh*. Cambridge: Cambridge University Press, 1986.
———. *Mending Bodies, Saving Souls: A History of Hospitals*. Oxford: Oxford University Press, 1999.
Schen, Claire S. "Constructing the Poor in Early Seventeenth-Century London." *Albion* 32 (3) (2000): 450–63.

Schleiner, Winfried. "Infection and Cure through Women: Renaissance Constructions of Syphilis." *Journal of Medieval and Renaissance Studies* 24 (1994) pp. 499–517.

———. *Medical Ethics in the Renaissance*. Washington, D.C.: Georgetown University Press, 1995.

———. "Moral Attitudes Towards Syphilis and Its Prevention in the Renaissance." *Bulletin of the History of Medicine* 68 (1994): 389–410.

Schwarz, Leonard. *London in the Age of Industrialisation: Entrepreneurs, Labour Force and Living Conditions*. Cambridge: Cambridge University Press, 1992.

Scott, James C. *Domination and the Arts of Resistance: Hidden Transcripts*. Yale University Press, 1990.

Sharpe, J. A. "Defamation and Sexual Slander in Early Modern England: The Church Courts at York." *Bothwick Papers* 58 (1980): 1–36.

Sheppard, Francis. *London: A History*. New York: Oxford University Press, 1998.

Shoemaker, Robert B. "Gendered Spaces: Patterns of Mobility and Perceptions of London's Geography, 1660–1750." In *Imagining Early Modern London: Perceptions and Portrayals of the City from Stow to Strype 1598–1720*, edited by J. F. Merritt. Cambridge: Cambridge University Press, 2001.

———. *Gender in English Society, 1650–1850: The Emergence of Separate Spheres?* London: Longman, 1998.

———. "Reforming the City: The Reformation of Manners Campaign in London, 1690–1738." In *Stilling the Grumbling Hive: The Response to Social and Economic Problems in England, 1689–1750*, edited by Lee Davison, Tim Hitchcock, Tim Keirn, and Robert B. Shoemaker. New York: St. Martin's Press, 1992.

———. "The Decline of Public Insult in London 1660–1800" *Past and Present* 169 (2000): 97–131.

Shorter, Edward. *Women's Bodies: A Social History of Women's Encounter with Health, Ill-Health, and Medicine*. New Brunswick, N.J.: Transactions Publications, 1991.

Siena, Kevin P. "Poverty and the Pox: Venereal Disease in London Hospitals, 1600–1800." Ph.D. diss., University of Toronto, 2001.

———. "Pollution, Promiscuity, and the Pox: English Venereology and the Early Modern Discourse on Social and Sexual Danger." *Journal of the History of Sexuality* 8 (1998): 553–74.

———. "The 'Foul' Disease and Privacy: The Effects of Venereal Disease and Patient Demand on the Medical Marketplace in Early Modern London." *Bulletin of the History of Medicine* 75 (2) (2001): 199–224.

Siraisi, Nancy. *The Clock and the Mirror: Girolamo Cardano and Renaissance Medicine*. Princeton: Princeton University Press, 1997.

Slack, Paul. *From Reformation to Improvement: Public Welfare in Early Modern England*. Oxford: Clarendon Press, 1999.

———. "Hospitals, Workhouses and the Relief of the Poor in Early Modern London." In *Health Care and the Poor Relief in Protestant Europe, 1500–1700*, edited by Ole Peter Grell and Andrew Cunningham. London: Routledge, 1997.

———. *Poverty and Policy in Tudor and Stuart England*. London: Longman, 1988.

———. *The English Poor Law, 1531–1782*. New York: Cambridge University Press, 1995.

———. *The Impact of Plague in Tudor and Stuart England*. London: Routledge, 1985.

Sokoll, Thomas, *Essex Pauper Letters, 1731–1837*. Oxford: Oxford University Press, 2001.

Sontag, Susan. *AIDS and its Metaphors*. New York: Straus and Giroux, 1989.

———. *Illness as Metaphor*. New York: Straus and Giroux, 1978.

Speck, W. A. "The Harlot's Progress in Eighteenth-Century England." *British Journal for Eighteenth-Century Studies* 3 (1980): 127–39.

Spongberg, Mary, *Feminizing Venereal Disease: The Body of the Prostitute in the Nineteenth-century*. London: Macmillan, 1997.

Staves, Susan. "British Seduced Maidens." *Eighteenth-Century Studies* 14 (2) (1980/81): 109–34.

Stewart., Mary Margaret. "'And Blights with Plagues the Marriage Hearse': Syphilis and Wives." In *The Secret Malady: Venereal Disease in Eighteenth-Century Britain and France*, edited by Linda Merians, Lexington: University of Kentucky Press, 1996.

Stone, Lawrence. *The Family, Sex and Marriage in England, 1500–1800*. London: Weidenfeld and Nicolson, 1977.

Suzuki, Akihito. "The Household and the Care of Lunatics in Eighteenth-Century London." In *The Locus of Care: Families, Communities, Institutions and the Provision of Welfare Since Antiquity*, edited by Peregrine Horden and Richard Smith. London: Routledge, 1998.

Taylor, James Stephen. *Jonas Hanway, Founder of the Marine Society: Charity and Policy in Eighteenth-Century Britain*. London: Scolar Press, 1985.

———. *Poverty, Migration, and Settlement in the Industrial Revolution: Sojourners' Narratives*. Palo Alto, Ca.: Society for the Promotion of Science and Scholarship, 1989.

———. "The Impact of Pauper Settlement 1691–1834." *Past and Present* 73 (1976): 42–74.

Temkin, Owsei. "On the History of 'Morality and Syphilis.'" In Owsei Temkin, *The Double Face of Janus and Other Essays in the History of Medicine*. Baltimore: Johns Hopkins University Press, 1977.

———. "The Role of Surgery in the Rise of Modern Medical Thought." *Bulletin of the History of Medicine* 25 (1951): 248–59.

———. "Therapeutic Trends and the Treatment of Syphilis before 1900." In Owsei Temkin, *The Double Face of Janus and Other Essays in the History of Medicine*, Baltimore: Johns Hopkins University Press, 1977.

Thomas, Keith. "Puritans and the English Adultery Act of 1650 Reconsidered." In *Puritans and Revolutionaries: Essays in Seventeenth-Century History Presented to Christopher Hill*, edited by Keith Thomas and Donald Pennington. Oxford: Clarendon Press, 1978.

Thompson, C. J. S. *The Quacks of Old London*. London: Brentano's, 1928.

Thompson, John D., and Grace Goldin. *The Hospital: A Social and Architectural History.* New Haven: Yale University Press, 1975.

Timken-Zinkann, R. F. "Some Aspects of Epidemics and German Art About 1500." *Medical History* 8 (4) (1969): 355–62.

Trumbach, Randolph. "Modern Prostitution and Gender in *Fanny Hill*: Libertine and Domesticated Fantasy." In *Sexual Underworlds of the Enlightenment*, edited by G. S. Rousseau and Roy Porter. Manchester: Manchester University Press, 1987.

———. *Sex and the Gender Revolution.* Vol. I, *Heterosexuality and the Third Gender in Enlightenment London.* Chicago: University of Chicago Press, 1998.

———. "Sex, Gender, and Sexual Identity in Modern Culture: Male Sodomy and Female Prostitution in Enlightenment London." *Journal of the History of Sexuality* 2 (2) (1991): 186–203.

Underdown, David. "The Taming of the Scold: The Enforcement of Patriarchal Authority in Early Modern England." In *Order and Disorder in Early Modern England*, edited by Anthony Fletcher and John Stephenson. Cambridge: Cambridge University Press, 1985.

Valenze, Deborah. "Charity, Custom, and Humanity: Changing Attitudes Towards the Poor in Eighteenth-Century England." In *Revival and Religion Since 1700: Essays for John Walsh*, edited by Jane Garnett and Colin Matthew. London: Hambledon Press, 1993.

Vickery, Amanda. "Golden Age to Separate Spheres? A Review of Categories and Chronology of English Women's History." *Historical Journal* 36 (1993): 383–414.

Villey, Raymond. *Histoire du Secret Médical.* Paris: Seghers, 1986.

Wagner, Peter. "The Discourse on Sex—or Sex as Discourse: Eighteenth-Century Medical and Paramedical Erotica." In *Sexual Underworlds of the Enlightenment*, edited by G. S. Rousseau and Roy Porter. Manchester: Manchester University Press, 1987.

———. "The Satire on Doctors in Hogarth's Graphic Works." In *Literature and Medicine During the Eighteenth Century*, edited by Marie Mulvey Roberts and Roy Porter. London: Routledge, 1993.

Walkowitz, Judith. *Prostitution and Victorian Society: Women, Class, and the State.* Cambridge: Cambridge University Press, 1980.

Wareing, John, "Changes in the Geographical Distribution of the Recruitment of Apprentices to the London Companies, 1486–1750." *Journal of Historical Geography* 6 (1980): 241–9.

Waugh, W. A. "Attitudes of Hospitals in London to Venereal Disease in the Eighteenth and Nineteenth Centuries." *British Journal of Venereal Disease* 47 (1971): 146–50.

Wear, Andrew. "Caring for the Sick Poor in the Parish of St. Bartholomew Exchange, 1580–1679." In *Living and Dying in London*, edited by W. F. Bynum and Roy Porter. *Medical History*, Supplement 11, London, 1991.

———. *Knowledge and Practice in English Medicine, 1550–1680.* Cambridge: Cambridge University press, 2001.

———. "Medical Ethics in Early Modern England." In *Doctors and Ethics: The Earlier Historical Setting of Professional Ethics,* edited by Andrew Wear, Johanna Geyer-Kordesch, and Roger French, 98–130. Amsterdam: Rodopi, 1993.

Weiner, Carol Z. "Sex Roles and Crime in Elizabethan Hertfordshire." *Journal of Social History* 8 (4), (1974/5): 38–60.

Whitteridge, Gweneth, and Veronica Stokes. *A Brief History of St. Bartholomew.* London: The Governors of the Hospital of St. Bartholomew, 1961.

Wild, Wayne, "Doctor-Patient Correspondence in Eighteenth-Century Britain: A Change in Rhetoric and Relationship." *Studies in Eighteenth-Century Culture* 29 (2000): 47–64.

Williams, David Innes, *The London Lock: A Charitable Hospital for Venereal Disease 1746–1952.* London, Royal Society of Medicine Press, 1996.

Williams, Gordon. "An Elizabethan Disease." *Trivium* 6 (1971): 43–58.

Wilson, Adrian. "Illegitimacy and Its Implications in Mid Eighteenth-Century London: The Evidence of the Foundling Hospital." *Continuity and Change* 4 (1)(1989): 103–64.

———. "Conflict, Consensus and Charity: Politics and the Provincial Voluntary Hospitals in the Eighteenth Century." *English Historical Review* 111 (442) (1996): 599–619.

Wilson, Phillip K. "Exposing the Secret Disease: Recognizing and Treating Syphilis in Daniel Turner's London." In *The Secret Malady: Venereal Disease in Eighteenth-Century Britain and France,* edited by Linda Merians, Lexington: University of Kentucky Press, 1996.

———. "'Sacred Sanctuaries of the Sick': Surgery at St. Thomas's Hospital 1725–26." *London Journal* 17 (1992): 136–53.

———. *Surgery, Skin and Syphilis: Daniel Turner's London (1667–1741).* Amsterdam: Rodopi, 1999.

Wood, Connie Shear. "Syphilis in Anthropological Perspective." *Social Science and Medicine* 12 (1978): 47–55.

Wrightson, Keith. "The Politics of the Parish in Early Modern England." In *The Experience of Authority in Early Modern England,* edited by Paul Griffiths and Steve Hindle. London: Macmillan, 1996.

Wrigley, E. A. "A Simple Model of London's Importance in Changing English Society and Economy, 1650–1750." *Past and Present* 37 (1967): 44–70.

———. "Marriage, Fertility and Population Growth in Eighteenth-Century England." In *Marriage and Society: Studies in the Social History of Marriage,* edited by R. B. Outhwaite. London: Europa Publications, 1981.

Wrigley, E. A., and Roger Schofield. *The Population History of England, 1541–1871.* Cambridge: Cambridge University Press, 1981.

Wunderli, Richard. *London Courts on the Eve of the Reformation,* Cambridge, Mass.: The Medieval Academy of America, 1981.

Wyke, T. J. "Hospital Facilities for, and Diagnosis and Treatment of, Venereal Disease in England, 1800–1870." *British Journal of Venereal Disease* 49 (1973): 78–85.

Wyman, A. L. "The Surgeoness: The Female Practitioner of Surgery, 1400–1800." *Medical History* 28 (1984): 22–41.

Zimbardo, Rose. "Satiric Representation of Venereal Disease: The Restoration Versus the Eighteenth-Century Model." In *The Secret Malady: Venereal Disease in Eighteenth-Century Britain and France*, edited by Linda Merians, Lexington: University of Kentucky Press, 1996.

INDEX

Acton, William, 127
advertisements, 41–42, 45–49, 51, 52–53, 55–59, 254, 279n57
alehouses, 26, 48–49, 66, 83, 86, 289n75, 291n97; bondsmen, 118–21, 255; Bird in Hand, 118–20; Cooper's Arms, 118; Guy's Head Tavern, 118–19; Ship and Shovel, 118; Spur Inn, 118
Allen, Peter Lewis, 77, 268n6
Amussen, Susan Dwyer, 91
Ancaster, Duke of, 185, 199
Anderson, Gilbert, 47
Andree, John, 26, 61
Andrew, Donna, 11, 184, 186–87, 203–4
Anselment, Raymond, 32, 34, 36
apothecaries, 27, 39, 48, 84, 138–39, 145, 186, 242, 300n16
apprenticeship, 56, 158, 173–74, 227, 283–84n131
Archer, John, 31
Ariès, Phillipe, 35–37
Arrizabalaga, Jon, 1, 64, 68
Atkins, Charles, 119–20

Baker, Robert, 49
Barber Surgeon's Company, 184
Bastardy. *See* illegitimacy, single mothers

Battersea, 153
Bayley, Lewis, 124
beauty, 34–35
Beckett, William, 122
Bethlem Hospital, 144, 285n2
Bills of Mortality, 32
birth of the clinic, 14, 125–30, 263–64, 298n88
Bisse, James, 87
Blanckaert, Stephan, 65
Blizzard, William, 207, 321n29
Boerhaave, Herman, 23
Bolton, Duke of, 27
booksellers, 48
Boswell, James, 16, 28, 45, 60, 273–74n68
bougies, 246, 326n95
Bridewell, 156–57, 167, 299–300n7
Bromfeild, Charles, 197, 245–46
Bromfeild, James, 186
Bromfeild, William, 181, 184–87, 196, 197–200, 205, 211, 218, 221, 228, 242, 313n80
Burrows, John, 34, 45
Bynum, William F., 289n73, 309n1

Cam, Joseph, 34, 43–44
Campbell, Pryse, 231–32
Carlson, Lawrence, 118
Case, John, 46

361

Chicoyneau, Francois, 43–44
Christ's Hospital, 285n2
Clapham Sect, 211, 212
Clare, Peter, 61, 284n151
Clark, Anna, 39, 213
Clark-Kennedy, A. E., 221
Clowes, William, 63–64, 66, 93
Coroner's Inquests, 24–27, 176–78, 242–43
Covent Garden, 46, 150

Dartmouth, Earl of, 199
Davidoff, Leonore, 211–12, 262
Dean, G. (empiric), 46
De Coetlogon, Edward Charles, 197–202, 211, 245
Desault, Pierre, 42
Despaignol, Samuel, 138–39
doctor-patient relationship, 7–8, 13, 36, 38, 50–55, 81, 125–30, 253–54, 282n120, 298n88
domestic service, 39–40, 60, 105, 168, 174–75, 176, 177, 225
Douglas, Mary, 36

Eden, Richard, 78
Elias, Nobert, 35–37
Ellis, William, 118
English Civil War, 72
Erasmus, 276n30
Evangelicals, 11, 183, 188, 200, 203–4, 211–13, 218, 238, 248–49

Fabricius, Johannes, 36
female healers, 14, 55–59, 254, 283–84nn131, 143
Field, Nicholas, 290–91n96
Fisher, Jalez, 203
Fissell, Mary, 59, 128–29, 263–64, 298n88, 302n36
Foot, Jesse, 216
Foucault, Michel, 127

Foul Disease, beliefs about contagion, 66–67, 89, 106, 201, 216, 291n109; issues related to diagnosis and terminology, 15, 78, 80–81, 170, 173, 226, 245, 264–65, 267n2, 306n115; symptoms, 16–22, 24–25, 31, 34, 65; treatment, 22–24 (*See also* mercury); theories on generation, 9, 215–16; incidence, 4, 10, 41–42, 78, 165–66, 219, 223–26, 270–71n26
Foundling Hospital, 40, 191
Freeman, Stephen, 48
French, Roger, 1, 64, 68

Garrick, David, 187
Gentleman's Magazine, 41
George II (king of England), 184
George, Dorothy, 149
Germany, 1, 64, 271n28, 289n66
Gillis, John, 47
Glorious Revolution, 76
Gonsale, Gerardts, 57
Goodman's Fields, 207
Gowing, Laura, 40, 156
Granshaw, Lindsay, 184
Graunt, John, 32
Gregory, John, 49–50, 51
Grubb, Robert, 47
guaiac, 45, 64, 83–84, 86, 290n91
Guy, Thomas, 220
Guy's Hospital, 10, 144, 152, 175, 220, 223, 224, 233, 293n14, 319n4

Habermas, Jürgen, 35
Hackney, 67
Hall, Catherine, 211–12, 262
Handel, George Frideric, 187
Hanway, Jonas, 205–10, 218, 223, 261
Harley, David, 15

Harris, Nicholas, 118
Harrison, John, 221
Harrison, Sarah, 119
Harvey, William, 67–68, 287n30
Hasselock, John, 78
Hawkins, Robert, 87
Henderson, John, 1, 64, 68
Henderson, Tony, 204, 235
Heusde, Sarah de, 56
Hippocratic Oath, 49
Hitchcock, Tim, 137–38, 140–41, 161, 164, 172, 223–24
Hogarth, William, 18–19, 204
Holborn, 58
Howles, John, 119–20
Hume, David, 49
Hyde Park Corner, 184

Ignatieff, Michael, 207
Illegitimacy, 224–26, 305nn96, 99, 321–22n38. *See also* single mothers
Incurabili Hospitals, 64
Ireland, 153
Italy, 1, 64, 68

Jones, Herbert, 207
Jütte, Robert, 1, 271n28, 289n66

Kent, John, 78, 289n75

Langham, W. (empiric), 57
Laverenst, Ann, 56
Lawrence, Susan, 125, 129
lazarettos, 12, 64–67, 80, 286n15. *See also* St. Bartholomew's Hospital
LeBarr, Lewis, 151–52, 154, 157, 160
Lisbon Diet Drink, 45
Lloyd, Stanley, 205
Lock Asylum for the Reception of Penitent Women, 9–11, 183, 207, 213–18, 261–64; discipline, 216–17, 249–50; foundation, 204, 213, 214; fundraising literature, 213–15, 250; governors, 211
Lock Hospital, 2, 8, 9, 11, 63, 168, 181–250; admissions process, 186, 237–40; chapel, 187–89, 196, 199, 201; children in, 118–19, 192–93, 195, 229, 235; discipline, 186, 188–89, 196, 247–49; foundation of, 181, 184–86; fundraising, 11, 185–87, 190–95, 202–3, 310n22; governors, 185, 198–99, 237–38; married women in, 195, 235; medical students, 185; ministers, 188, 200–202 (*See also* Martin Madan, Edward Charles de Coetlogon, and Thomas Scott); moral reformation, 195–96, 197–206, 314n94; patients, 19–22, 119–20, 168, 195, 196, 201, 235–50, 324n69; patient's strategies when applying, 237–39; policy on settled parish patients, 228–35; population, 195, 235–36; religious instruction, 188–90, 197–206; therapeutics, 245–47; waiting lists, 240–44, 324–25n78
London, City of, 64, 67, 101, 133, 156, 158, 234
London Hospital, 207, 220, 221–24, 319–20nn15, 19
Lucas, Joyce, 56
Lying-In Hospital, 191

MacDonald, Michael, 177, 273n66
Madan, Martin, 187–89, 191–93, 195, 196, 197–200; *Every Man our Neighbour*, 191; *Thelypthora, or a treatise on Female Ruin*, 204–5

Magdalen Hospital, 205–6, 207–10, 211, 216, 217, 315–16n107
Manchester, Duke of, 199
Maris, Elizabeth, 56, 57–59
Maris, Peter, 52–53, 57–58
marriage 40, 52–54, 90, 111, 166, 307n120
Marten, John, 45, 53, 282n115
Mayo, Thomas, 230
McAllister, Marie, 45
McCullough, Laurence B., 49
Medical conditions, ague, 171, 175; broken limbs, 143; consumption, 172; dropsy, 143; fever, 68, 171, 143, 270n26; gonorrhea, 15–17; impotence 276n23; itch, 140, 143, 175, 320n24; leprosy, 65–67, 286n15; lying-in, 140, 146, 157, 167; looseness, 143; madness, 140, 144, 146, 276n23; measles, 143; plague, 10, 270n26, 276n23, 287n32; pox (*See* Foul Disease); rheumatism, 143, 145, 169; scrofula, 60; sore leg, 18, 170, 171–72; smallpox, 68, 271n26, 320n24; syphilis, 15, 95, 173, 226, 264–65; vomiting, 143
medical confidentiality. *See* privacy
medical education, 13, 125–30
medical ethics, 49–51, 53–54, 281n106
medical marketplace, 30, 40–61, 121, 253–54
merchants, 40–41, 43, 48, 52
mercury. *See* salivation
Merians, Linda E., 184, 185, 203, 214
Middleton, Charles, 211
midwives, 139, 156
migration, 82, 227–28
Misericordia Dispensary, 239–40
Misericordia Hospital, 183, 206–10, 211, 216, 217, 218, 223, 224, 261, 264, 321n29

Montpellier method, 42–45, 99, 151, 253, 279n63
Moore, Norman, 2
Murphy, Terrance, 177, 273n66

Nash, Stanley, 205
Nicholson, I. F., 45
Norwich, 64
nurses, 14, 26, 27, 38, 39, 89, 102, 118–19, 139, 144, 177

Ogborn, Miles, 207–8
Old Bailey Sessions Papers, 16, 23–24, 25, 37, 38–39, 57, 60–61, 193–95, 251–53

Padua, 68
Palmer, Samuel, 99–100
Paracelsus, 64–65
parish relief, 12, 82, 97, 105, 113–14, 116, 133–34, 135–80, 227–35, 256, 289n72; entitlement to, 141–42, 149, 153, 160–61, 172, 176, 178–79, 233, 258; officers' attitudes, 152–53, 158–61, 170–71, 178–79; passing, 153, 155–56, 304n83, 305n90, 306nn105, 106; settlement enquiries, 153–58, 170–71, 257, 306n103; strategies for applying for, 145, 161, 169–72, 176, 307–8n144; Abbington, 230–31, 232; Airencester, 232; Arsley, 232; Clapham, 232; Clewer, 232; Darenth, 232; Dartford, 232; Deal, 232; Great Missenden, 232; Ham, 232; Hammersmith, 229, 232; Hayes, 232; Hurst, 229, 232; Henden, 232; Horton Kirby, 232; Kingston, 232; Leatherhead, 232; Long Ditton, 232;

Newland, 232; Putney, 232; Richmond, 231, 232; St. Andrew's Holborn, 144, 146, 149, 151, 162–65; St. Ann's Westminster, 119, 252, 258; St. Clement's Dane, 26; St. Giles in the Field, 146–47, 152, 252, 258; St. George Bloomsbury, 146–47, 152, 154; St. George Hanover Square, 138–39, 140, 144, 148, 153, 177, 229; 323n57; St. George the Martyr, 144, 151; St. James Clerkenwell, 155; St. James, Westminster, 144, 147, 151; St. Luke's Chelsea, 140–41, 152–53, 162–64, 168–69, 172, 173–76, 233; St. Margaret's Westminster, 135, 140–46, 149, 150, 152, 157–58, 162, 167, 293n14; St. Martin's in the Fields, 148, 151, 154; St. Sepulchre (London Division), 140–44, 146, 147, 149, 151, 152–53, 154–55, 157–59, 162, 167, 173; Sutton, 232; Upton, 232; Wimbledon, 232; Walton St. Lawrence, 232; Windsor, 232
Parsons, F. G., 2, 72, 86
Patient strategies, 13–14, 24, 25–27, 47–49, 52–54, 54–59, 79–82, 104–5, 107, 113–21, 123, 126–27, 128–32, 145–46, 149, 161, 169–76, 228, 236–38, 241, 249, 253–60, 263. *See also* Lock Hospital, parish relief, St. Bartholomew's Hospital, St. Thomas's Hospital
Pearson, John, 16–18, 235, 245–47
Peckwell, Henry, 211
Pelling, Margaret, 41, 50, 282–83n107

penitentiary movement, 183, 207, 210–11, 248, 261
penny post, 47–48, 254
Pepys, Elizabeth, 38–39
Pepys, Samuel, 19, 32, 34, 38, 50–51, 275n12
Pepys, Thomas, 32
Percival, Thomas, 50
Perry, Ruth, 213
Peter, Charles, 17, 31, 62, 83, 282n120
physicians, 34, 67, 83, 86, 139, 140, 186
Pitt, Charles Morton, 211
Plenck, Joseph, 34, 50, 54
Pollack, Linda, 37–38
Poor Law, New, 136
Poor Law, Old. *See* parish relief, workhouses
Porter, Roy, 47, 50, 59, 88
privacy, 13, 16, 35–39, 40, 45–54, 56, 113, 116, 121–22, 169, 171, 237, 255, 281n100, 281–82n107
Profily, John, 40–41
prostitutes, 9, 34, 129, 165, 204–6, 213–15, 224–26, 235–36, 321n36
Pyecroft, Robert, 108

quackery, 17, 44, 50, 63, 279n67
Quaiffe, G. G., 238
Quétel, Claude, 32, 89

Ranby, John, 184
rape, 16, 37, 39, 57, 119, 193–95, 271n32, 312nn48, 54, 55, 56, 320n25
reputation, 38–41, 43, 54–55, 172, 177. *See also* privacy, shame
Reynolds, John, 189
Ritchie, John, 266–37, 246
Robinson, Nicholas, 23–24, 43

Rogers, Nicholas, 321–22n38
Rose, Craig, 92
royal hospitals. *See* Bethlem Hospital, Christ's Hospital, St. Bartholomew's Hospital, St. Thomas's Hospital
Royal Society, 65

St. Bartholomew's Hospital, 2, 39, 62–64, 144, 160, 221, 234, 255–56, 315–16n107; admissions, 79–81, 107, 287n30; discipline, 86–89, 291n99; effects of fire, 73–76, 96–97; fees, 75–77, 78, 96, 102, 166, 255, 288n51; finances, 69–71, 73, 75–76, 84, 93, 96–100, 106–7, 108–9; governors, 92–93; Kingsland Outhouse, 67–68, 76, 77–78, 84, 87, 99, 144, 167, 217, 290n80; Lock Outhouse, 67–68, 76, 77–78, 84, 86–87, 99, 144; medical staff, 83; patients, 79–93, 105–6; outhouses, 64–66, 67–68, 69–71, 74, 76–81, 83, 96–100; closure of, 73, 106–9; religious instruction, 87–88, 124; therapeutics, 83–86, 94–95; VD operations, scope of, 69–73, 75, 77, 96–102, 109; wards, 67, 109
St. George's Hospital, 27, 184–86, 190–91, 242–43, 309n9
St. Thomas's Hospital, 2, 10, 33, 62–63, 68, 144, 151, 169–70, 173, 174, 220, 233, 255–56, 308n149, 315–16n107; admissions, 79–81, 82, 107, 111, 113–21; corporal punishment, 10, 89–92, 122, 125, 256; discipline, 86, 89–93, 122–25; effects of fire, 73–76, 84–85;
fees, 102–4, 115–16, 123–24, 166, 255, 323n56; finances, 73–74, 82, 93; governors, 92–93, 114–15; matron, 118–19; medical staff, 83; medical students, 126–27; nurses, 102, 118–19, 293n13; patients, 79–93, 105–6, 111–32, 170, 174, 324n69; policies, 69, 73–74, 82, 289n72; religious instruction, 123–25; therapeutics, 83–86, 131; VD operations, scope, 71–73, 75, 109–13; wards, 68, 72–73, 110–12, 295nn41, 46; salivation, 22–24, 25, 33, 61, 62, 85–86, 102–3, 124, 131, 150–51, 158, 172, 245–47, 325nn90–93; alternatives to, 42–45, 83, 151–52; as punishment, 85–86, 90–91, 124; opposition to, 42–45; resistance to, 24, 40–41, 43; satirized, 33–34. *See also* guaiac
satire, 31–33
Saunders, William, 54
Sawrey, Solomon, 34
Schleiner, Winfried, 36, 50, 54, 193, 274n6, 281n106
Scott, James, 8, 324n70
Scott, Thomas, 203, 211
segregation, 12, 66–69, 89, 106–7, 110, 111–12, 123, 150–51, 201
separate spheres, 211–13, 262
sexuality, patterns of, 223–26
shame, 16, 31–37, 52, 115, 148, 239, 262, 278n53
Short Title Catalogue, The, 41
Shorter, Edward, 269n13
Sigogne, Bouez de, 40
single mothers, 39–40, 156–57, 160, 165, 166–68, 187, 305nn96, 99, 320n24. *See also* Illegitimacy
slander, 36–37, 38, 54

Society for the Promotion of Christian Knowledge (SPCK), 137, 140, 148, 182
Society for the Reformation of Manners, 182
Soeburgh, Willemina van, 56
soldiers, 24–25, 26, 60, 71, 76, 116, 269nn12, 13, 288nn42, 53, 296n61, 309n9
Souberg, Abraham, 57
Southwark, 67–68, 86, 118, 150
Spongberg, Mary, 215
Stafford, Sydney, 232
Staves, Susan, 204–5
Strand, 239
suicide, 24–28, 176–78, 242–43, 273n66
surgeons, 16–17, 26, 34, 43–44, 45, 50, 53, 61, 74, 75, 77–78, 82, 88, 122, 138, 140, 143, 158, 186, 216, 229, 236, 245–47, 289n73, 290n89
Swain, Hadley, 211
Swediur, Franz, 279n63
Sydenham, Thomas, 23

Temkin, Owsei, 85–86, 90
theft, 25, 39, 60–61
Thornton, Henry, 211
Thornton, Robert, 211
Tilbourn, Cornelius, 57
Topliffe, John, 83
Tories, 124
Trumbach, Randolph, 39, 193, 203, 224–26, 235–36, 296n61, 321n36, 322n38
Tucker, John, 147
Tuff, John, 106–108
Turner, Daniel, 43–44, 99, 279n67

vagrancy, 155–56, 227, 234
voluntary hospitals, 114, 132–33, 185–86, 225, 240, 260, 302n38

Wales, Prince of, 184–85
Walkowitz, Judith, 182
Wall, W. (medical practitioner), 62–63
Wallington, John, 108
Warden, Thomas, 118
Wastell, Henry, 54
Wear, Andrew, 50
Westbrook, John, 140
Westminster Infirmary, 1, 3, 104–5, 132, 190–91
wet nursing, 17, 90, 111, 126–27, 129, 156, 192–93, 195, 238
Whigs, 123
widows, 119
Wilberforce, William, 204, 211
Williams, David Innes, 184, 188, 197, 200, 203
Williams, Thomas, 186
Wilson, Adrian, 5
Wilson, Phillip, 17, 44, 61, 85
Wood, Basil, 211
Wood, Loftus, 239
workhouses, 26, 256–57; age and gender of inmates, 161–67; children in, 163, 168; discipline, 147–48; illness in, 140–141, 149–51; incidence of VD in, 149–50; infirmaries, 140–46, 150, 298n3; female inmates, 135–36, 151–52, 153, 154–61, 167–69, 170, 172, 172–78; living conditions in, 147; male inmates, 153, 169–70, 172; masters, 147–48, 154, 230; medical personnel, 138–39, 151–52; pauper opinion of, 147–49, 172–76, 256–57; population, 140–41; therapeutics, 150–52. *See also* parish relief
Workhouse Test Act (1727), 137
Wynell, John, 31